Operator Theory: Advances and Applications
Vol. 175

Operator Theory in Inner Product Spaces

Karl-Heinz Förster
Peter Jonas
Heinz Langer
Carsten Trunk
Editors

Birkhäuser
Basel · Boston · Berlin

Editors:

Karl-Heinz Förster
Peter Jonas
Carsten Trunk
Institut für Mathematik, MA 6-4
Technische Universität Berlin
Straße des 17. Juni 136
D-10623 Berlin

e-mail: foerster@math.tu-berlin.de
jonas@math.tu-berlin.de
trunk@math.tu-berlin.de

Heinz Langer
Mathematik
Technische Universität Wien
Wiedner Hauptstrasse 8–10/1411
A-1040 Wien

e-mail: hlanger@mail.zserv.tuwien.ac.at

2000 Mathematics Subject Classification 12D10, 15A22, 15A57, 34B20, 34B40, 46C20, 47Axx, 47B25, 47B50, 47E05

Library of Congress Control Number: 2007922256

Bibliographic information published by Die Deutsche Bibliothek
Die Deutsche Bibliothek lists this publication in the Deutsche Nationalbibliografie; detailed bibliographic data is available in the Internet at <http://dnb.ddb.de>.

ISBN 978-3-7643-8269-8 Birkhäuser Verlag, Basel – Boston – Berlin

Part of Springer Science+Business Media
Printed on acid-free paper produced from chlorine-free pulp. TCF ∞
Cover design: Heinz Hiltbrunner, Basel
Printed in Germany
ISBN-10: 3-7643-8269-3
ISBN-13: 978-3-7643-8269-8

e-ISBN-10: 3-7643-8270-8
e-ISBN-13: 978-3-7643-8270-4

9 8 7 6 5 4 3 2 1

www.birkhauser.ch

Contents

Preface

This volume contains papers written by the participants of the 4th Workshop on Operator Theory in Krein Spaces and Applications, which was held at the Technische Universität Berlin, Germany, December 17 to 19, 2004.

The workshop covered topics from spectral, perturbation and extension theory of linear operators and relations in inner product spaces. They included spectral analysis of differential operators, the theory of generalized Nevanlinna functions and related classes of functions, spectral theory of matrix polynomials and problems from scattering theory. All these topics are reflected in the present volume. The workshop was attended by 58 participants from 12 countries.

It is a pleasure to acknowledge the substantial financial support received from the

- Deutsche Forschungsgemeinschaft (DFG),
- DFG-Forschungszentrum MATHEON "Mathematik für Schlüsseltechnologien",
- Institute of Mathematics of the Technische Universität Berlin.

We would also like to thank Petra Grimberger for her great help. Last but not least, special thanks are due to Jussi Behrndt and Christian Mehl for their excellent work in the organisation of the workshop and the preparation of this volume. Without their assistance the workshop might not have taken place.

The Editors

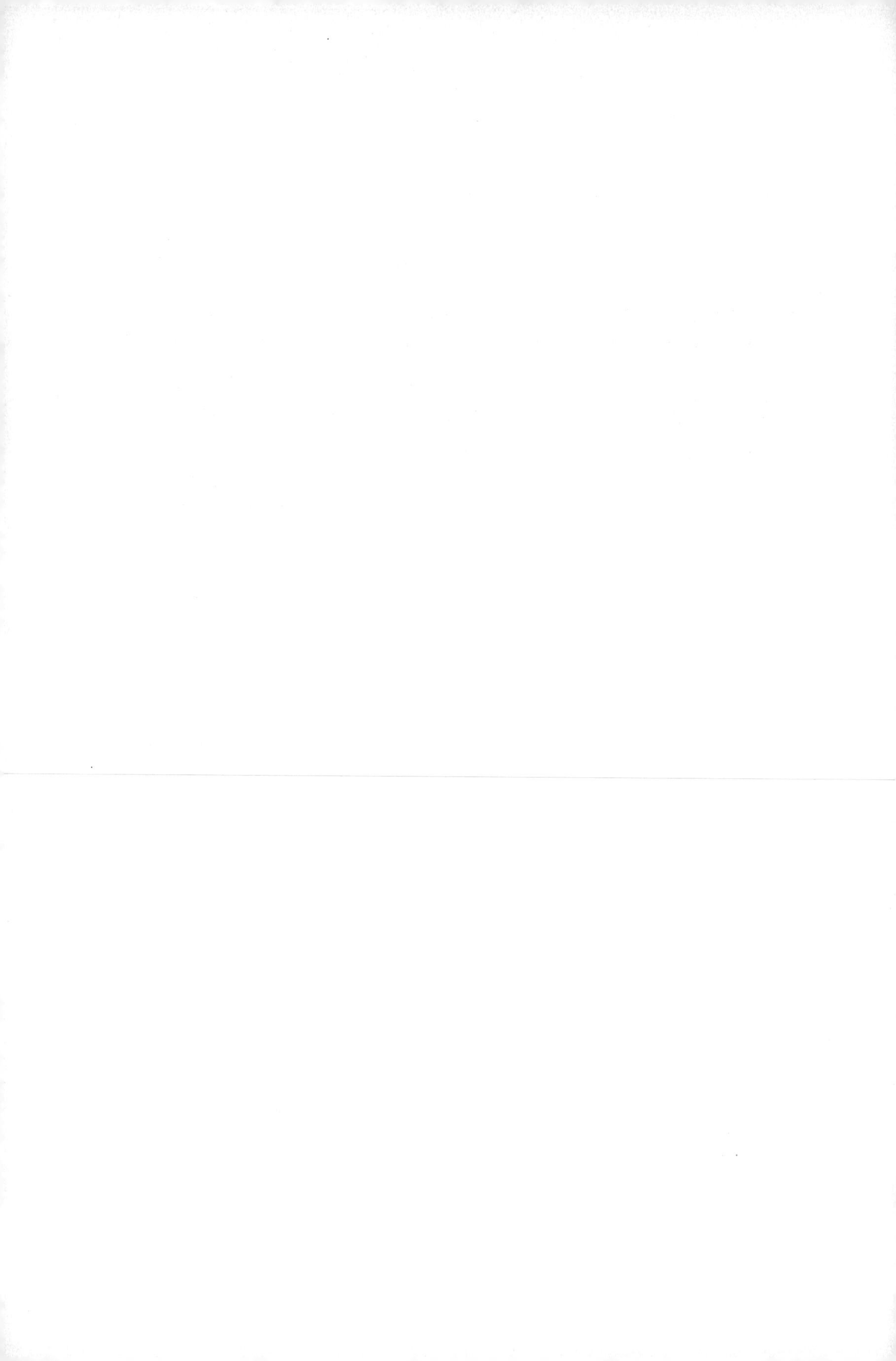

Operator Theory:
Advances and Applications, Vol. 175, 1–11

Linear Operators in Almost Krein Spaces

Tomas Ya. Azizov and Lioudmila I. Soukhotcheva

Abstract. The aim of this paper is to study the completeness and basicity problems for selfadjoint operators of the class $\mathbf{K}(\mathbf{H})$ in almost Krein spaces and prove criteria for the basicity and completeness of root vectors of linear pencils.

Mathematics Subject Classification (2000). Primary 47B50; Secondary 46C50.

Keywords. Krein space, operator pencil, completeness and basicity problem.

1. Introduction

Let $\mathcal{H}$ be a Hilbert space, let A and B be compact operators and let A be additionally a positive operator. Consider the linear operator pencil

$$L(\lambda) = A^{-1} - \lambda(I + B). \tag{1}$$

Such a pencil appears, for instance, if the following spectral problem in $L_2(0, \pi)$ is considered:

$$\begin{cases} -\dfrac{d^2 f}{dt^2} + q(t)f(t) = \lambda(f(t) + \displaystyle\int_0^\pi K(t,s)f(s)ds) \\ f(0) = f(\pi) = 0. \end{cases} \tag{2}$$

Assume q is a continuous real function with $q(t) > -1$, the kernel $K(t, s)$ is symmetric and continuous on $[0, \pi] \times [0, \pi]$. It remains to define A and B in the following way:

$$A^{-1} f = -\frac{d^2 f}{dt^2} + qf,$$

$$\operatorname{dom} A^{-1} = \Big\{ f \in L_2(0, \pi) \mid f, f' \text{ absolutely continuous on } (0, \pi), \\ -\frac{d^2 f}{dt^2} + qf \in L_2(0, \pi), f(0) = f(\pi) = 0 \Big\}$$

This research is supported partially by the grant RFBR 05-01-00203.

$$(Bf)(t) = \int_0^{\pi} K(t,s)f(s)ds.$$

The completeness and basicity problems for the pencil (1) and for the operator

$$H = A(I + B) \tag{3}$$

are closely related. The completeness problem for operators (3) under the assumption $\lambda = 0$ is not an eigenvalue of H ($0 \notin \sigma_p(H)$) was considered for the first time by M.V. Keldysh (see, for instance, [6, Theorem 8.1]). The case when B is also selfadjoint and $\ker H = \{0\}$ was studied by I.Ts. Gokhberg and M.G. Krein in [6, p. 322]. There it was shown that there is a Riesz basis in $\mathcal{H}$ which consists of root vectors of H. The proof is based on the selfadjointness of the operator H with respect to the indefinite inner product

$$[\cdot,\cdot] = ((I + B)\cdot,\cdot), \tag{4}$$

which turns $\{\mathcal{H}, [\cdot,\cdot]\}$ into a Pontryagin space. General criteria for the completeness and basicity of root vectors of selfadjoint operators in Pontryagin spaces can be found in [2], and, more generally, in [3, Theorem IV.2.12].

If $\ker(I + B) \neq 0$, the inner product (4) is degenerate and, due to the compactness of the operator B, we have

(a) the isotropic part

$$\mathcal{H}^0 := \{x \in \mathcal{H} \mid [x,y] = 0, \quad \text{for all } y \in \mathcal{H}\}$$

of $\mathcal{H}$ is finite-dimensional, and

(b) the factor-space $\widehat{\mathcal{H}} = \mathcal{H}/\mathcal{H}_0$ is a Pontryagin space.

Indefinite inner product spaces $\mathcal{H}$ with the properties (a) and (b) are called almost Pontryagin spaces. The operator H in (3) is selfadjoint in the degenerate almost Pontryagin space $\{\mathcal{H}, [\cdot,\cdot]\}$. First results about the completeness and basicity problems for compact selfadjoint operators in almost Pontryagin space were obtained in [1].

Recently, spectral properties of operators acting in almost Pontryagin and almost Krein spaces (for a definition of almost Krein spaces we refer to Definition 1 below) and their applications were studied in, for example, [4], [5], [7]–[13].[1]

The main aim of this paper is to study the completeness and basicity problems for selfadjoint operators of the class $\mathbf{K}(\mathbf{H})$ in almost Krein spaces (a definition of the class $\mathbf{K}(\mathbf{H})$ see below on p. 3). Moreover, we will prove criteria for the basicity and completeness of root vectors of the pencil (1).

[1]The authors thank Chr. Mehl (TU Berlin) for his help with the bibliography.

2. Main definitions

We shortly recall some definitions and notions related to Krein spaces. For more details we refer to [3].

A linear space $\mathcal{K}$ equipped with an indefinite inner product $[\cdot,\cdot]$ is called a Krein space if it admits a canonical decomposition

$$\mathcal{K} = \mathcal{K}^{+}[+]\mathcal{K}^{-},$$

where $\mathcal{K}^{+}$ is $[\cdot,\cdot]$-orthogonal to $\mathcal{K}^{-}$, and $\mathcal{K}^{\pm}$ is a Hilbert space with respect to the scalar product $\pm[\cdot,\cdot]$, respectively.

Definition 1. A space $\mathcal{K}$ with an indefinite inner product $[\cdot,\cdot]$ is called an almost Krein space if its isotropic part $\mathcal{K}^0$ is finite-dimensional (we do not exclude the case $\mathcal{K}^0 = \{0\}$) and the factor-space

$$\widehat{\mathcal{K}} = \mathcal{K}/\mathcal{K}^0, \qquad [\hat{x},\hat{y}]\widehat{\ } = [x,y],\ \hat{x},\hat{y} \in \hat{\mathcal{K}},\ x \in \hat{x},\ y \in \hat{y}, \tag{5}$$

is a Krein space with respect to the naturally reduced indefinite inner product $[\cdot,\cdot]\widehat{\ }$.

Let us note that each almost Pontryagin space is an almost Krein space.

As in the Krein space case we say that a nonnegative/nonpositive subspace $\mathcal{L}_{\pm}$ belongs to the class $\mathbf{h}^{\pm}$, if it is represented as a sum of a finite-dimensional neutral and a uniformly positive/negative subspace.

Remark 2. From the definitions it follows immediately that nonnegative/nonpositive subspace $\mathcal{L}_{\pm}$ in an almost Pontryagin space belongs to the class $\mathbf{h}^{\pm}$.

All operators below are considered to be linear, everywhere defined and bounded.

An operator A in an almost Krein space is called selfadjoint, if $[Ax,y] = [x,Ay]$ for all $x,y \in \mathcal{K}$.

By definition, a selfadjoint operator A belongs to the class $\mathbf{H}$, if it has a maximal nonpositive and a maximal nonnegative invariant subspaces $\mathcal{L}_{\pm}$ and each such a subspace belongs to $\mathbf{h}^{\pm}$, respectively.

We say that a selfadjoint operator A belongs to $\mathbf{K}(\mathbf{H})$, if there is a selfadjoint operator $B \in \mathbf{H}$, which commutes with A.

Remark 3. Let $\mathcal{K}$ be an almost Pontryagin space. From Remark 2 and the definition of the class $\mathbf{H}$ it follows that $I \in \mathbf{H}$. Hence each selfadjoint operator in an almost Pontryagin space belongs to the class $\mathbf{K}(\mathbf{H})$. Moreover, from Theorem 4 below we have that every selfadjoint operator in an almost Pontryagin space belongs to the class $\mathbf{H}$.

3. Invariant dual pairs

$\{\mathcal{L}_+, \mathcal{L}_-\}$ is said to be a dual pair, if $\mathcal{L}_\pm$ is a nonnegative/nonpositive subspace and $[x_+, x_-] = 0$ for all $x_\pm \in \mathcal{L}_\pm$. We prove an analog of a known result about invariant subspaces of operators of the class $\mathbf{K}(\mathbf{H})$ in a Krein space (see [3, § III.5]) for the case of almost Krein spaces.

Theorem 4. *Let $A \in \mathbf{K}(\mathbf{H})$ be a selfadjoint operator in an almost Krein space $\mathcal{K}$ and let B be a selfadjoint operator in $\mathbf{H}$ which commutes with A. Assume $\{\mathcal{L}_+, \mathcal{L}_-\}$ is an A-invariant dual pair (it is not excluded that $\mathcal{L}_+ = \mathcal{L}_- = \{0\}$).*
Then there exists an A-invariant maximal dual pair $\{\mathcal{L}^+, \mathcal{L}^-\}$ such that $\mathcal{L}_\pm \subset \mathcal{L}^\pm$.

If $\{\mathcal{L}_+, \mathcal{L}_-\}$ is also B-invariant, then we can choose a maximal dual pair which is simultaneously A- and B-invariant.

In particular, there exists an A-invariant maximal dual pair $\{\mathcal{L}^+, \mathcal{L}^-\}$ such that $\mathcal{L}^\pm \in h^\pm$.

Proof. Since the isotropic part $\mathcal{K}^0$ of $\mathcal{K}$ is a part of each maximal semidefinite subspace and $\{\mathcal{L}_+ + \mathcal{K}^0, \mathcal{L}_- + \mathcal{K}^0\}$ is A-invariant, without loss of generality we assume that $\mathcal{K}^0 \subset \mathcal{L}_+ \cap \mathcal{L}_-$.

Consider the factor-space (5). Let us note that the isotropic part $\mathcal{K}^0$ of $\mathcal{K}$ is invariant under all selfadjoint operators in $\mathcal{K}$, in particular, it is A- and B-invariant. Therefore the selfadjoint operators $\widehat{A}$ and $\widehat{B}$ in $\widehat{\mathcal{K}}$, generated by A and B, are well defined. Since $\widehat{A}\widehat{B} = \widehat{B}\widehat{A}$ and $\widehat{B} \in \mathbf{H}$, we have $\widehat{A} \in \mathbf{K}(\mathbf{H})$.

Denote $\widehat{\mathcal{L}}_\pm = \mathcal{L}_\pm/\mathcal{K}^0$. The dual pair $\{\widehat{\mathcal{L}}_+, \widehat{\mathcal{L}}_-\}$ is $\widehat{A}$-invariant. By [3, Corollary III.5.13] and [3, Theorem III.1.13] there is an $\widehat{A}$-invariant maximal dual pair $\{\widehat{\mathcal{L}}^+, \widehat{\mathcal{L}}^-\}$ in $\widehat{\mathcal{K}}$. Then $\{\mathcal{L}^+, \mathcal{L}^-\}$, where

$$\mathcal{L}^\pm = \{x \in \mathcal{K} \mid \hat{x} \in \widehat{\mathcal{L}}^\pm\},$$

is a desired A-invariant dual pair.

If $\{\mathcal{L}_+, \mathcal{L}_-\}$ is an A- and B-invariant dual pair then by [3, Theorem III.5.12], [3, Theorem III.1.13] and the same arguments as above there is an A- and B-invariant maximal dual pair which is an extension of $\{\mathcal{L}_+, \mathcal{L}_-\}$.

In particular, if $\mathcal{L}_+ = \mathcal{L}_- = \{0\}$, then, by the above statement, there exists an A- and B-invariant maximal dual pair $\{\mathcal{L}^+, \mathcal{L}^-\}$. $B \in \mathbf{H}$ implies $\mathcal{L}^\pm \in h^\pm$. □

From Definition 1 it follows that a space $\mathcal{K}$ with an indefinite inner product $[\cdot,\cdot]$ is an almost Krein space if and only if it admits a decomposition in a direct sum

$$\mathcal{K} = \mathcal{K}^0 \dotplus \mathcal{K}_1, \tag{6}$$

with finite-dimensional isotropic part $\mathcal{K}^0$ and a subspace $\mathcal{K}_1$ which is a Krein space with respect to $[\cdot,\cdot]$.

Let $\{\mathcal{L}^+, \mathcal{L}^-\}$ be an A-invariant maximal dual pair in $\mathcal{K}$. Then $\mathcal{L}^\pm = \mathcal{K}^0 \dotplus \mathcal{L}_1^\pm$, where $\mathcal{L}_1^\pm = \mathcal{L}^\pm \cap \mathcal{K}_1$. The dual pair $\{\mathcal{L}_1^+, \mathcal{L}_1^-\}$ is maximal in the Krein space $\mathcal{K}_1$. Let $\mathcal{L}_1^0 = \mathcal{L}_1^+ \cap \mathcal{L}_1^-$ and let $\mathcal{M}_1^0 \subset \mathcal{K}_1$ be a subspace skew-linked with $\mathcal{L}_1^0$. Then:

$$\mathcal{K} = \mathcal{K}^0 \dotplus \mathcal{L}_1^0 \dotplus \mathcal{L}_1^+ \dotplus \mathcal{L}_1^- \dotplus \mathcal{M}_1^0. \tag{7}$$

Introduce in $\mathcal{K}$ a Hilbert scalar product $(\cdot,\cdot)$ such that all subspaces in (7) are orthogonal to each others with respect to $(\cdot,\cdot)$ and $(\cdot,\cdot)|\mathcal{L}_1^{\pm} = \pm[\cdot,\cdot]|\mathcal{L}_1^{\pm}$. Then with respect to (7) the operator A has a triangular representation

$$A = \begin{bmatrix} A_{00} & A_{01} & A_{02} & A_{03} & A_{04} \\ 0 & A_{11} & A_{12} & A_{13} & A_{14} \\ 0 & 0 & A_{22} & 0 & A_{24} \\ 0 & 0 & 0 & A_{33} & A_{34} \\ 0 & 0 & 0 & 0 & A_{44} \end{bmatrix}, \tag{8}$$

where A_{0k}, $k = 0,1,2,3,4$, are bounded operators, A_{11} and A_{44}^* are similar, and A_{22} and A_{33} are selfadjoint with respect to the introduced scalar product. We omit information about other components of (8) since it is not essential in this paper.

Corollary 5. *The spectrum of a selfadjoint operator $A \in \mathbf{K}(\mathbf{H})$ in an almost Krein space is real, except for at most finite number of non-real normal eigenvalues.*

There are not more than a finite number of real eigenvalues λ of A with degenerate $\ker(A-\lambda)$. In particular, the set of eigenvalues with nontrivial Jordan chains is finite.

Proof. The triangular representation (8) implies $\sigma(A) = \bigcup_{k=0}^{4} \sigma(A_{kk})$. Since A_{22} and A_{33} are Hilbert space selfadjoint operators, the non-real spectrum of A coincides with the non-real spectra of A_{00}, A_{11} and A_{44}. Hence the statement about the non-real spectrum of A follows immediately from the finite dimensionality of $\mathcal{K}^0 + \mathcal{L}_1^0 + \mathcal{M}_1^0$.

Assume that A has an infinite number of real eigenvalues λ such that $\ker(A-\lambda)$ is degenerate. Let $\mathcal{L}$ be the closed linear span of the isotropic parts of these kernels. Let $B \in \mathbf{H}$ be an operator which commutes with A. Then $B\mathcal{L} \subset \mathcal{L}$. This contradicts the assumption $B \in \mathbf{H}$.

The last statement of the corollary follows from the fact that kernels with non-trivial Jordan chains are degenerate. □

4. Criteria for the completeness and basicity

In the sequel we will need the following lemma.

Lemma 6. *Assume $\mathcal{K}$ is an almost Krein space and $A \in \mathbf{K}(\mathbf{H})$ is a selfadjoint operator in $\mathcal{K}$ with $\sigma(A) = \{0\}$. Then A is a nilpotent operator of finite rank.*

Proof. We consider the decomposition (7) of $\mathcal{K}$ and the corresponding decomposition (8) of A. Note that $\sigma(A) = \{0\}$ implies

$$\sigma(A_{00}) = \sigma(A_{11}) = \sigma(A_{22}) = \sigma(A_{33}) = \sigma(A_{44}) = \{0\}$$

and the selfadjointness of A_{22} and A_{33} implies $A_{22} = 0$ and $A_{33} = 0$. □

Below we use the following notations:

- $\mathcal{L}_\lambda(A)$ is the root subspace of A related to the eigenvalue $\lambda:\ \lambda \in \sigma_p(A)$;
- $\mathfrak{E}(A)$ is the closed linear span (c.l.s.) of the root vectors of A: $\mathfrak{E}(A) = \text{c.l.s.}\,\{\mathcal{L}_\lambda(A) \mid \lambda \in \sigma_p(A)\}$;
- $\mathfrak{E}_0(A) = \text{c.l.s.}\,\{\ker(A-\lambda) \mid \lambda \in \sigma_p(A)\}$;
- $\mathfrak{E}_{\mathbb{R}\setminus 0}(A) = \text{c.l.s.}\,\{\mathcal{L}_\lambda(A) \mid \lambda \in \mathbb{R}\setminus\{0\}\}$.

A vector system $\{e_k\}$ is called almost orthonormal if it is a union of two systems $\{e_k\} = \{e'_k\} \cup \{e_k''\}$, where $\{e'_k\}$ is orthonormal: $[e'_k, e'_k] \neq 0$, $[e'_k, e'_j] = \delta_{kj}\,\text{sign}\,[e'_k, e'_k]$, and $\{e_k''\}$ is a finite system with $[e'_k, e_j''] = 0$.

Corollary 7. *Let $A \in \mathbf{K}(\mathbf{H})$ be a compact selfadjoint operator in a Krein space $\mathcal{K}$. Then $\mathfrak{E}(A) = \mathcal{K}$ if and only if $\mathcal{L}_0(A) \cap \mathfrak{E}_{\mathbb{R}\setminus 0}(A) = \{0\}$. Moreover, if $\mathfrak{E}(A) = \mathcal{K}$, there exists in $\mathcal{K}$ an almost orthonormal Riesz basis composed of root vectors of A.*

Proof. Indeed, let $\mathcal{L} = \mathcal{L}_0(A) \cap \mathfrak{E}_{\mathbb{R}\setminus 0}(A)$. Then $\mathcal{L}$ is orthogonal to all root subspaces of A. If $\mathfrak{E}(A) = \mathcal{K}$, we have the orthogonality of $\mathcal{L}$ to the whole Krein space which is non-degenerate. Hence $\mathcal{L} = \{0\}$.

Assume $\mathcal{L} = \{0\}$. Let $B \in \mathbf{H}$ and $AB = BA$. Since $\mathfrak{E}_{\mathbb{R}\setminus 0}(A)$ is A- and B-invariant, the isotropic part $\mathfrak{E}_{\mathbb{R}\setminus 0}(A)^0$ of $\mathfrak{E}_{\mathbb{R}\setminus 0}(A)$ is also A- and B-invariant. Hence, taking into account that $B \in \mathbf{H}$ and $\mathfrak{E}_{\mathbb{R}\setminus 0}(A)^0$ is B-invariant neutral subspace, we obtain from Theorem 4 that $\dim \mathfrak{E}_{\mathbb{R}\setminus 0}(A)^0 < \infty$. Since the root subspaces $\mathcal{L}_\lambda(A)$, $\lambda \neq 0$ are non-degenerate, we have $\mathfrak{E}_{\mathbb{R}\setminus 0}(A)^0 \subset \mathcal{L}_0(A)$, and hence $\mathfrak{E}_{\mathbb{R}\setminus 0}(A)^0 \subset \mathcal{L}$, that is, $\mathfrak{E}_{\mathbb{R}\setminus 0}(A)^0 = \{0\}$. Therefore $\mathfrak{E}_{\mathbb{R}\setminus 0}(A)$ is a Krein space. The spectrum of the restriction of A to the orthogonal complement $\mathfrak{E}_{\mathbb{R}\setminus 0}(A)^{[\perp]}$ to $\mathfrak{E}_{\mathbb{R}\setminus 0}(A)$ consists of a unique point $\lambda = 0$. Hence, it follows from Lemma 6 that $\mathcal{L}_0(A) = \mathfrak{E}_{\mathbb{R}\setminus 0}(A)^{[\perp]}$. The latter implies $\mathfrak{E}(A) = \mathcal{K}$.

The existence of an almost orthonormal basis is proved in [3, Thm. IV.2.12]. □

Below we give necessary and sufficient conditions for the basicity and completeness of the set of root vectors of a compact selfadjoint operator in an almost Krein space. We show that in almost Krein spaces, in contrast to the Pontryagin or Krein space case, the completeness of the root vector system is not sufficient for its basicity.

Theorem 8. *Let $A \in \mathbf{K}(\mathbf{H})$ be a compact selfadjoint operator in an almost Krein space $\mathcal{K}$ and $\mathfrak{E}(A) = \mathcal{K}$. Then the following assertions are equivalent:*

(i) $\mathcal{K}^0 \cap \mathcal{L}_0(A) \cap \mathfrak{E}_{\mathbb{R}\setminus 0}(A) = \{0\}$;

(ii) $\mathcal{L}_0(A) \cap \mathfrak{E}_{\mathbb{R}\setminus 0}(A) = \{0\}$;

(iii) *there exists in $\mathcal{K}$ an almost orthonormal Riesz basis consisting of root vectors of A.*

Proof. Corollary 5 implies that the non-real spectrum of A and the real nonzero eigenvalues $\lambda \in \sigma_p(A)$ with degenerate kernels $\ker(A-\lambda)$ is a finite subset of the normal eigenvalues. Let P be the Riesz projector related to this part of $\sigma(A)$. Since

$\dim P\mathcal{K} < \infty$ and $P\mathcal{K}$ is both A- and B-invariant, where $B \in \mathbf{H}$ commutes with A, we have that $A|(I-P)\mathcal{K}$ satisfies all conditions of the theorem and the basicity property for the root systems of A and $A|(I-P)\mathcal{K}$ is equivalent. Moreover,

$$\mathcal{K}^0 \cap \mathcal{L}_0(A) \cap \mathfrak{E}_{\mathbb{R}\setminus 0}(A) = (I-P)\mathcal{K}^0 \cap \mathcal{L}_0(A|(I-P)\mathcal{K}) \cap \mathfrak{E}_{\mathbb{R}\setminus 0}(A|(I-P)\mathcal{K})$$

and

$$\mathcal{L}_0(A) \cap \mathfrak{E}_{\mathbb{R}\setminus 0}(A) = \{0\} \Longleftrightarrow \mathcal{L}_0(A|(I-P)\mathcal{K}) \cap \mathfrak{E}_{\mathbb{R}\setminus 0}(A|(I-P)\mathcal{K}) = \{0\}.$$

Hence without loss of generality we can assume $P = 0$, that is, both the set of the non-real eigenvalues and the set of nonzero real eigenvalues with degenerate kernels are empty.

(i) $\Rightarrow$ (ii). Let $\mathcal{K}^0 \cap \mathcal{L}_0(A) \cap \mathfrak{E}_{\mathbb{R}\setminus 0}(A) = \{0\}$. Then $\mathfrak{E}_{\mathbb{R}\setminus 0}(A)$ is a nondegenerate subspace. Indeed, let us suppose the opposite, i.e., that this subspace is degenerate and its isotropic part $\mathfrak{E}_{\mathbb{R}\setminus 0}(A)^0$ is nonzero. The subspace $\mathfrak{E}_{\mathbb{R}\setminus 0}(A)^0$ is finite-dimensional and A-invariant. Therefore, there exists an eigenvalue λ_0 and a corresponding eigenvector $x_0 \in \mathfrak{E}_{\mathbb{R}\setminus 0}(A)^0$. Since all kernels $\ker(A-\lambda)$ with $\lambda \neq 0$ are nondegenerate, we obtain $\lambda_0 = 0$. So, $x_0 \in \ker A$, it is orthogonal to $\mathcal{L}_0(A)$ and thus is orthogonal to $\mathfrak{E}(A)$. Hence, $x_0 \in \mathcal{K}^0 \cap \mathcal{L}_0(A) \cap \mathfrak{E}_{\mathbb{R}\setminus 0}(A)$, and $x_0 = 0$. This contradicts the assumption $x \neq 0$. We have that $\mathfrak{E}_{\mathbb{R}\setminus 0}(A)$ is nondegenerate and therefore $\mathcal{L}_0(A) \cap \mathfrak{E}_{\mathbb{R}\setminus 0}(A) = \{0\}$.

(ii) $\Rightarrow$ (i) is trivial.

(ii) $\Rightarrow$ (iii). According to the assumption all eigenvalues of A are real and all $\ker(A-\lambda)$, $\lambda \neq 0$, are nondegenerate. Hence the subspace $\mathfrak{E}_{\mathbb{R}\setminus 0}(A)$ is nondegenerate. Let us prove that this subspace is a Krein space. Really, it is B-invariant, where $B \in \mathbf{H}$ is an operator commuting with A. In each subspace $\ker(A-\lambda)$ the operator B has an invariant maximal dual pair $\{\mathcal{L}_\lambda^+, \mathcal{L}_\lambda^-\}$, and $\{\mathcal{L}^+, \mathcal{L}^-\}$ with $\mathcal{L}^\pm = \text{c.l.s.}\{\mathcal{L}_\lambda^\pm \mid \lambda \in \sigma(A)\}$, is a B-invariant maximal dual pair in $\mathfrak{E}_{\mathbb{R}\setminus 0}(A)$. The assumption $B \in \mathbf{H}$ implies $\mathcal{L}^\pm \in h^\pm$. Hence $\mathcal{L}^+ + \mathcal{L}^-$ is an almost Krein space. By construction, the defect of this subspace in $\mathfrak{E}_{\mathbb{R}\setminus 0}(A)$ is finite. Hence $\mathfrak{E}_{\mathbb{R}\setminus 0}(A)$ is also an almost Krein space and, by assumption, it is nondegenerate. So, $\mathfrak{E}_{\mathbb{R}\setminus 0}(A)$ is a Krein space. Therefore the orthogonal decomposition

$$\mathcal{K} = \mathfrak{E}_{\mathbb{R}\setminus 0}(A)[+]\mathfrak{E}_{\mathbb{R}\setminus 0}(A)^{[\perp]}$$

holds. Since $B \in \mathbf{H}$ has in $\mathfrak{E}_{\mathbb{R}\setminus 0}(A)$ an invariant maximal dual pair, we have $B|\mathfrak{E}_{\mathbb{R}\setminus 0}(A) \in \mathbf{H}$ and $B|\mathfrak{E}_{\mathbb{R}\setminus 0}(A)^{[\perp]} \in \mathbf{H}$. Hence,

$$A|\mathfrak{E}_{\mathbb{R}\setminus 0}(A) \in \mathbf{K}(\mathbf{H}) \quad \text{and} \quad \mathbf{A}|\mathfrak{E}_{\mathbb{R}\setminus \mathbf{0}}(\mathbf{A})^{[\perp]} \in \mathbf{K}(\mathbf{H}).$$

By Lemma 6, the subspace $\mathfrak{E}_{\mathbb{R}\setminus 0}(A)^{[\perp]} \supset \mathcal{L}_0(A)$ coincides with $\mathcal{L}_0(A|\mathfrak{E}_{\mathbb{R}\setminus 0}(A)^{[\perp]}) \subset \mathcal{L}_0(A)$. This implies $\mathfrak{E}_{\mathbb{R}\setminus 0}(A)^{[\perp]}) = \mathcal{L}_0(A)$ and

$$\mathcal{K} = \mathfrak{E}_{\mathbb{R}\setminus 0}(A)[+]\mathcal{L}_0(A). \tag{9}$$

Since $\mathfrak{E}_{\mathbb{R}\setminus 0}(A)$ is a Krein space and $A|\mathfrak{E}_{\mathbb{R}\setminus 0}(A) \in \mathbf{K}(\mathbf{H})$ it follows from [3, Theorem IV.2.12] that there exists in $\mathfrak{E}_{\mathbb{R}\setminus 0}(A)$ an orthonormal Riesz basis constructed from eigenvectors of A. If we add to this basis an arbitrary almost orthonormal Riesz basis composed of vectors of $\mathcal{L}_0(A)$, we obtain desired basis in $\mathcal{K}$.

(iii) $\Rightarrow$ (ii). Assume there exists an almost orthonormal Riesz basis in $\mathcal{K}$ composed of root vectors of A. Let $\{e_k\}$ be the part of this basis contained in $\mathfrak{E}_{\mathbb{R}\setminus 0}(A)$. Using the same arguments as above we can assume, without loss of generality, that all vectors e_k are definite and orthogonal to each other. If $x_0 \in \mathcal{L}_0(A) \cap \mathfrak{E}_{\mathbb{R}\setminus 0}(A)$, then $x_0 \in$ c.l.s. $\{e_k\} : x_0 = \sum \alpha_k e_k$. On the other hand it is orthogonal to all these vectors. Hence, $\alpha_k = [x_0, e_k] \operatorname{sign} [e_k, e_k] = 0$, that is, $x_0 = 0$. □

Remark 9. In the proof of Theorem 8 we have used the assumption that the root vector system of A is complete only in the proof of the implication (i) $\Rightarrow$ (ii).

We can rewrite the assumptions (i) and (ii) as $0 \notin \sigma_p(A|\mathcal{K}^0 \cap \mathfrak{E}_{\mathbb{R}\setminus 0})$ and $0 \notin \sigma_p(A|\mathcal{L}_0(A) \cap \mathfrak{E}_{\mathbb{R}\setminus 0})$, respectively.

Theorem 10. *Let $A \in \mathbf{K}(\mathbf{H})$ be a compact selfadjoint operator in an almost Krein space $\mathcal{K}$. Then*

$$\{\mathfrak{E}(A) = \mathcal{K}\} \iff \{\mathcal{L}_0(A) \cap \mathfrak{E}_{\mathbb{R}\setminus 0}(A) \subset \mathcal{K}^0\}. \tag{10}$$

Proof. $\{\mathfrak{E}(A) = \mathcal{K}\} \Longrightarrow \{\mathcal{L}_0(A) \cap \mathfrak{E}_{\mathbb{R}\setminus 0}(A) \subset \mathcal{K}^0\}$.

Let $x_0 \in \mathcal{L}_0(A) \cap \mathfrak{E}_{\mathbb{R}\setminus 0}(A)$. Since $\mathcal{L}_0(A)$ is orthogonal to c.l.s. $\{\mathcal{L}_\lambda(A) \mid \lambda \neq 0\}$, the vector x_0 is orthogonal to $\mathfrak{E}(A)$, that is, $x_0 \in \mathcal{K}^0$.

$\{\mathfrak{E}(A) = \mathcal{K}\} \Longleftarrow \{\mathcal{L}_0(A) \cap \mathfrak{E}_{\mathbb{R}\setminus 0}(A) \subset \mathcal{K}^0\}$.

With respect to decomposition (6) the operator A admits the following matrix representation:

$$A = \begin{bmatrix} A_{00} & A_{01} \\ 0 & A_{11} \end{bmatrix}. \tag{11}$$

Consider also the operator $\widehat{A}$ induced by the operator A in the factor space $\widehat{\mathcal{K}} = \mathcal{K}/\mathcal{K}^0$. It follows from $\mathcal{L}_0(A) \cap \mathfrak{E}_{\mathbb{R}\setminus 0}(A) \subset \mathcal{K}^0$ that $\mathcal{L}_0(\widehat{A}) \cap \mathfrak{E}_{\mathbb{R}\setminus 0}(\widehat{A}) = \{0\}$. By Theorem 8, there is in $\widehat{\mathcal{K}}$ a Riesz basis composed of root vectors of $\widehat{A}$. Since the operators $\widehat{A}$ and A_{11} are similar, there is in $\mathcal{K}_1$ a Riesz basis composed of root vectors of A_{11}. Hence, $\mathfrak{E}(A_{11}) = \mathcal{K}_1$. It remains to use that $\mathfrak{E}(A) = \mathcal{K}^0[+]\mathfrak{E}(A_{11}) = \mathcal{K}$ (see (11)). □

5. Keldysh type operators

Let $\mathcal{H}$ be a Hilbert space with the scalar product $(\cdot, \cdot)$, let A and B be compact selfadjoint operators acting in this space and $A > 0$. Consider an operator H as in (3) and introduce in $\mathcal{H}$ an inner product $[\cdot, \cdot]$ given by

$$[\cdot, \cdot] = ((I + B)\cdot, \cdot).$$

Denote $\mathcal{K} = \{\mathcal{H}, [\cdot, \cdot]\}$. Then $\mathcal{K}$ is an almost Krein space, moreover, $\mathcal{K}$ is an almost Pontryagin space. The operator H is selfadjoint in $\mathcal{K}$. Thus, we can apply to H Theorems 8 and 10.

Theorem 11. *Let an operator H as in (3) satisfy the above-mentioned properties. Then*

(a) $\mathfrak{E}(H) = \mathcal{K}$;

(b) *There is an almost orthonormal Riesz basis in the almost Krein space $\mathcal{K}$ composed of root vectors of H if and only if one of the following equivalent assumptions holds.*

(b1) $\dim\ker A(I+B) = \dim\ker(I+B)A$;

(b2) $\ker(I+B) \subset \operatorname{ran} A$.

Proof. **(a)** The isotropic part $\mathcal{K}^0$ of the almost Krein space $\mathcal{K}$ coincides with $\ker(I+B)$. Since $A > 0$, we have $\mathcal{L}_0(H) = \ker H = \ker(I+B)$. Hence, $\mathcal{L}_0(H) \cap \mathfrak{E}_{\mathbb{R}\setminus 0}(H) \subset \ker(I+B) = \mathcal{K}^0$. Now **(a)** follows directly from Theorem 10.

(b) Let us note that the assumptions **(b1)** and **(b2)** are equivalent since the operator A has a trivial kernel.

Suppose that there exists an almost orthonormal Riesz basis in $\mathcal{K}$ composed of root vectors of H. By Theorem 8(ii) we have

$$\mathcal{K} = \ker H[+]\mathfrak{E}_{\mathbb{R}\setminus 0}(H). \tag{12}$$

Since the operator H is diagonal with respect to (12) and the operator $H|\mathfrak{E}_{\mathbb{R}\setminus 0}(H)$ is a Pontryagin space selfadjoint, $H|\mathfrak{E}_{\mathbb{R}\setminus 0}(H)$ is similar to its Hilbert space adjoint. Hence $H = A(I+B)$ and its Hilbert space adjoint $H^* = (I+B)A$ are similar too. This implies **(b1)**.

Assume **(b2)**. Consider the linear pencil

$$L(\lambda) = A^{-1} - \lambda(I+B). \tag{13}$$

Since the set of root vectors of this pencil and of the operator H are the same, it is sufficient to check that there is in $\mathcal{K}$ an almost orthonormal Riesz basis composed of root vectors of the pencil (13). Consider the orthogonal decomposition of $\mathcal{H}$:

$$\mathcal{H} = \ker(I+B) \oplus \operatorname{ran}(I+B). \tag{14}$$

Let us rewrite the operator A^{-1} and the pencil (13) in the matrix form with respect to the decomposition (14):

$$A^{-1} = \begin{bmatrix} C_{00} & C_{01} \\ C_{10} & C_{11} \end{bmatrix},$$

$$L(\lambda) = \begin{bmatrix} C_{00} & C_{01} \\ C_{10} & C_{11} - \lambda(I+B_{11}) \end{bmatrix}. \tag{15}$$

From **(b2)** it follows that C_{00} is a finite-dimensional operator and positive with respect to the Hilbert space scalar product, that C_{01} and C_{10} are bounded operators of finite ranks and that the operator C_{11} is positive with respect to the Hilbert space scalar product and has a compact inverse. Hence, $x = \begin{pmatrix} x_0 \\ x_1 \end{pmatrix}$ is an eigenvector of (15), related to $\lambda \neq 0$, if and only if, $x_0 = -C_{00}^{-1}C_{01}x_1$, where x_1 is an eigenvector of the pencil

$$L_1(\lambda) = C_{11} - C_{10}C_{00}^{-1}C_{01} - \lambda(I + B_{11}), \tag{16}$$

corresponding to the same eigenvalue. As in the proof of Theorem 8 we assume without loss of generality that the root subspaces and kernels corresponding to the same eigenvalue coincides. Since $C_{11} - C_{10}C_{00}^{-1}C_{01}$ is a Hilbert space positive operator in $\operatorname{ran}(I + B)$, its inverse is compact, and $\ker(I + B_{11}) = \{0\}$, the sets of eigenvectors of (16) coincide with the sets of eigenvectors of the operator

$$H_1 = (C_{11} - C_{10}C_{00}^{-1}C_{01})^{-1}(I + B_{11}). \tag{17}$$

Consider $\operatorname{ran}(I + B)$ as a Pontryagin space $\mathcal{K}_1$ with indefinite inner product $[\cdot,\cdot] = ((I + B_{11})\cdot,\cdot)$. Because the compact operator H_1 is positive in $\mathcal{K}_1$, the assumption (ii) of Theorem 8 is fulfilled. Hence, there is an almost orthonormal Riesz basis in $\mathcal{K}_1$ composed of eigenvectors of H_1. Taking into account the relation between eigenvectors of H_1 and H and the equality $\ker(I + B) = \ker H$, we obtain the existence of an almost orthonormal Riesz basis in $\mathcal{K}$ composed of eigenvectors vectors of H. The latter is true if there are no associated vectors. In the general case, there is an almost orthonormal Riesz basis in $\mathcal{K}$ composed of root vectors of H. □

References

[1] Azizov T.Ya. *On completely continuous operators that are selfadjoint with respect to a degenerate indefinite metric*, Matem. issled., 7 (1972), 4, 237–240. (Russ.)

[2] Azizov T.Ya., Iokhvidov I.S. *A criterion of the completeness and basicity of root vectors of a completely continuous J-selfadjoint operator in a Pontryagin space Π_κ.* Matem. issled., 6 (1971), 1, 158–161. (Russ.)

[3] T.Ya. Azizov, I.S. Iokhvidov, *Foundation of the theory of linear operators in spaces with an indefinite metric*, Nauka, Moscow, 1986 (Russ.); English transl.: *Linear operators in spaces with an indefinite metric*, Wiley, New York, 1989.

[4] P. Binding and R. Hryniv. *Full and partial range completeness.* Oper. Theory Adv. Appl. 130:121–133, 2002.

[5] Vladimir Bolotnikov, Chi-Kwong Li, Patrick Meade, Christian Mehl, Leiba Rodman. *Shells of matrices in indefinite inner product spaces.* Electron. J. Linear Algebra, 9: 67–92, 2002.

[6] I.Ts. Gokhberg, M.G. Krein. *Introduction to the theory of linear non-selfadjoint operators in a Hilbert space.* Nauka, Moscow, 1965 (Russian).

[7] M. Kaltenbäck and H. Woracek. *Selfadjoint extensions of symmetric operators in degenerated inner product spaces.* Integral Equations Operator Theory, 28 (1997), 289–320.

[8] P. Lancaster, A.S. Markus, and P. Zizler. *The order of neutrality for linear operators on inner product spaces.* LAA 259 (1997), 25–29.

[9] H. Langer, R. Mennicken, and C. Tretter. *A self-adjoint linear pencil $Q - \lambda P$ of ordinary differential operators*. Methods Funct. anal. Topology, 2 (1996), 38–54.

[10] A. Luger. *A factorization of regular generalized Nevanlinna functions*. Integral Equations Operator Theory 43: 326–345, 2002.

[11] Christian Mehl, Leiba Rodman. *Symmetric matrices with respect to sesquilinear forms*. Linear Algebra Appl., 349: 55–75, 2002.

[12] Christian Mehl, Andre C.M. Ran, Leiba Rodman. *Semidefinite invariant subspaces: degenerate inner products*, to appear in Oper. Theory Adv. Appl.

[13] H. Woracek. *Resolvent matrices in degenerate inner product spaces*. Math. Nachr. 213 (2000), 155–175.

Tomas Ya. Azizov and Lioudmila I. Soukhotcheva
Department of Mathematics
Voronezh State University
Universitetskaya pl., 1 394022,
Voronezh, Russia
e-mail: azizov@math.vsu.ru
e-mail: lsuh@math.vsu.ru

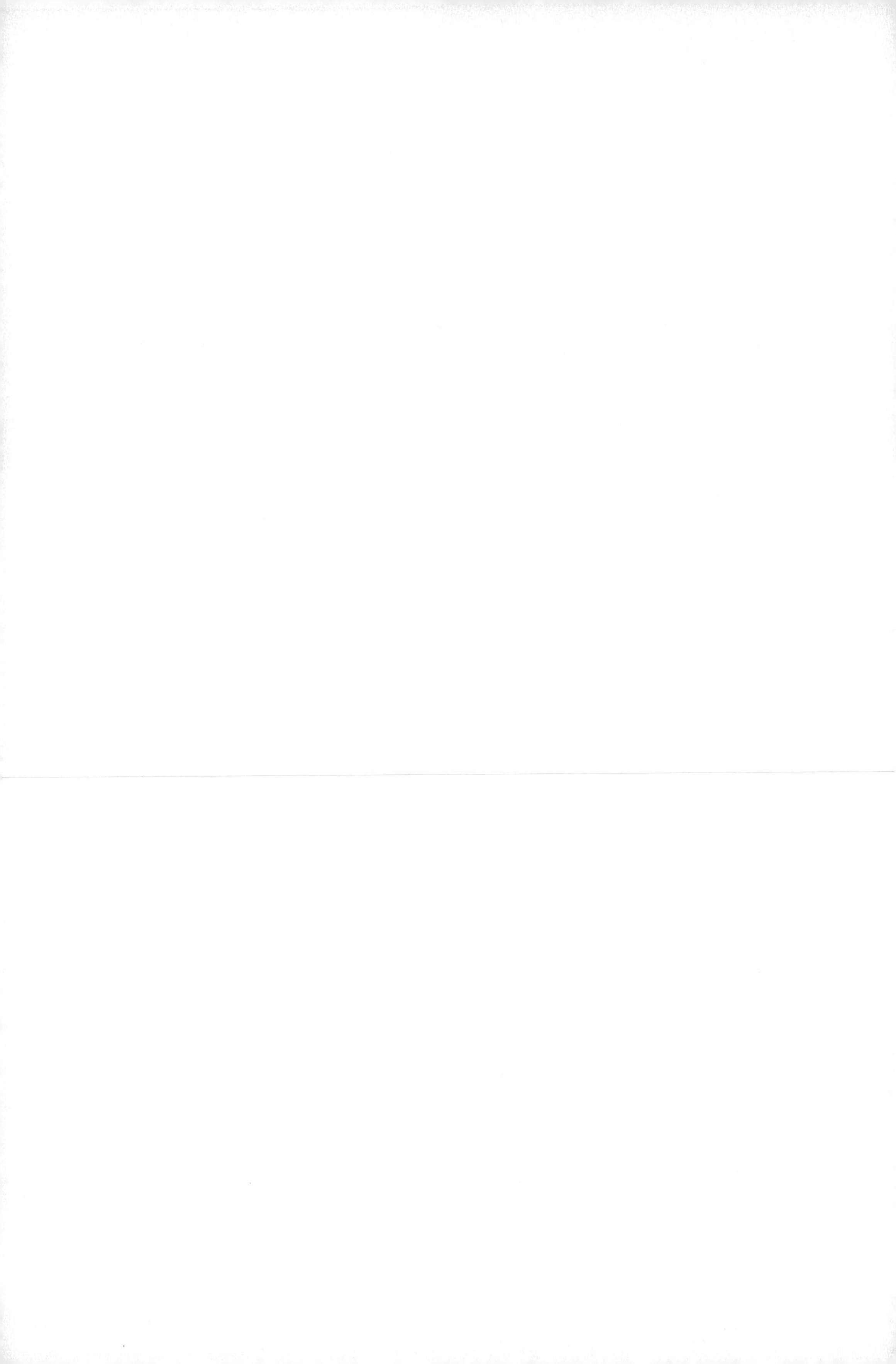

Operator Theory:
Advances and Applications, Vol. 175, 13–32

Generalized Resolvents of a Class of Symmetric Operators in Krein Spaces

Jussi Behrndt, Annemarie Luger and Carsten Trunk

Abstract. Let A be a closed symmetric operator of defect one in a Krein space $\mathcal{K}$ and assume that A possesses a self-adjoint extension in $\mathcal{K}$ which locally has the same spectral properties as a definitizable operator. We show that the Krein-Naimark formula establishes a bijective correspondence between the compressed resolvents of locally definitizable self-adjoint extensions $\widetilde{A}$ of A acting in Krein spaces $\mathcal{K}\times\mathcal{H}$ and a special subclass of meromorphic functions.

Mathematics Subject Classification (2000). Primary: 47B50; Secondary: 47B25.

Keywords. Generalized resolvents, Krein-Naimark formula, self-adjoint extensions, locally definitizable operators, locally definitizable functions, boundary value spaces, Weyl functions, Krein spaces.

1. Introduction

Let A be a densely defined closed symmetric operator with defect one in a Hilbert space $\mathcal{K}$ and let $\{\mathbb{C},\Gamma_0,\Gamma_1\}$ be a boundary value space for the adjoint operator A^*. Let A_0 be the self-adjoint extension $A^*\upharpoonright\ker\Gamma_0$ of A in $\mathcal{K}$ and denote the γ-field and Weyl function corresponding to the boundary value space $\{\mathbb{C},\Gamma_0,\Gamma_1\}$ by γ and M, respectively. Here M is a scalar Nevanlinna function, that is, it maps the upper half-plane $\mathbb{C}^+$ holomorphically into $\mathbb{C}^+\cup\mathbb{R}$ and is symmetric with respect to the real axis. It is well known that in this case the Krein-Naimark formula

$$P_{\mathcal{K}}(\widetilde{A}-\lambda)^{-1}|_{\mathcal{K}}=(A_0-\lambda)^{-1}-\gamma(\lambda)\big(M(\lambda)+\tau(\lambda)\big)^{-1}\gamma(\overline{\lambda})^* \tag{1.1}$$

establishes a bijective correspondence between the class of Nevanlinna functions τ (including the constant ∞) and the compressed resolvents of self-adjoint extensions $\widetilde{A}$ of A in $\mathcal{K}\times\mathcal{H}$, where $\mathcal{H}$ is a Hilbert space, cf. [22, 32]. The compressed resolvent on the left-hand side of (1.1) is said to be a generalized resolvent of A. We note that if A has equal deficiency indices >1 the generalized resolvents of A can still be described with formula (1.1), where the parameters τ are so-called Nevanlinna families, cf. [11, 23, 30, 31].

Various generalizations of the Krein-Naimark formula in an indefinite setting have been proved in the last decades. The case that A is a symmetric operator in a Pontryagin space $\mathcal{K}$ and $\mathcal{H}$ is a Hilbert space was investigated by M.G. Krein and H. Langer in [24]. Later V. Derkach considered both $\mathcal{K}$ and $\mathcal{H}$ to be Pontryagin or even Krein spaces, cf. [10]. In the general situation of Krein spaces $\mathcal{K}$ and $\mathcal{H}$ one obtains a correspondence between locally holomorphic relation-valued functions τ and self-adjoint extensions $\widetilde{A}$ of A with a non-empty resolvent set. Under additional assumptions other variants of (1.1) were proved in [6, 7, 8, 9, 10, 14, 27].

If, e.g., $\mathcal{H}$ is a Pontryagin space, then the parameters τ belong to the class of $\mathcal{N}_\kappa$-families, a class of relation-valued functions which includes the generalized Nevanlinna functions. If $\mathcal{H}$ is a Krein space and the hermitian forms $[A\cdot,\cdot]$ and $[\widetilde{A}\cdot,\cdot]$ both have finitely many negative squares, then τ belongs to a special subclass of the definitizable functions, cf. [7, 19].

It is the aim of this paper to prove a new variant of formula (1.1). Here we allow both $\mathcal{K}$ and $\mathcal{H}$ to be Krein spaces and we assume that A is of defect one and possesses a self-adjoint extension A_0 in $\mathcal{K}$ which locally has the same spectral properties as a definitizable operator or relation, cf. [20, 28]. Under the assumption that $\widetilde{A}$ is also locally definitizable and that its sign types coincide "in essence" (i.e., with the exception of a discrete set, see Definition 2.6) with the sign types of A_0 we prove in Theorem 3.2 that there exists a so-called locally definitizable function τ such that (1.1) holds. The proof is based on a coupling method developed in [11, §5] and a recent perturbation result from [4].

One of the main difficulties here is to show that the symmetric relation $\widetilde{A}\cap\mathcal{H}^2$ possesses a self-adjoint extension in the Krein space $\mathcal{H}$ with a non-empty resolvent set and to choose a boundary value space for the adjoint of $\widetilde{A}\cap\mathcal{H}^2$ in $\mathcal{H}$ such that (1.1) holds with the corresponding Weyl function τ. In connection with a class of abstract λ-dependent boundary value problems the converse direction was already proved in [3], i.e., for a given locally definitizable function τ a self-adjoint extension $\widetilde{A}$ of A in $\mathcal{K}\times\mathcal{H}$ such that (1.1) holds was constructed.

The paper is organized as follows:

In Section 2 we recall the definitions and basic properties of locally definitizable self-adjoint operators and relations and the class of locally definitizable functions introduced and studied by P. Jonas, see, e.g., [20, 21]. The notion of d-compatibility of sign types of locally definitizable relations and functions is defined in the end of Section 2.3. In the beginning of Section 3 we recall some basics on boundary value spaces and associated Weyl functions.

Section 3.2 contains our main result. We prove in Theorem 3.2 that formula (1.1) establishes a bijective correspondence between an appropriate subclass of the locally definitizable functions and the compressed resolvents of locally definitizable $\mathcal{K}$-minimal self-adjoint exit space extensions $\widetilde{A}$ of A in a Krein space $\mathcal{K}\times\mathcal{H}$ with spectral sign types d-compatible to those of A_0. Finally, in the end of Section 3.2, we formulate a variant of the Krein-Naimark formula for self-adjoint extensions A_0 and $\widetilde{A}$ of A in $\mathcal{K}$ and $\mathcal{K}\times\mathcal{H}$, respectively, which locally have the same spectral properties as self-adjoint operators or relations in Pontryagin spaces and functions τ from the local generalized Nevanlinna class.

2. Locally definitizable self-adjoint relations and locally definitizable functions

2.1. Notations and definitions

Let $(\mathcal{K}, [\cdot,\cdot])$ be a separable Krein space with a corresponding fundamental symmetry J. The linear space of bounded linear operators defined on a Krein space $\mathcal{K}_1$ with values in a Krein space $\mathcal{K}_2$ is denoted by $\mathcal{L}(\mathcal{K}_1, \mathcal{K}_2)$. If $\mathcal{K} := \mathcal{K}_1 = \mathcal{K}_2$ we simply write $\mathcal{L}(\mathcal{K})$. We study linear relations in $\mathcal{K}$, that is, linear subspaces of $\mathcal{K}^2$. The set of all closed linear relations in $\mathcal{K}$ is denoted by $\widetilde{\mathcal{C}}(\mathcal{K})$. Linear operators in $\mathcal{K}$ are viewed as linear relations via their graphs. For the usual definitions of the linear operations with relations, the inverse etc., we refer to [15] and [16].

The sum and the direct sum of subspaces in $\mathcal{K}^2$ are denoted by $\boldsymbol{+}$ and $\dot{\boldsymbol{+}}$. We define an indefinite inner product on $\mathcal{K}^2$ by

$$[\![\hat{f}, \hat{g}]\!] := i([f, g'] - [f', g]), \quad \hat{f} = \begin{pmatrix} f \\ f' \end{pmatrix}, \; \hat{g} = \begin{pmatrix} g \\ g' \end{pmatrix} \in \mathcal{K}^2.$$

Then $(\mathcal{K}^2, [\![\cdot,\cdot]\!])$ is a Krein space and $\mathcal{J} = \left(\begin{smallmatrix} 0 & -iJ \\ iJ & 0 \end{smallmatrix}\right) \in \mathcal{L}(\mathcal{K}^2)$ is a corresponding fundamental symmetry. If necessary we will indicate the underlying space by subscripts, e.g., $[\![\cdot,\cdot]\!]_{\mathcal{K}^2}$.

Let A be a linear relation in $\mathcal{K}$. The adjoint relation $A^+ \in \widetilde{\mathcal{C}}(\mathcal{K})$ is defined as

$$A^+ := A^{[\![\perp]\!]} = \{\hat{h} \in \mathcal{K}^2 \mid [\![\hat{h}, \hat{f}]\!] = 0 \text{ for all } \hat{f} \in A\},$$

where $A^{[\![\perp]\!]}$ denotes the orthogonal companion of A with respect to $[\![\cdot,\cdot]\!]$. A is said to be *symmetric* (*self-adjoint*) if $A \subset A^+$ (resp. $A = A^+$).

Let S be a closed linear relation in $\mathcal{K}$. The resolvent set $\rho(S)$ of $S \in \widetilde{\mathcal{C}}(\mathcal{K})$ is the set of all $\lambda \in \mathbb{C}$ such that $(S - \lambda)^{-1} \in \mathcal{L}(\mathcal{K})$, the spectrum $\sigma(S)$ of S is the complement of $\rho(S)$ in $\mathbb{C}$. The extended spectrum $\widetilde{\sigma}(S)$ of S is defined by $\widetilde{\sigma}(S) = \sigma(S)$ if $S \in \mathcal{L}(\mathcal{K})$ and $\widetilde{\sigma}(S) = \sigma(S) \cup \{\infty\}$ otherwise. A point $\lambda \in \mathbb{C}$ is called a *point of regular type* of S, $\lambda \in r(S)$, if $(S - \lambda)^{-1}$ is a bounded operator. We say that $\lambda \in \mathbb{C}$ belongs to the *approximate point spectrum of* S, denoted by $\sigma_{\mathrm{ap}}(S)$, if there exists a sequence $\left(\begin{smallmatrix} x_n \\ y_n \end{smallmatrix}\right) \in S$, $n = 1, 2, \ldots$, such that $\|x_n\| = 1$ and $\lim_{n\to\infty} \|y_n - \lambda x_n\| = 0$. The *extended approximate point spectrum* $\widetilde{\sigma}_{\mathrm{ap}}(S)$ *of* S is defined by

$$\widetilde{\sigma}_{\mathrm{ap}}(S) := \begin{cases} \sigma_{\mathrm{ap}}(S) \cup \{\infty\} & \text{if } \; 0 \in \sigma_{\mathrm{ap}}(S^{-1}) \\ \sigma_{\mathrm{ap}}(S) & \text{if } \; 0 \notin \sigma_{\mathrm{ap}}(S^{-1}) \end{cases}.$$

We remark, that the boundary points of $\widetilde{\sigma}(S)$ in $\overline{\mathbb{C}}$ belong to $\widetilde{\sigma}_{\mathrm{ap}}(S)$.

Next we recall the definitions of the spectra of positive and negative type of self-adjoint relations, cf. [20] (for bounded self-adjoint operators see [29]). For equivalent descriptions of the spectra of positive and negative type we refer to [20, Theorem 3.18].

Definition 2.1. Let A_0 be a self-adjoint relation in $\mathcal{K}$. A point $\lambda \in \sigma_{ap}(A_0)$ is said to be of *positive type* (*negative type*) with respect to A_0, if for every sequence $\binom{x_n}{y_n} \in A_0$, $n = 1, 2, \ldots$, with $\|x_n\| = 1$, $\lim_{n\to\infty} \|y_n - \lambda x_n\| = 0$ we have

$$\liminf_{n\to\infty} [x_n, x_n] > 0 \quad \Big(\text{resp. } \limsup_{n\to\infty} [x_n, x_n] < 0\Big).$$

If $\infty \in \widetilde{\sigma}_{ap}(A_0)$, then ∞ is said to be of *positive type* (*negative type*) with respect to A_0 if 0 is of positive type (resp. negative type) with respect to A_0^{-1}. We denote the set of all points of $\widetilde{\sigma}_{ap}(A_0)$ of positive type (negative type) by $\sigma_{++}(A_0)$ (resp. $\sigma_{--}(A_0)$).

We remark that the self-adjointness of the relation A_0 yields that the points of positive and negative type introduced in Definition 2.1 belong to $\overline{\mathbb{R}}$.

An open subset Δ of $\overline{\mathbb{R}}$ is said to be of *positive type* (*negative type*) with respect to A_0 if each point $\lambda \in \Delta \cap \widetilde{\sigma}(A_0)$ is of positive type (resp. negative type) with respect to A_0. An open subset Δ of $\overline{\mathbb{R}}$ is called of *definite type* with respect to A_0 if it is either of positive or of negative type with respect to A_0.

For each $\lambda \in \sigma_{++}(A_0)$ $(\sigma_{--}(A_0))$ there exists an open neighborhood $\mathcal{U}_\lambda$ in $\overline{\mathbb{C}}$ such that $(\mathcal{U}_\lambda \cap \widetilde{\sigma}(A_0) \cap \overline{\mathbb{R}}) \subset \sigma_{++}(A_0)$ (resp. $(\mathcal{U}_\lambda \cap \widetilde{\sigma}(A_0) \cap \overline{\mathbb{R}}) \subset \sigma_{--}(A_0)$), $\mathcal{U}_\lambda \backslash \overline{\mathbb{R}} \subset \rho(A_0)$ and

$$\|(A_0 - \lambda)^{-1}\| \leq M |\mathrm{Im}\,\lambda|^{-1}$$

holds for some $M > 0$ and all $\lambda \in \mathcal{U}_\lambda \backslash \overline{\mathbb{R}}$, cf. [1], [20] (and [29] for bounded operators).

2.2. Locally definitizable self-adjoint relations

In this section we briefly recall the notion of locally definitizable self-adjoint relations and intervals of type π_+ and type π_- from [20].

Let Ω be some domain in $\overline{\mathbb{C}}$ symmetric with respect to the real axis such that $\Omega \cap \overline{\mathbb{R}} \neq \emptyset$ and the intersections of Ω with the upper and lower open half-planes are simply connected.

Definition 2.2. Let A_0 be a self-adjoint relation in the Krein space $\mathcal{K}$ such that $\sigma(A_0) \cap (\Omega \backslash \overline{\mathbb{R}})$ consists of isolated points which are poles of the resolvent of A_0, and no point of $\Omega \cap \overline{\mathbb{R}}$ is an accumulation point of the non-real spectrum of A_0 in Ω. The relation A_0 is said to be *definitizable over* Ω, if the following holds.

(i) Every point $\mu \in \Omega \cap \overline{\mathbb{R}}$ has an open connected neighborhood I_μ in $\overline{\mathbb{R}}$ such that both components of $I_\mu \backslash \{\mu\}$ are of definite type with respect to A_0.

(ii) For every finite union Δ of open connected subsets of $\overline{\mathbb{R}}$, $\overline{\Delta} \subset \Omega \cap \overline{\mathbb{R}}$, there exists $m \geq 1$, $M > 0$ and an open neighborhood $\mathcal{U}$ of $\overline{\Delta}$ in $\overline{\mathbb{C}}$ such that

$$\|(A_0 - \lambda)^{-1}\| \leq M(1 + |\lambda|)^{2m-2} \, |\mathrm{Im}\,\lambda|^{-m}$$

holds for all $\lambda \in \mathcal{U} \backslash \overline{\mathbb{R}}$.

By [20, Theorem 4.7] a self-adjoint relation A_0 in $\mathcal{K}$ is definitizable over $\overline{\mathbb{C}}$ if and only if A_0 is *definitizable*, that is, the resolvent set of A_0 is non-empty and there exists a rational function $r \neq 0$ with poles only in $\rho(A_0)$ such that $r(A_0) \in \mathcal{L}(\mathcal{K})$ is a nonnegative operator in $\mathcal{K}$, that is

$$[r(A_0)x, x] \geq 0$$

holds for all $x \in \mathcal{K}$ (see [28] and [16, §4 and §5]).

Let $A_0 = A_0^+$ be definitizable over Ω and let $\delta \mapsto E(\delta)$ be the local spectral function of A_0 on $\Omega \cap \mathbb{R}$. Recall that $E(\delta)$ is defined for all finite unions δ of connected subsets of $\Omega \cap \overline{\mathbb{R}}$, $\overline{\delta} \subset \Omega \cap \overline{\mathbb{R}}$, the endpoints of which belong to $\Omega \cap \overline{\mathbb{R}}$ and are of definite type with respect to A_0 (see [20, Section 3.4 and Remark 4.9]). With the help of the local spectral function $E(\cdot)$ the open subsets of definite type in $\Omega \cap \overline{\mathbb{R}}$ can be characterized in the following way. An open subset Δ, $\overline{\Delta} \subset \Omega \cap \overline{\mathbb{R}}$, is of positive type (negative type) with respect to A_0 if and only if for every finite union δ of open connected subsets of Δ, $\overline{\delta} \subset \Delta$, such that the boundary points of δ in $\overline{\mathbb{R}}$ are of definite type with respect to A_0, the spectral subspace $(E(\delta)\mathcal{K}, [\cdot,\cdot])$ (resp. $(E(\delta)\mathcal{K}, -[\cdot,\cdot])$) is a Hilbert space (cf. [20, Theorem 3.18]).

We say that an open subset Δ, $\overline{\Delta} \subset \Omega \cap \overline{\mathbb{R}}$, is of *type* π_+ (*type* π_-) with respect to A_0 if for every finite union δ of open connected subsets of Δ, $\overline{\delta} \subset \Delta$, such that the boundary points of δ in $\overline{\mathbb{R}}$ are of definite type with respect to A_0 the spectral subspace $(E(\delta)\mathcal{K}, [\cdot,\cdot])$ is a Pontryagin space with finite rank of negativity (resp. positivity). We shall say that A_0 is of *type* π_+ *over* Ω (*type* π_- *over* Ω) if $\Omega \cap \overline{\mathbb{R}}$ is of type π_+ (resp. type π_-) with respect to A_0 and $\sigma(A_0) \cap \Omega \backslash \overline{\mathbb{R}}$ consists of eigenvalues with finite algebraic multiplicity.

We remark, that spectral points in sets of type π_+ and type π_- can also be characterized with the help of approximative eigensequences (see [1, 2]).

2.3. Matrix-valued locally definitizable functions

In this section we recall the definition of matrix-valued locally definitizable functions from [21]. Although in the formulation of the main theorem in Section 3.2 below only scalar locally definitizable functions appear, matrix-valued functions will be used within the proof.

Let Ω be a domain as in the beginning of Section 2.2 and let τ be an $\mathcal{L}(\mathbb{C}^n)$-valued piecewise meromorphic function in $\Omega \backslash \overline{\mathbb{R}}$ which is symmetric with respect to the real axis, that is $\tau(\overline{\lambda}) = \tau(\lambda)^*$ for all points λ of holomorphy of τ. If, in addition, no point of $\Omega \cap \overline{\mathbb{R}}$ is an accumulation point of nonreal poles of τ we write $\tau \in M^{n \times n}(\Omega)$. The set of the points of holomorphy of τ in $\Omega \backslash \overline{\mathbb{R}}$ and all points $\mu \in \Omega \cap \mathbb{R}$ such that τ can be analytically continued to μ and the continuations from $\Omega \cap \mathbb{C}^+$ and $\Omega \cap \mathbb{C}^-$ coincide, is denoted by $\mathfrak{h}(\tau)$.

The following definition of sets of positive and negative type with respect to matrix functions and Definition 2.4 below of locally definitizable matrix functions can be found in [21].

Definition 2.3. Let $\tau \in M^{n\times n}(\Omega)$. An open subset $\Delta \subset \Omega \cap \overline{\mathbb{R}}$ is said to be of *positive type* with respect to τ if for every $x \in \mathbb{C}^n$ and every sequence (μ_k) of points in $\Omega \cap \mathbb{C}^+ \cap \mathfrak{h}(\tau)$ which converges in $\overline{\mathbb{C}}$ to a point of Δ we have

$$\liminf_{k\to\infty} \operatorname{Im} \big(\tau(\mu_k)x, x\big) \geq 0.$$

An open subset $\Delta \subset \Omega \cap \overline{\mathbb{R}}$ is said to be of *negative type* with respect to τ if Δ is of positive type with respect to $-\tau$. Δ is said to be of *definite type* with respect to τ if Δ is of positive or of negative type with respect to τ.

Definition 2.4. A function $\tau \in M^{n\times n}(\Omega)$ is called *definitizable in* Ω if the following holds.

(i) Every point $\mu \in \Omega \cap \overline{\mathbb{R}}$ has an open connected neighborhood I_μ in $\overline{\mathbb{R}}$ such that both components of $I_\mu \backslash \{\mu\}$ are of definite type with respect to τ.

(ii) For every finite union Δ of open connected subsets in $\overline{\mathbb{R}}$, $\overline{\Delta} \subset \Omega \cap \overline{\mathbb{R}}$, there exist $m \geq 1$, $M > 0$ and an open neighborhood $\mathcal{U}$ of $\overline{\Delta}$ in $\overline{\mathbb{C}}$ such that

$$\|\tau(\lambda)\| \leq M(1 + |\lambda|)^{2m} \, |\operatorname{Im} \lambda|^{-m}$$

holds for all $\lambda \in \mathcal{U} \backslash \overline{\mathbb{R}}$.

The class of $\mathcal{L}(\mathbb{C}^n)$-valued definitizable functions in Ω will be denoted by $\mathcal{D}^{n\times n}(\Omega)$. In the case $n = 1$ we write $\mathcal{D}(\Omega)$ instead of $\mathcal{D}^{1\times 1}(\Omega)$ and we set

$$\widetilde{\mathcal{D}}(\Omega) := \mathcal{D}(\Omega) \cup \{d_\infty\},$$

where d_∞ denotes the relation $\left\{ \binom{0}{c} \,|\, c \in \mathbb{C} \right\} \in \widetilde{\mathcal{C}}(\mathbb{C})$.

A function $\tau \in M^{n\times n}(\overline{\mathbb{C}})$ which is definitizable in $\overline{\mathbb{C}}$ is called *definitizable*, see [20]. We note that $\tau \in M^{n\times n}(\overline{\mathbb{C}})$ is definitizable if and only if there exists a rational function g symmetric with respect to the real axis such that the poles of g belong to $\mathfrak{h}(\tau) \cup \{\infty\}$ and $g\tau$ is the sum of a Nevanlinna function and a meromorphic function in $\overline{\mathbb{C}}$ (cf. [20]). For a comprehensive study of definitizable functions we refer to the papers [18, 19] of P. Jonas. We mention only that the generalized Nevanlinna class is a subclass of the definitizable functions. Recall that a function $\tau \in M^{n\times n}(\overline{\mathbb{C}})$ is said to be a *generalized Nevanlinna function* if the kernel K_τ,

$$K_\tau(\lambda, \mu) = \frac{\tau(\lambda) - \tau(\overline{\mu})}{\lambda - \overline{\mu}},$$

has finitely many negative squares (see [25] and [26]).

In [21] it is shown that a function $\tau \in M^{n\times n}(\Omega)$ is definitizable in Ω if and only if for every finite union Δ of open connected subsets of $\overline{\mathbb{R}}$ such that $\overline{\Delta} \subset \Omega \cap \overline{\mathbb{R}}$, τ can be written as the sum

$$\tau = \tau_0 + \tau_{(0)} \tag{2.1}$$

of an $\mathcal{L}(\mathbb{C}^n)$-valued definitizable function τ_0 and an $\mathcal{L}(\mathbb{C}^n)$-valued function $\tau_{(0)}$ which is locally holomorphic on $\overline{\Delta}$. We say that a locally definitizable function $\tau \in \mathcal{D}^{n\times n}(\Omega)$ is a *generalized Nevanlinna function in* Ω if the function τ_0 in (2.1) can be chosen as a generalized Nevanlinna function.

The class of $\mathcal{L}(\mathbb{C}^n)$-valued generalized Nevanlinna functions in Ω will be denoted by $\mathcal{N}^{n\times n}(\Omega)$. In the case $n = 1$ we write $\mathcal{N}(\Omega)$ instead of $\mathcal{N}^{1\times 1}(\Omega)$ and we set

$$\widetilde{\mathcal{N}}(\Omega) := \mathcal{N}(\Omega) \cup \{d_\infty\},$$

where d_∞ denotes the relation $\{\binom{0}{c} \mid c \in \mathbb{C}\} \in \widetilde{\mathcal{C}}(\mathbb{C})$.

The following theorem is a consequence of [21, Propositions 2.8 and 3.4]. It establishes a connection between self-adjoint relations which are locally definitizable (locally of type π_+) and $\mathcal{L}(\mathbb{C}^n)$-valued locally definitizable functions (resp. local generalized Nevanlinna functions).

Theorem 2.5. *Let Ω be a domain as above and let A_0 be a self-adjoint relation in the Krein space $\mathcal{K}$ which is definitizable over Ω. Let $\gamma \in \mathcal{L}(\mathbb{C}^n, \mathcal{K})$ and $S = S^* \in \mathcal{L}(\mathbb{C}^n)$, fix some point $\lambda_0 \in \rho(A_0) \cap \Omega$ and define*

$$\tau(\lambda) := S + \gamma^+\big((\lambda - \operatorname{Re}\lambda_0) + (\lambda - \lambda_0)(\lambda - \overline{\lambda}_0)(A_0 - \lambda)^{-1}\big)\gamma$$

for all $\lambda \in \rho(A_0) \cap \Omega$. Then the following holds.

(i) *The function τ is definitizable in Ω, $\tau \in \mathcal{D}^{n\times n}(\Omega)$.*

(ii) *If A_0 is of type π_+ over Ω, then τ belongs to $\mathcal{N}^{n\times n}(\Omega)$.*

(iii) *An open subset Δ of $\Omega \cap \overline{\mathbb{R}}$ which is of positive type (negative type) with respect to A_0 is of positive type (resp. negative type) with respect to τ.*

In the sequel we shall often assume that the sign types of self-adjoint relations which are definitizable over Ω, and definitizable functions in Ω coincide outside of a discrete set in $\Omega \cap \overline{\mathbb{R}}$. A notion for this concept is introduced in the next definition, cf. [3, Definition 2.8].

Definition 2.6. Let A_0 and A_1 be self-adjoint relations which are definitizable over Ω and let τ and $\widetilde{\tau}$ be $\mathcal{L}(\mathbb{C}^n)$-valued definitizable functions in Ω. We shall say that *the sign types of A_0 and A_1 (A_0 and τ, τ and $\widetilde{\tau}$) are d-compatible in Ω* if for every $\mu \in \Omega \cap \overline{\mathbb{R}}$ there exists an open connected neighborhood $I_\mu \subset \Omega \cap \overline{\mathbb{R}}$ of μ such that each component of $I_\mu \backslash \{\mu\}$ is either of positive type with respect to A_0 and A_1 (resp. A_0 and τ, τ and $\widetilde{\tau}$) or of negative type with respect to A_0 and A_1 (resp. A_0 and τ, τ and $\widetilde{\tau}$).

If A_0 is definitizable over Ω and the function $\tau \in \mathcal{D}^{n\times n}(\Omega)$ is defined as in Theorem 2.5, then obviously the sign types of A_0 and τ are d-compatible in Ω. A typical nontrivial situation where d-compatibility of sign types of locally definitizable self-adjoint relations appears is shown in Theorem 2.7 below. For a proof we refer to [4, Theorem 3.2].

Theorem 2.7. *Let A_0 and A_1 be self-adjoint relations in $\mathcal{K}$, assume that the set $\rho(A_0) \cap \rho(A_1) \cap \Omega$ is non-empty and that A_0 is definitizable over Ω. If*

$$(A_1 - \mu)^{-1} - (A_0 - \mu)^{-1}$$

is a finite rank operator for some (and hence for all) $\mu \in \rho(A_0) \cap \rho(A_1) \cap \Omega$, then A_1 is definitizable over Ω and the sign types of A_0 and A_1 are d-compatible in Ω.

3. Generalized resolvents of a class of symmetric operators

3.1. Boundary value spaces and Weyl functions associated with symmetric relations in Krein spaces

Let $(\mathcal{K}, [\cdot,\cdot])$ be a separable Krein space, let J be a corresponding fundamental symmetry and let A be a closed symmetric relation in $\mathcal{K}$. We say that A has *defect* $m \in \mathbb{N} \cup \{\infty\}$, if both deficiency indices

$$n_\pm(JA) = \dim \ker((JA)^* - \overline{\lambda}), \qquad \lambda \in \mathbb{C}^\pm,$$

of the symmetric relation JA in the Hilbert space $(\mathcal{K}, [J\cdot,\cdot])$ are equal to m. Here * denotes the Hilbert space adjoint. We remark, that this is equivalent to the fact that there exists a self-adjoint extension of A in $\mathcal{K}$ and that each self-adjoint extension $\hat{A}$ of A in $\mathcal{K}$ satisfies $\dim(\hat{A}/A) = m$.

We shall use the so-called boundary value spaces for the description of the self-adjoint extensions of closed symmetric relations in Krein spaces. The following definition is taken from [10].

Definition 3.1. Let A be a closed symmetric relation in the Krein space $\mathcal{K}$. We say that $\{\mathcal{G}, \Gamma_0, \Gamma_1\}$ is a *boundary value space* for A^+ if $\mathcal{G}$ is a Hilbert space and $\Gamma_0, \Gamma_1 : A^+ \to \mathcal{G}$ are mappings such that $\Gamma := \binom{\Gamma_0}{\Gamma_1} : A^+ \to \mathcal{G}^2$ is surjective, and the relation

$$[\![\Gamma\hat{f}, \Gamma\hat{g}]\!]_{\mathcal{G}^2} = [\![\hat{f}, \hat{g}]\!]_{\mathcal{K}^2}$$

holds for all $\hat{f}, \hat{g} \in A^+$.

In the following we recall some basic facts on boundary value spaces which can be found in, e.g., [8] and [10]. For the Hilbert space case we refer to [17], [12] and [13]. Let A be a closed symmetric relation in $\mathcal{K}$. Then

$$\mathcal{N}_{\lambda,A^+} := \ker(A^+ - \lambda) = \operatorname{ran}(A - \overline{\lambda})^{[\perp]}$$

denotes the defect subspace of A at the point $\lambda \in r(A)$ and we set

$$\hat{\mathcal{N}}_{\lambda,A^+} := \left\{ \binom{f_\lambda}{\lambda f_\lambda} \,\middle|\, f_\lambda \in \mathcal{N}_{\lambda,A^+} \right\}.$$

When no confusion can arise we write $\mathcal{N}_\lambda$ and $\hat{\mathcal{N}}_\lambda$ instead of $\mathcal{N}_{\lambda,A^+}$ and $\hat{\mathcal{N}}_{\lambda,A^+}$.

If there exists a self-adjoint extension A' of A such that $\rho(A') \neq \emptyset$ then we have

$$A^+ = A' \dotplus \hat{\mathcal{N}}_\lambda \quad \text{for all} \quad \lambda \in \rho(A').$$

In this case there exists a boundary value space $\{\mathcal{G}, \Gamma_0, \Gamma_1\}$ for A^+ such that $\ker \Gamma_0 = A'$ (cf. [10]).

Let in the following A, $\{\mathcal{G}, \Gamma_0, \Gamma_1\}$ and Γ be as in Definition 3.1. It follows that the mappings Γ_0 and Γ_1 are continuous. The self-adjoint extensions

$$A_0 := \ker \Gamma_0 \qquad \text{and} \qquad A_1 := \ker \Gamma_1$$

of A are transversal, i.e., $A_0 \cap A_1 = A$ and $A_0 \,\widehat{+}\, A_1 = A^+$. The mapping Γ induces, via

$$A_\Theta := \Gamma^{-1}\Theta = \{\hat{f} \in A^+ \mid \Gamma\hat{f} \in \Theta\}, \quad \Theta \in \widetilde{\mathcal{C}}(\mathcal{G}), \tag{3.1}$$

a bijective correspondence $\Theta \mapsto A_\Theta$ between the set of all closed linear relations $\widetilde{\mathcal{C}}(\mathcal{G})$ in $\mathcal{G}$ and the set of closed extensions $A_\Theta \subset A^+$ of A. In particular (3.1) gives a one-to-one correspondence between the closed symmetric (self-adjoint) extensions of A and the closed symmetric (resp. self-adjoint) relations in $\mathcal{G}$. If Θ is a closed operator in $\mathcal{G}$, then the corresponding extension A_Θ of A is determined by

$$A_\Theta = \ker\big(\Gamma_1 - \Theta\Gamma_0\big). \tag{3.2}$$

Assume that $\rho(A_0) \neq \emptyset$ and denote by π_1 the orthogonal projection onto the first component of $\mathcal{K} \times \mathcal{K}$. For every $\lambda \in \rho(A_0)$ we define the operators

$$\gamma(\lambda) := \pi_1(\Gamma_0|\hat{\mathcal{N}_\lambda})^{-1} \in \mathcal{L}(\mathcal{G},\mathcal{K}) \quad\text{and}\quad M(\lambda) := \Gamma_1(\Gamma_0|\hat{\mathcal{N}_\lambda})^{-1} \in \mathcal{L}(\mathcal{G}).$$

The functions $\lambda \mapsto \gamma(\lambda)$ and $\lambda \mapsto M(\lambda)$ are called the *γ-field* and *Weyl function* corresponding to $\{\mathcal{G},\Gamma_0,\Gamma_1\}$. They are holomorphic on $\rho(A_0)$ and the relations

$$\gamma(\zeta) = \big(1 + (\zeta-\lambda)(A_0-\zeta)^{-1}\big)\gamma(\lambda) \tag{3.3}$$

and

$$M(\lambda) - M(\zeta)^* = (\lambda - \overline{\zeta})\gamma(\zeta)^+\gamma(\lambda) \tag{3.4}$$

hold for all $\lambda, \zeta \in \rho(A_0)$ (cf. [10]). It follows that

$$\begin{aligned} M(\lambda) = \operatorname{Re} M(\lambda_0) + \gamma(\lambda_0)^+\big((\lambda - \operatorname{Re}\lambda_0) \\ + (\lambda-\lambda_0)(\lambda-\overline{\lambda}_0)(A_0-\lambda)^{-1}\big)\gamma(\lambda_0) \end{aligned} \tag{3.5}$$

holds for a fixed $\lambda_0 \in \rho(A_0)$ and all $\lambda \in \rho(A_0)$. If, in addition, the condition $\mathcal{K} = \operatorname{clsp}\{\mathcal{N}_\lambda \,|\, \lambda \in \rho(A_0)\}$ is fulfilled, then it follows from (3.3) and (3.4) that the function M is *strict*, that is

$$\bigcap_{\lambda\in\mathfrak{h}(M)} \ker\left(\frac{M(\lambda)-M(\mu)^*}{\lambda-\overline{\mu}}\right) = \{0\} \tag{3.6}$$

holds for some (and hence for all) $\mu \in \mathfrak{h}(\tau)$.

If $\Theta \in \widetilde{\mathcal{C}}(\mathcal{G})$ and A_Θ is the corresponding extension of A (see (3.1)), then for every point $\lambda \in \rho(A_0)$ we have

$$\lambda \in \rho(A_\Theta) \qquad \text{if and only if} \qquad 0 \in \rho(\Theta - M(\lambda)). \tag{3.7}$$

For $\lambda \in \rho(A_\Theta) \cap \rho(A_0)$ the well-known resolvent formula

$$(A_\Theta - \lambda)^{-1} = (A_0-\lambda)^{-1} + \gamma(\lambda)\big(\Theta - M(\lambda)\big)^{-1}\gamma(\overline{\lambda})^+ \tag{3.8}$$

holds (for a proof see, e.g., [10]).

3.2. A variant of the Krein-Naimark formula

We choose a domain Ω as in the beginning of Section 2.2. Let A be a (not necessarily densely defined) closed symmetric operator in the Krein space $\mathcal{K}$, let $\{\mathcal{G},\Gamma_0,\Gamma_1\}$ be a boundary value space for A^+ and let $\mathcal{H}$ be a further Krein space. A self-adjoint extension $\widetilde{A}$ of A in $\mathcal{K} \times \mathcal{H}$ is said to be an *exit space extension of* A and $\mathcal{H}$ is called the *exit space*. The exit space extension $\widetilde{A}$ of A is said to be $\mathcal{K}$*-minimal* if $\rho(\widetilde{A}) \cap \Omega$ is non-empty and

$$\mathcal{K} \times \mathcal{H} = \operatorname{clsp}\left\{\mathcal{K}, (\widetilde{A} - \lambda)^{-1}\mathcal{K} \,|\, \lambda \in \rho(\widetilde{A}) \cap \Omega\right\}$$

holds. Note, that the definition of $\mathcal{K}$-minimality depends on the domain Ω. The elements of $\mathcal{K} \times \mathcal{H}$ will be written in the form $\{k, h\}$, $k \in \mathcal{K}$, $h \in \mathcal{H}$. Let $P_{\mathcal{K}} : \mathcal{K} \times \mathcal{H} \to \mathcal{H}$, $\{k, h\} \mapsto k$, be the projection onto the first component of $\mathcal{K} \times \mathcal{H}$. Then the compression

$$P_{\mathcal{K}}(\widetilde{A} - \lambda)^{-1}|_{\mathcal{K}}, \qquad \lambda \in \rho(\widetilde{A}),$$

of the resolvent of $\widetilde{A}$ to $\mathcal{K}$ is called a *generalized resolvent* of A.

Theorem 3.2. *Let A be a closed symmetric operator of defect one in the Krein space $\mathcal{K}$ and let $\{\mathbb{C}, \Gamma_0, \Gamma_1\}$, $A_0 = \ker \Gamma_0$, be a boundary value space for A^+ with corresponding γ-field γ and Weyl function M. Assume that A_0 is definitizable over Ω and that the condition $\mathcal{K} = \operatorname{clsp}\{\mathcal{N}_{\lambda,A^+} \,|\, \lambda \in \rho(A_0) \cap \Omega\}$ is fulfilled. Then the following assertions hold.*

(i) *Let $\widetilde{A}$ be a $\mathcal{K}$-minimal self-adjoint exit space extension of A in $\mathcal{K} \times \mathcal{H}$ which is definitizable over Ω and assume that the sign types of $\widetilde{A}$ and A_0 are d-compatible in Ω. Then there exists a locally definitizable function $\tau \in \widetilde{\mathcal{D}}(\Omega)$ such that the sign types of τ, $\widetilde{A}$ and A_0 are d-compatible in Ω,*

$$\rho(\widetilde{A}) \cap \rho(A_0) \cap \mathfrak{h}(\tau) \cap \Omega$$

is a subset of $\mathfrak{h}((M + \tau)^{-1})$ and the formula

$$P_{\mathcal{K}}(\widetilde{A} - \lambda)^{-1}|_{\mathcal{K}} = (A_0 - \lambda)^{-1} - \gamma(\lambda)\big(M(\lambda) + \tau(\lambda)\big)^{-1}\gamma(\overline{\lambda})^+ \tag{3.9}$$

holds for all $\lambda \in \rho(\widetilde{A}) \cap \rho(A_0) \cap \mathfrak{h}(\tau) \cap \Omega$.

(ii) *Let $\tau \in \widetilde{\mathcal{D}}(\Omega)$ be a locally definitizable function such that $M(\mu) + \tau(\mu) \neq 0$ for some $\mu \in \Omega$, assume that the sign types of τ and A_0 are d-compatible in Ω and let Ω' be a domain with the same properties as Ω, $\overline{\Omega'} \subset \Omega$. Then there exists a Krein space $\mathcal{H}$ and a $\mathcal{K}$-minimal self-adjoint exit space extension $\widetilde{A}$ of A in $\mathcal{K} \times \mathcal{H}$ which is definitizable over Ω', such that the sign types of $\widetilde{A}$, τ and A_0 are d-compatible in Ω',*

$$\rho(A_0) \cap \mathfrak{h}(\tau) \cap \mathfrak{h}\big((M + \tau)^{-1}\big) \cap \Omega'$$

is a subset of $\rho(\widetilde{A})$ and formula (3.9) holds for all points λ belonging to $\rho(A_0) \cap \mathfrak{h}(\tau) \cap \mathfrak{h}((M + \tau)^{-1}) \cap \Omega'$.

Proof. The proof of assertion (i) consists of four steps. Let $\widetilde{A}$ be a $\mathcal{K}$-minimal self-adjoint exit space extension of A in $\mathcal{K} \times \mathcal{H}$ which is definitizable over Ω such that the sign types of $\widetilde{A}$ and A_0 are d-compatible in Ω.

1. In this first step we prove assertion (i) for the case $\mathcal{H} = \{0\}$. Here $\widetilde{A}$ is a canonical extension of A and therefore, by (3.1), there exists a self-adjoint constant $\tau \in \mathbb{R} \cup \{d_\infty\}$ such that

$$(\widetilde{A} - \lambda)^{-1} = (A_0 - \lambda)^{-1} - \gamma(\lambda)\big(M(\lambda) + \tau\big)^{-1}\gamma(\overline{\lambda})^+$$

holds (cf. (3.8)), that is, $\widetilde{A}$ coincides with the canonical self-adjoint extension $A_{-\tau}$ of A and by (3.7) we have $\rho(A_{-\tau}) \cap \rho(A_0) \subset \mathfrak{h}((M+\tau)^{-1})$. Here each point in $\Omega \cap \overline{\mathbb{R}}$ is of positive as well as of negative type with respect to τ and hence assertion (i) follows.

2. In the following we assume $\mathcal{H} \neq \{0\}$. Following the lines of [11, §5] we define in this step a symmetric relation T in $\mathcal{H}$ and a special boundary value space for the adjoint T^+.

Below we will deal with direct products of linear relations. The following notation will be used. If U is a relation in $\mathcal{K}$ and V is a relation in $\mathcal{H}$ we shall write $U \times V$ for the direct product of U and V which is a relation in $\mathcal{K} \times \mathcal{H}$,

$$U \times V = \left\{ \begin{pmatrix} \{f_1, f_2\} \\ \{f_1', f_2'\} \end{pmatrix} \,\middle|\, \begin{pmatrix} f_1 \\ f_1' \end{pmatrix} \in U,\ \begin{pmatrix} f_2 \\ f_2' \end{pmatrix} \in V \right\}.$$

For the pair $\left(\begin{smallmatrix} \{f_1, f_2\} \\ \{f_1', f_2'\} \end{smallmatrix}\right)$ we shall also write $\{\hat{f}_1, \hat{f}_2\}$, where $\hat{f}_1 = \left(\begin{smallmatrix} f_1 \\ f_1' \end{smallmatrix}\right)$ and $\hat{f}_2 = \left(\begin{smallmatrix} f_2 \\ f_2' \end{smallmatrix}\right)$.

The linear relations

$$S := \widetilde{A} \cap \mathcal{K}^2 = \left\{ \begin{pmatrix} k \\ k' \end{pmatrix} \,\middle|\, \begin{pmatrix} \{k, 0\} \\ \{k', 0\} \end{pmatrix} \in \widetilde{A} \right\}$$

and

$$T := \widetilde{A} \cap \mathcal{H}^2 = \left\{ \begin{pmatrix} h \\ h' \end{pmatrix} \,\middle|\, \begin{pmatrix} \{0, h\} \\ \{0, h'\} \end{pmatrix} \in \widetilde{A} \right\}$$

are closed and symmetric in $\mathcal{K}$ and $\mathcal{H}$, respectively, and we have $A \subset S$. Let $J_\mathcal{K}$ and $J_\mathcal{H}$ be fundamental symmetries in the Krein spaces $\mathcal{K}$ and $\mathcal{H}$, respectively, and choose

$$J := \begin{pmatrix} J_\mathcal{K} & 0 \\ 0 & J_\mathcal{H} \end{pmatrix} \in \mathcal{L}(\mathcal{K} \times \mathcal{H})$$

as a fundamental symmetry in the Krein space $\mathcal{K} \times \mathcal{H}$. Then $J_\mathcal{K} S = J\widetilde{A} \cap \mathcal{K}^2$ and $J_\mathcal{H} T = J\widetilde{A} \cap \mathcal{H}^2$ are symmetric relations in the Hilbert spaces $(\mathcal{K}, [J_\mathcal{K}\cdot, \cdot])$ and $(\mathcal{H}, [J_\mathcal{H}\cdot, \cdot])$, respectively. It follows from [11, §5] that the deficiency indices of $J_\mathcal{K} S$ and $-J_\mathcal{H} T$ coincide. As $J_\mathcal{K} S$ is a symmetric extension of the symmetric operator $J_\mathcal{K} A$ in the Hilbert space $(\mathcal{K}, [J_\mathcal{K}\cdot, \cdot])$ the deficiency indices $n_\pm(J_\mathcal{K} S)$ of $J_\mathcal{K} S$ are $(1,1)$ or $(0,0)$.

The case $n_\pm(J_\mathcal{K} S) = 0$ is impossible here as otherwise also the relation $J_\mathcal{H} T$ would be self-adjoint in $(\mathcal{H}, [J_\mathcal{H}\cdot, \cdot])$ and therefore $J\widetilde{A}$ would coincide with $J_\mathcal{K} S \times J_\mathcal{H} T$. But as $\widetilde{A} = S \times T$ is by assumption a $\mathcal{K}$-minimal exit space extension of A we would obtain $\mathcal{H} = \{0\}$, a contradiction.

This implies $m_{11}(\lambda) = m_{12}(\lambda)m_{21}(\lambda) = 0$ and since m_{12} and m_{21} are piecewise meromorphic functions in $\Omega\backslash\overline{\mathbb{R}}$ and $\widetilde{M}$ is symmetric with respect to the real axis we conclude $m_{12}(\lambda) = m_{21}(\lambda) = 0$, $\lambda \in \rho(\widetilde{A}) \cap \Omega$, which contradicts the strictness of $\widetilde{M}$.

It is straightforward to check that the matrix

$$V := \begin{pmatrix} 0 & 0 & 1 & 0 \\ 0 & -\alpha & \alpha & 1 \\ -1 & 0 & 0 & 1 \\ 0 & -1 & 1 & 0 \end{pmatrix} \in \mathcal{L}(\mathbb{C}^4) \tag{3.17}$$

is unitary in $(\mathbb{C}^4, [\![\cdot,\cdot]\!]_{\mathbb{C}^4})$, cf. (3.14). Let $\{\mathbb{C}^2, \widehat{\Gamma}_0, \widehat{\Gamma}_1\}$ be the boundary value space for $A^+ \times T^+$ defined by

$$\begin{pmatrix} \widehat{\Gamma}_0 \\ \widehat{\Gamma}_1 \end{pmatrix} := V \begin{pmatrix} \widetilde{\Gamma}_0 \\ \widetilde{\Gamma}_1 \end{pmatrix} = VW \begin{pmatrix} \Gamma_0'' \\ \Gamma_1'' \end{pmatrix}, \tag{3.18}$$

(see (3.13)). From

$$VW = \begin{pmatrix} 1 & 0 & 0 & 0 \\ 0 & -\alpha & 0 & 1 \\ 0 & 0 & 1 & 0 \\ 0 & -1 & 0 & 0 \end{pmatrix}$$

we obtain

$$\widehat{\Gamma}_0\{\hat{f}_1, \hat{f}_2\} = \begin{pmatrix} \Gamma_0 \hat{f}_1 \\ \Gamma_1' \hat{f}_2 - \alpha \Gamma_0' \hat{f}_2 \end{pmatrix}, \qquad \hat{f}_1 \in A^+, \hat{f}_2 \in T^+,$$

and

$$\widehat{\Gamma}_1\{\hat{f}_1, \hat{f}_2\} = \begin{pmatrix} \Gamma_1 \hat{f}_1 \\ -\Gamma_0' \hat{f}_2 \end{pmatrix}, \qquad \hat{f}_1 \in A^+, \hat{f}_2 \in T^+.$$

We denote the self-adjoint extension $\ker(\Gamma_1' - \alpha\Gamma_0') \in \widetilde{\mathcal{C}}(\mathcal{H})$ of T in $\mathcal{H}$ by T_α. Then the self-adjoint extension $\ker \widehat{\Gamma}_0$ of $A \times T$ in $\mathcal{K} \times \mathcal{H}$ coincides with $A_0 \times T_\alpha$.

Since (3.18) and (3.17) imply

$$\begin{aligned} A_0 \times T_\alpha = \ker \widehat{\Gamma}_0 &= \ker \left(\begin{pmatrix} 0 & 0 \\ 0 & -\alpha \end{pmatrix} \widetilde{\Gamma}_0 + \begin{pmatrix} 1 & 0 \\ \alpha & 1 \end{pmatrix} \widetilde{\Gamma}_1 \right) \\ &= \ker \left(\widetilde{\Gamma}_1 - \begin{pmatrix} 0 & 0 \\ 0 & \alpha \end{pmatrix} \widetilde{\Gamma}_0 \right), \end{aligned}$$

we find from (3.15), (3.2) and (3.7) that a point $\lambda \in \rho(\widetilde{A})$ belongs to the set $\rho(A_0 \times T_\alpha)$ if and only if 0 belongs to the resolvent set of

$$\widetilde{M}(\lambda) - \begin{pmatrix} 0 & 0 \\ 0 & \alpha \end{pmatrix}.$$

But we have chosen α such that this function is invertible for some $\lambda' \in \rho(\widetilde{A}) \cap \Omega$, therefore λ' belongs to $\rho(A_0 \times T_\alpha)$. In particular $\lambda' \in \rho(A_0) \cap \rho(T_\alpha)$ and

$$\rho(T_\alpha) \cap \rho(A_0) \cap \rho(\widetilde{A}) \cap \Omega \neq \emptyset.$$

As $A \times T$ is a symmetric relation of defect two and $\widetilde{A}$ and $A_0 \times T_\alpha$ are self-adjoint extensions of $A \times T$ in $\mathcal{K} \times \mathcal{H}$ we have

$$\dim\big(\operatorname{ran}\,\big((\widetilde{A} - \lambda)^{-1} - ((A_0 \times T_\alpha) - \lambda)^{-1}\big)\big) \leq 2$$

for all $\lambda \in \rho(\widetilde{A}) \cap \rho(A_0) \cap \rho(T_\alpha) \cap \Omega$. Since $\widetilde{A}$ is definitizable over Ω we obtain from Theorem 2.7 that also the self-adjoint relation $A_0 \times T_\alpha$ is definitizable over Ω and that the sign types of $\widetilde{A}$ and the sign types of $A_0 \times T_\alpha$ are d-compatible in Ω.

It is a simple consequence from Definition 2.1 that

$$\big(\sigma_{++}(A_0 \times T_\alpha) \cap \sigma_{\mathrm{ap}}(T_\alpha)\big) \subset \sigma_{++}(T_\alpha)$$

and

$$\big(\sigma_{--}(A_0 \times T_\alpha) \cap \sigma_{\mathrm{ap}}(T_\alpha)\big) \subset \sigma_{--}(T_\alpha)$$

holds. Hence, real points from $\sigma_{++}(A_0 \times T_\alpha)$ ($\sigma_{--}(A_0 \times T_\alpha)$) belong to $\rho(T_\alpha)$ or to $\sigma_{++}(T_\alpha)$ (resp. $\sigma_{--}(T_\alpha)$). Therefore T_α is definitizable over Ω and the sign types of T_α in Ω are d-compatible with the sign types of $A_0 \times T_\alpha$ and, hence, with the sign types of $\widetilde{A}$ and A_0 in Ω.

4. In this step we show that also T_0 in (3.10) has a non-empty resolvent set and that formula (3.9) holds with the Weyl function τ corresponding to the boundary value space $\{\mathbb{C}, \Gamma_0', \Gamma_1'\}$. Moreover, we show that τ is locally definitizable and that its sign types are d-compatible with the sign types of A_0 and $\widetilde{A}$ in Ω.

It is straightforward to verify that $\{\mathbb{C}, \Gamma_1' - \alpha\Gamma_0', -\Gamma_0'\}$ is a boundary value space for T^+ and we have $T_\alpha = \ker(\Gamma_1' - \alpha\Gamma_0')$ and $T_0 = \ker(-\Gamma_0')$. The corresponding Weyl function τ_α is defined for all $\lambda \in \rho(T_\alpha)$. As T_α is definitizable over Ω the function τ_α belongs to the class $\mathcal{D}(\Omega)$ and the sign types of τ_α are d-compatible with the sign types of T_α, $\widetilde{A}$ and A_0 in Ω (cf. Theorem 2.5 and (3.5) or [3, Proposition 3.2]). Relation (3.16) implies that τ_α is strict and in particular τ_α is not identically equal to zero.

Then, by (3.8), for $\lambda \in \rho(T_\alpha) \cap \mathfrak{h}(\tau_\alpha^{-1})$ we have

$$(T_0 - \lambda)^{-1} = (T_\alpha - \lambda)^{-1} - \gamma_\alpha'(\lambda)\frac{1}{\tau_\alpha(\lambda)}\gamma_\alpha'(\overline{\lambda})^+,$$

where γ_α' is the γ-field of the boundary value space $\{\mathbb{C}, \Gamma_1' - \alpha\Gamma_0', -\Gamma_0'\}$. Therefore the set $\rho(T_\alpha) \cap \rho(T_0) \cap \Omega$ is non-empty and by Theorem 2.7 the self-adjoint relation T_0 is definitizable over Ω and the sign types of T_0 and T_α are d-compatible in Ω. The Weyl function τ corresponding to the boundary value space $\{\mathbb{C}, \Gamma_0', \Gamma_1'\}$ satisfies

$$\tau(\lambda) = -\tau_\alpha(\lambda)^{-1} + \alpha, \qquad \lambda \in \mathfrak{h}(\tau_\alpha^{-1}) \cap \rho(T_\alpha).$$

and is holomorphic on $\rho(T_0)$.

It follows from Theorem 2.5 and (3.5) that τ belongs to the class $\mathcal{D}(\Omega)$ and that its sign types are d-compatible with the sign types of T_0 and T_α and hence also with the sign types of $\widetilde{A}$ and A_0. The γ-field corresponding to $\{\mathbb{C}, \Gamma'_0, \Gamma'_1\}$ will be denoted by γ'.

Since A_0 and T_0 are both definitizable over Ω the set $\rho(A_0)\cap\rho(T_0)\cap\Omega$ is non-empty. The γ-field γ'' and the Weyl function M'' corresponding to the boundary value space $\{\mathbb{C}^2, \Gamma''_0, \Gamma''_1\}$ defined in (3.12) are given by

$$\lambda \mapsto \gamma''(\lambda) = \begin{pmatrix} \gamma(\lambda) & 0 \\ 0 & \gamma'(\lambda) \end{pmatrix}, \quad \lambda \in \rho(A_0)\cap\rho(T_0)\cap\Omega, \tag{3.19}$$

and

$$\lambda \mapsto M''(\lambda) = \begin{pmatrix} M(\lambda) & 0 \\ 0 & \tau(\lambda) \end{pmatrix}, \quad \lambda \in \rho(A_0)\cap\rho(T_0)\cap\Omega, \tag{3.20}$$

respectively. The relation

$$\Theta := \left\{ \begin{pmatrix} \{u, -u\} \\ \{v, v\} \end{pmatrix} \,\middle|\, u, v \in \mathbb{C} \right\} \in \widetilde{\mathcal{C}}(\mathbb{C}^2) \tag{3.21}$$

is self-adjoint and the corresponding self-adjoint extension of $A \times T$ is given by

$$\begin{pmatrix} \Gamma''_0 \\ \Gamma''_1 \end{pmatrix}^{-1} \Theta = \left\{ \{\hat{f}_1, \hat{f}_2\} \in A^+ \times T^+ \,|\, \Gamma_0 \hat{f}_1 + \Gamma'_0 \hat{f}_2 = \Gamma_1 \hat{f}_1 - \Gamma'_1 \hat{f}_2 = 0 \right\} \tag{3.22}$$

and coincides with $\widetilde{A}$ (see (3.11)).

By (3.7) a point $\lambda \in \rho(A_0 \times T_0)$ belongs to $\rho(\widetilde{A})$ if and only if

$$0 \in \rho(\Theta - M''(\lambda)).$$

Hence, for $\lambda \in \rho(A_0 \times T_0)\cap\rho(\widetilde{A})\cap\Omega = \rho(A_0)\cap\mathfrak{h}(\tau)\cap\rho(\widetilde{A})\cap\Omega$

$$\big(\Theta - M''(\lambda)\big)^{-1} = \left\{ \begin{pmatrix} \{v - M(\lambda)u, v + \tau(\lambda)u\} \\ \{u, -u\} \end{pmatrix} \,\middle|\, u, v \in \mathbb{C} \right\}$$

is an operator. Therefore $(M(\lambda) + \tau(\lambda))u = 0$ implies $u = 0$ and we conclude that the set $\rho(\widetilde{A})\cap\rho(A_0)\cap\mathfrak{h}(\tau)\cap\Omega$ is a subset of $\mathfrak{h}((M+\tau)^{-1})$. Setting $x = v - M(\lambda)u$ and $y = v + \tau(\lambda)u$ we obtain

$$u = -\big(M(\lambda) + \tau(\lambda)\big)^{-1} x + \big(M(\lambda) + \tau(\lambda)\big)^{-1} y$$

for $\lambda \in \rho(A_0 \times T_0)\cap\rho(\widetilde{A})\cap\Omega$. This implies

$$\big(\Theta - M''(\lambda)\big)^{-1} = \begin{pmatrix} -(M(\lambda) + \tau(\lambda))^{-1} & (M(\lambda) + \tau(\lambda))^{-1} \\ (M(\lambda) + \tau(\lambda))^{-1} & -(M(\lambda) + \tau(\lambda))^{-1} \end{pmatrix}. \tag{3.23}$$

For all $\lambda \in \rho(A_0 \times T_0)\cap\rho(\widetilde{A})\cap\Omega$ the relation

$$(\widetilde{A} - \lambda)^{-1} = \big((A_0 \times T_0) - \lambda\big)^{-1} + \gamma''(\lambda)\big(\Theta - M''(\lambda)\big)^{-1}\gamma''(\overline{\lambda})^+ \tag{3.24}$$

holds (cf. (3.8)) and it follows from (3.24), (3.19) and (3.23) that the formula

$$P_{\mathcal{K}}(\widetilde{A} - \lambda)^{-1}|_{\mathcal{K}} = (A_0 - \lambda)^{-1} - \gamma(\lambda)\big(M(\lambda) + \tau(\lambda)\big)^{-1}\gamma(\overline{\lambda})^+$$

holds. This completes the proof of assertion (i).

5. Assertion (ii) was already proved in [3] in a slightly different form. For the convenience of the reader we sketch the proof.

If τ is identically equal to a real constant, then $A_{-\tau} := \ker(\Gamma_1 + \tau\Gamma_0)$ is a canonical self-adjoint extension of A. As the Weyl function M corresponding to A^+ and $\{\mathbb{C}, \Gamma_0, \Gamma_1\}$ is strict we obtain $\rho(A_{-\tau}) \cap \Omega \neq \emptyset$ and Theorem 2.7 implies that $A_{-\tau}$ is definitizable over Ω and that the sign types of A_0, $A_{-\tau}$ and $\tau \in \mathbb{R}$ are d-compatible. By (3.8)

$$(A_{-\tau} - \lambda)^{-1} = (A_0 - \lambda)^{-1} - \gamma(\lambda)\big(M(\lambda) + \tau\big)^{-1}\gamma(\overline{\lambda})^+$$

holds for all $\lambda \in \rho(A_0) \cap ((M+\tau)^{-1})$. In the case $\tau = d_\infty = \big\{\big(\begin{smallmatrix}0\\c\end{smallmatrix}\big) \,|\, c \in \mathbb{C}\big\}$ we have $A_{-\tau} = A_0$.

Assume now that $\tau \in \mathcal{D}(\Omega)$ is not equal to a constant and let Ω' be a domain with the same properties as Ω, $\overline{\Omega'} \subset \Omega$. With the help of [21, Theorem 3.8] it was shown in [3, Theorem 3.3] that there exists a Krein space $\mathcal{H}$, a closed symmetric operator T of defect one in $\mathcal{H}$ and a boundary value space $\{\mathbb{C}, \Gamma_0', \Gamma_1'\}$ for T^+ such that $T_0 := \ker \Gamma_0'$ is definitizable over Ω', the sign types of τ and T_0 are d-compatible and τ coincides with the Weyl function corresponding to $\{\mathbb{C}, \Gamma_0', \Gamma_1'\}$ on $\Omega' \cap \rho(T_0)$. Moreover the condition

$$\mathcal{H} = \operatorname{clsp}\big\{\gamma'(\lambda) \,|\, \lambda \in \rho(T_0) \cap \Omega'\big\} \tag{3.25}$$

is fulfilled. We choose the boundary value space $\{\mathbb{C}^2, \Gamma_0'', \Gamma_1''\}$ for $A^+ \times T^+$ as in (3.12) with γ-field and Weyl function given by (3.19) and (3.20), respectively. The self-adjoint extension corresponding to Θ in (3.21) via (3.1) is denoted by $\widetilde{A}$. Then $\widetilde{A}$ has the form (3.22) and the relation (3.24) holds for all $\lambda \in \Omega'$ which belong to $\rho(A_0 \times T_0)$ and fulfil $0 \in \rho(\Theta - M''(\lambda))$. From (3.23) we conclude

$$\rho(A_0) \cap \mathfrak{h}(\tau) \cap \mathfrak{h}\big((M+\tau)^{-1}\big) \cap \Omega' \subset \rho(\widetilde{A})$$

and (3.24) implies that the formula (3.9) holds. Since the minimality condition (3.25) is fulfilled it follows from (3.24) that $\widetilde{A}$ is a $\mathcal{K}$-minimal exit space extension of A. As $A_0 \times T_0$ is definitizable over Ω' the relation (3.24) and Theorem 2.7 imply that $\widetilde{A}$ is also definitizable over Ω' and the sign types of $\widetilde{A}$, A_0 and τ are d-compatible. □

The next theorem is a variant of the Krein-Naimark formula for the case that A_0 and $\widetilde{A}$ are locally of type π_+ and τ is a local generalized Nevanlinna function. The proof of Theorem 3.3 below is essentially the same as the proof of Theorem 3.2. Instead of the result on finite rank perturbations of locally definitizable self-adjoint relations from [4], cf. Theorem 2.7, one has to use [5, Theorem 2.4] on the stability of self-adjoint operators and relations locally of type π_+ under compact perturbations in resolvent sense. We leave the details to the reader.

Theorem 3.3. *Let A be a closed symmetric operator of defect one in the Krein space $\mathcal{K}$ and let $\{\mathbb{C}, \Gamma_0, \Gamma_1\}$ be a boundary value space for A^+ with corresponding γ-field γ and Weyl function M. Assume that $A_0 = \ker \Gamma_0$ is of type π_+ over Ω and that the condition $\mathcal{K} = \operatorname{clsp}\{\mathcal{N}_{\lambda,A^+} \,|\, \lambda \in \rho(A_0) \cap \Omega\}$ is fulfilled.*

Then the following assertions hold:

(i) *For every $\mathcal{K}$-minimal self-adjoint exit space extension $\widetilde{A}$ of A in $\mathcal{K}\times\mathcal{H}$ which is of type π_+ over Ω there exists a function $\tau\in\widetilde{\mathcal{N}}(\Omega)$ such that*

$$\rho(\widetilde{A})\cap\rho(A_0)\cap\mathfrak{h}(\tau)\cap\Omega$$

is a subset of $\mathfrak{h}((M+\tau)^{-1})$ *and the formula*

$$P_{\mathcal{K}}(\widetilde{A}-\lambda)^{-1}|_{\mathcal{K}}=(A_0-\lambda)^{-1}-\gamma(\lambda)\bigl(M(\lambda)+\tau(\lambda)\bigr)^{-1}\gamma(\overline{\lambda})^+ \tag{3.26}$$

holds for all $\lambda\in\rho(\widetilde{A})\cap\rho(A_0)\cap\mathfrak{h}(\tau)\cap\Omega$.

(ii) *Let $\tau\in\widetilde{\mathcal{N}}(\Omega)$ be a local generalized Nevanlinna function such that $M(\mu)+\tau(\mu)\neq 0$ for some $\mu\in\Omega$ and let Ω' be a domain with the same properties as Ω, $\overline{\Omega'}\subset\Omega$. Then there exists a Krein space $\mathcal{H}$ and a $\mathcal{K}$-minimal self-adjoint exit space extension $\widetilde{A}$ of A in $\mathcal{K}\times\mathcal{H}$ which is of type π_+ over Ω', such that*

$$\rho(A_0)\cap\mathfrak{h}(\tau)\cap\mathfrak{h}\bigl((M+\tau)^{-1}\bigr)\cap\Omega$$

is a subset of $\rho(\widetilde{A})$ and formula (3.26) *holds for all points λ belonging to* $\rho(A_0)\cap\mathfrak{h}(\tau)\cap\mathfrak{h}((M+\tau)^{-1})\cap\Omega$.

References

[1] T.Ya. Azizov, J. Behrndt, P. Jonas, C. Trunk: Spectral Points of Type π_+ and Type π_- for Closed Linear Relations in Krein Spaces, *submitted.*

[2] T.Ya. Azizov, P. Jonas, C. Trunk: Spectral Points of Type π_+ and Type π_- of Selfadjoint Operators in Krein Spaces, *J. Funct. Anal.* **226** (2005), 114–137.

[3] J. Behrndt: A Class of Abstract Boundary Value Problems with Locally Definitizable Functions in the Boundary Condition, Operator Theory: Advances and Applications **163** (2005), 55–73.

[4] J. Behrndt: Finite Rank Perturbations of Locally Definitizable Operators in Krein Spaces, *to appear in* J. Operator Theory.

[5] J. Behrndt, P. Jonas: On Compact Perturbations of Locally Definitizable Selfadjoint Relations in Krein Spaces, Integral Equations Operator Theory **52** (2005), 17–44.

[6] J. Behrndt, H.C. Kreusler: Boundary Relations and Generalized Resolvents of Symmetric Relations in Krein Spaces, *submitted.*

[7] J. Behrndt, C. Trunk: On Generalized Resolvents of Symmetric Operators of Defect One with Finitely many Negative Squares, Proceedings of the Algorithmic Information Theory Conference, Vaasan Yliop. Julk. Selvityksiä Rap., 124, <http://www.ams.org/mathscinet/search/series.html?cn=Vaasan_Yliop_Julk_Selvityksia_Rap> Vaasan Yliopisto, Vaasa, (2005), 21–30.

[8] V.A. Derkach: On Weyl Function and Generalized Resolvents of a Hermitian Operator in a Krein Space, Integral Equations Operator Theory **23** (1995), 387–415.

[9] V.A. Derkach: On Krein Space Symmetric Linear Relations with Gaps, Methods of Funct. Anal. Topology **4** (1998), 16–40.

[10] V.A. Derkach: On Generalized Resolvents of Hermitian Relations in Krein Spaces, J. Math. Sci. (New York) **97** (1999), 4420–4460.

[11] V.A. Derkach, S. Hassi, M.M. Malamud, H.S.V. de Snoo: Generalized Resolvents of Symmetric Operators and Admissibility, Methods Funct. Anal. Topology **6** (2000), 24–53.

[12] V.A. Derkach, M.M. Malamud: Generalized Resolvents and the Boundary Value Problems for Hermitian Operators with Gaps, J. Funct. Anal. **95** (1991), 1–95.

[13] V.A. Derkach, M.M. Malamud: The Extension Theory of Hermitian Operators and the Moment Problem, J. Math. Sci. (New York) **73** (1995), 141–242.

[14] V.A. Derkach, M.M. Malamud: On some Classes of Holomorphic Operator Functions with Nonnegative Imaginary Part, Operator Theory, Operator Algebras and related Topics (Timişoara, 1996), Theta Found., Bucharest (1997), 113–147.

[15] A. Dijksma, H.S.V. de Snoo: Symmetric and Selfadjoint Relations in Krein Spaces I, Operator Theory: Advances and Applications **24**, Birkhäuser Verlag Basel (1987), 145–166.

[16] A. Dijksma, H.S.V. de Snoo: Symmetric and Selfadjoint Relations in Krein Spaces II, Ann. Acad. Sci. Fenn. Math. **12**, (1987), 199–216.

[17] V.I. Gorbachuk, M.L. Gorbachuk: Boundary Value Problems for Operator Differential Equations, Kluwer Academic Publishers, Dordrecht (1991).

[18] P. Jonas: A Class of Operator-valued Meromorphic Functions on the Unit Disc, Ann. Acad. Sci. Fenn. Math. **17** (1992), 257–284.

[19] P. Jonas: Operator Representations of Definitizable Functions, Ann. Acad. Sci. Fenn. Math. **25** (2000), 41–72.

[20] P. Jonas: On Locally Definite Operators in Krein Spaces, in: Spectral Theory and Applications, Theta Foundation 2003, 95–127.

[21] P. Jonas: On Operator Representations of Locally Definitizable Functions, Operator Theory: Advances and Applications **162**, Birkhäuser Verlag Basel (2005), 165–190.

[22] M.G. Krein: On Hermitian Operators with Defect-indices equal to Unity, Dokl. Akad. Nauk SSSR, **43** (1944), 339–342.

[23] M.G. Krein: On the Resolvents of an Hermitian Operator with Defect-index (m, m), Dokl. Akad. Nauk SSSR, **52** (1946), 657–660.

[24] M.G. Krein, H. Langer: On Defect Subspaces and Generalized Resolvents of Hermitian Operators in Pontryagin Spaces, Funktsional. Anal. i Prilozhen. **5** No. 2 (1971) 59-71; **5** No. 3 (1971) 54-69 (Russian); English transl.: Funct. Anal. Appl. **5** (1971/1972), 139–146, 217–228.

[25] M.G. Krein, H. Langer: Über einige Fortsetzungsprobleme, die eng mit der Theorie hermitescher Operatoren im Raume Π_κ zusammenhängen. I. Einige Funktionenklassen und ihre Darstellungen, Math. Nachr. **77** (1977), 187–236.

[26] M.G. Krein, H. Langer: Some Propositions on Analytic Matrix Functions Related to the Theory of Operators in the Space Π_κ, Acta Sci. Math. (Szeged), **43** (1981), 181–205.

[27] H. Langer: Verallgemeinerte Resolventen eines J-nichtnegativen Operators mit endlichem Defekt, J. Funct. Anal. **8** (1971), 287–320.

[28] H. Langer: Spectral Functions of Definitizable Operators in Krein Spaces, Functional Analysis Proceedings of a Conference held at Dubrovnik, Yugoslavia, November 2–14 (1981), Lecture Notes in Mathematics **948**, Springer Verlag Berlin-Heidelberg-New York (1982), 1–46.

[29] H. Langer, A. Markus, V. Matsaev: Locally Definite Operators in Indefinite Inner Product Spaces, Math. Ann. **308** (1997), 405–424.

[30] H. Langer, B. Textorius: On Generalized Resolvents and Q-functions of Symmetric Linear Relations (Subspaces) in Hilbert Space, Pacific J. Math. **72** (1977), 135–165.

[31] M.M. Malamud: On a Formula for the Generalized Resolvents of a Non-densely defined Hermitian Operator (*Russian*), Ukrain. Mat. Zh. **44** (1992), 1658–1688; *translation in* Ukrainian Math. J. **44** (1993), 1522–1547.

[32] M.A. Naimark: On Spectral Functions of a Symmetric Operator, Izv. Akad. Nauk SSSR, Ser. Matem. **7** (1943), 373–375.

Jussi Behrndt
Institut für Mathematik, MA 6-4
Technische Universität Berlin
Straße des 17. Juni 136
D-10623 Berlin, Germany
e-mail: `behrndt@math.tu-berlin.de`

Annemarie Luger
Institut für Analysis und Scientific Computing
Technische Universität Wien
Wiedner Hauptstraße 8-10
A-1040 Wien, Austria
e-mail: `aluger@mail.zserv.tuwien.ac.at`

Carsten Trunk
Institut für Mathematik, MA 6-3
Technische Universität Berlin
Straße des 17. Juni 136
D-10623 Berlin, Germany
e-mail: `trunk@math.tu-berlin.de`

Operator Theory:
Advances and Applications, Vol. 175, 33–49

Block Operator Matrices, Optical Potentials, Trace Class Perturbations and Scattering

Jussi Behrndt, Hagen Neidhardt and Joachim Rehberg

Abstract. For an operator-valued block-matrix model, which is called in quantum physics a Feshbach decomposition, a scattering theory is considered. Under trace class perturbations the channel scattering matrices are calculated. Using Feshbach's optical potential it is shown that for a given spectral parameter the channel scattering matrices can be recovered either from a dissipative or from a Lax-Phillips scattering theory.

Mathematics Subject Classification (2000). Primary 47A40; Secondary 47A55, 47B44.

Keywords. Feshbach decomposition, optical potential, Lax-Phillips scattering theory, dissipative scattering theory, scattering matrix, characteristic function, dissipative operators.

1. Introduction

Let L and L_0 be self-adjoint operators in a separable Hilbert space $\mathfrak{L}$ and denote by $P^{ac}(L_0)$ the orthogonal projection onto the absolutely continuous subspace $\mathfrak{L}^{ac}(L_0)$ of L_0. The pair $\{L, L_0\}$ of self-adjoint operators is said to perform a scattering system if the wave operators $W_\pm(L, L_0)$,

$$W_\pm(L, L_0) := s - \lim_{t\to\pm\infty} e^{itL}e^{-itL_0}P^{ac}(L_0), \tag{1.1}$$

exist, cf. [5]. If the wave operators exist, then they are isometries from the absolutely continuous subspace $\mathfrak{L}^{ac}(L_0)$ into the absolutely continuous subspace $\mathfrak{L}^{ac}(L)$, i.e., $\operatorname{ran}(W_\pm(L, L_0)) \subseteq \mathfrak{L}^{ac}(L)$. The scattering system $\{L, L_0\}$ is called complete if the ranges of the wave operators $W_\pm(L, L_0)$ coincide with $\mathfrak{L}^{ac}(L)$, cf. [5]. The operator

$$S(L, L_0) := W_+(L, L_0)^*W_-(L, L_0)$$

is called the scattering operator of the scattering system $\{L, L_0\}$. The scattering operator regarded as an operator in $\mathfrak{L}^{ac}(L_0)$ commutes with L_0^{ac}. If the scattering system $\{L, L_0\}$ is complete, then $S(L, L_0)$ is a unitary operator in $\mathfrak{L}^{ac}(L_0)$. In physical applications L_0 is usually called the unperturbed or free Hamiltonian while L is called the perturbed or full Hamiltonian. Since $S(L, L_0)$ commutes with the free Hamiltonian L_0 the scattering operator is unitarily equivalent to a multiplication operator induced by a family $\{S(\lambda)\}_{\lambda\in\mathbb{R}}$ of unitary operators in the spectral representation of L_0. This family is called the scattering matrix of the complete scattering system $\{L, L_0\}$ and is the most important quantity in the analysis of scattering processes.

In this paper we investigate the special case that the Hilbert space $\mathfrak{L}$ splits into two subspaces $\mathfrak{H}_1$ and $\mathfrak{H}_2$,

$$\mathfrak{L} = \begin{array}{c}\mathfrak{H}_1\\ \oplus\\ \mathfrak{H}_2\end{array},$$

and the unperturbed Hamiltonian L_0 is of the form

$$L_0 = \begin{pmatrix} H_1 & 0 \\ 0 & H_2 \end{pmatrix} : \begin{array}{c}\mathfrak{H}_1\\ \oplus\\ \mathfrak{H}_2\end{array} \longrightarrow \begin{array}{c}\mathfrak{H}_1\\ \oplus\\ \mathfrak{H}_2\end{array}. \tag{1.2}$$

In physics the subspaces $\mathfrak{H}_j$ and the self-adjoint operators H_j, $j = 1, 2$, are often called scattering channels and channel Hamiltonians, respectively. With respect to the decomposition (1.2) one introduces the channel wave operators

$$W_\pm(L, H_j) := s - \lim_{t\to\pm\infty} e^{itL} J_j e^{-itH_j} P^{ac}(H_j)$$

where $J_j : \mathfrak{H}_j \longrightarrow \mathfrak{L}$ is the natural embedding operator. Introducing the channel scattering operators

$$S_{ij} = W_+(L, H_i)^* W_-(L, H_j) : \mathfrak{H}_j \longrightarrow \mathfrak{H}_i, \quad i, j = 1, 2,$$

one obtains a channel decomposition of the scattering operator

$$S(L, L_0) = \begin{pmatrix} S_{11} & S_{12} \\ S_{21} & S_{22} \end{pmatrix}. \tag{1.3}$$

In physics the decomposition (1.2) is often motivated either by the exclusive interest to scattering data in a certain channel or by the limited measuring process which allows to measure the scattering data only of a certain channel, say $\mathfrak{H}_1$. Thus, let us assume that only the channel scattering operator $S_{11} : \mathfrak{H}_1^{ac} \longrightarrow \mathfrak{H}_1^{ac}$ in the scattering channel $\mathfrak{H}_1$ is known. This gives rise to the following problem: Is it possible to replace the full Hamiltonian L by an effective one H acting only in $\mathfrak{H}_1$ such that the scattering operator of the scattering system $\{H, H_1\}$ coincides with S_{11}? Since S_{11} is a contraction, in general, this implies that either the scattering system $\{H, H_1\}$ cannot be complete or H is not self-adjoint.

The problem has a solution within the scope of dissipative scattering systems developed in [18, 19, 20] for pairs $\{H, H_1\}$ of dissipative and self-adjoint operators in some separable Hilbert space. For such pairs the wave operators $W^D_\pm(H, H_1)$ are defined by

$$W^D_+(H, H_1) := s - \lim_{t\to+\infty} e^{itH^*} e^{-itH_1} P^{ac}(H_1)$$

and

$$W^D_-(H, H_1) := s - \lim_{t\to+\infty} e^{-itH} e^{itH_1} P^{ac}(H_1),$$

and the notion of completeness is generalized, cf. [18, 19, 20]. The scattering operator of a dissipative scattering system $\{H, H_1\}$ is defined by

$$S_D := W^D_+(H, H_1)^* W^D_-(H, H_1).$$

It turns out that S_D is a contraction acting on the absolutely continuous subspace $\mathfrak{H}^{ac}_1$ of H_1 which commutes with H_1. In [17, 18] it was shown that for any self-adjoint operator H_1 in $\mathfrak{H}_1$ and any contraction S_D acting on the absolutely continuous subspace $\mathfrak{H}^{ac}_1$ and commuting with H_1 there is a maximal dissipative operator H on $\mathfrak{H}_1$ such that $\{H, H_1\}$ performs a complete scattering system with scattering operator given by S_D. In particular, this holds for the self-adjoint operator H_1 and the channel scattering operator S_{11}. That means, there is a maximal dissipative operator H on $\mathfrak{H}_1$ such that the channel scattering operator S_{11} is the scattering operator of the complete dissipative scattering system $\{H, H_1\}$. Hence, roughly speaking, the scattering operator S_{11} can be always viewed as the scattering operator of a suitable chosen dissipative scattering system on $\mathfrak{H}_1$. The disadvantage of this fact is that H is not known explicitly.

Another approach to this problem was suggested by Feshbach in [10, 11], see also [6, 9]. He proposes a concrete dissipative perturbation V_1 of the channel Hamiltonian H_1, called “optical potential”, such that the scattering operator S_1 of the dissipative scattering system $\{H_1 + V_1, H_1\}$ approximates S_{11} with a certain accuracy. To explain this approach in more detail let us assume that the full Hamiltonian L is obtained from L_0 by an additive perturbation, $L = L_0 + V$, where V is given by

$$V = \begin{pmatrix} 0 & G \\ G^* & 0 \end{pmatrix} : \begin{matrix} \mathfrak{H}_1 \\ \oplus \\ \mathfrak{H}_2 \end{matrix} \longrightarrow \begin{matrix} \mathfrak{H}_1 \\ \oplus \\ \mathfrak{H}_2 \end{matrix} . \tag{1.4}$$

Introducing the “optical potential”

$$V_1(\lambda) := -G(H_2 - \lambda - i0)^{-1} G^*, \quad \lambda \in \mathbb{R}, \tag{1.5}$$

it was shown in [8, Theorem 4.4.4] that under strong assumptions indeed the scattering operator $S_1[\lambda]$ of the (in general dissipative) scattering system $\{H_1(\lambda), H_1\}$,

$$H_1(\lambda) := H_1 + V_1(\lambda), \quad \lambda \in \mathbb{R}, \tag{1.6}$$

coincides with the scattering operator S_{11} with an error of second order in the coupling constant.

We show that Feshbach's proposal can be made precise in another sense. Note first that the decomposition (1.2) leads not only to the decomposition (1.3) of the scattering operator S but also to a decomposition of the scattering matrix $\{S(\mu)\}_{\mu\in\mathbb{R}}$,

$$S(\mu) := \begin{pmatrix} S_{11}(\mu) & S_{12}(\mu) \\ S_{21}(\mu) & S_{22}(\mu) \end{pmatrix},$$

where $\{S_{ij}(\mu)\}_{\mu\in\mathbb{R}}$ are called the channel scattering matrices. Denoting the scattering matrix of the dissipative scattering system $\{H_1(\lambda), H_1\}$ by $\{S_1[\lambda](\mu)\}_{\mu\in\mathbb{R}}$ we prove that

$$S_{11}(\lambda) = S_1[\lambda](\lambda) \tag{1.7}$$

holds for a.e. $\lambda \in \mathbb{R}$. This shows, that Feshbach's proposal gives in fact a good approximation of the channel scattering matrix $\{S_{11}(\mu)\}_{\mu\in\mathbb{R}}$ in a neighborhood of the chosen spectral parameter λ of the optical potential $V_1(\lambda)$.

Moreover, Feshbach's proposal implies a second problem. Similarly to the optical potential $V_1(\lambda)$ in the first channel $\mathfrak{H}_1$ one can introduce an optical potential $V_2(\lambda)$ in the second channel,

$$V_2(\lambda) := -G^*(H_1 - \lambda - i0)^{-1}G, \quad \lambda \in \mathbb{R}, \tag{1.8}$$

and define a perturbed operator $H_2(\lambda)$,

$$H_2(\lambda) := H_2 + V_2(\lambda), \quad \lambda \in \mathbb{R}, \tag{1.9}$$

in $\mathfrak{H}_2$. We show below that the characteristic function $\Theta_2[\lambda](\xi)$, $\xi \in \mathbb{C}_-$, of the dissipative operator $H_2(\lambda)$ and the scattering matrix $\{S_{11}(\lambda)\}_{\lambda\in\mathbb{R}}$ are related by

$$S_{11}(\lambda) = \Theta_2[\lambda](\lambda)^* \tag{1.10}$$

for a.e. $\lambda \in \mathbb{R}$. By [1]–[4] the last relation also yields that the scattering matrix $\{S_{11}(\lambda)\}_{\lambda\in\mathbb{R}}$ can be regarded as the scattering matrix $S_{LP}[\lambda](\mu)$ of a Lax-Phillips scattering system at the point λ.

Below we restrict ourself to a complete scattering system $\{L, L_0\}$, $L = L_0 + V$, where the perturbation V is a self-adjoint trace class operator. The assumption that V is a trace class operator is made for simplicity. Indeed, it would be sufficient to assume that the resolvent difference $(L - z)^{-p} - (L_0 - z)^{-p}$ is nuclear for a certain $p \in \mathbb{N}$ or, more generally, that the conditions of the so-called "stationary" scattering theory are satisfied, cf. [5, Section 14]. However, we emphasize that in contrast to [8] the smallness of the perturbation V is not assumed. Following the lines of [5] we show in Section 2 how the scattering matrix of the scattering system $\{L, L_0\}$ can be calculated. Under the additional assumptions (1.2) and (1.4) we find in Section 3 the channel scattering matrices $\{S_{ij}(\lambda)\}_{\lambda\in\mathbb{R}}$. In Section 4 we prove relation (1.7). Section 5 is devoted to the proof of (1.10). Moreover, the Lax-Phillips scattering theory for which $\{\Theta_2[\lambda](\mu)^*\}_{\mu\in\mathbb{R}}$ is the scattering matrix is indicated.

2. Scattering matrix

In this section we briefly recall the notion of the scattering matrix $\{S(\lambda)\}_{\lambda\in\mathbb{R}}$ of a scattering system $\{L, L_0\}$, where it is assumed that the unperturbed operator L_0 is self-adjoint in the separable Hilbert space $\mathfrak{L}$ and the perturbed operator L differs from L_0 by a self-adjoint trace class operator $V \in \mathcal{B}_1(\mathfrak{L})$,

$$L = L_0 + V, \qquad V = V^* \in \mathcal{B}_1(\mathfrak{L}). \tag{2.1}$$

Let $E_0(\cdot)$ be the spectral measure of L_0 and denote by $\mathfrak{B}(\mathbb{R})$ the set of all Borel subsets of the real axis $\mathbb{R}$. Without loss of generality we assume throughout the paper that the condition

$$\mathfrak{L} = \text{clospan}\{E_0(\Delta)\text{ran}\,(|V|) : \Delta \in \mathfrak{B}(\mathbb{R})\} \tag{2.2}$$

is satisfied, where $|V| := (V^*V)^{1/2}$. By Theorem X.4.4 of [13] the scattering system $\{L, L_0\}$ is complete, that is, the ranges of the wave operators $W_\pm(L, L_0)$ in (1.1) coincide with the absolutely continuous subspace $\mathfrak{L}^{ac}(L)$ of L. The operator V admits the representation

$$V = |V|^{1/2}C|V|^{1/2}, \quad |V| = (V^*V)^{1/2}, \quad C = \text{sgn}(V), \tag{2.3}$$

where $|V|^{1/2}$ belongs to the Hilbert-Schmidt class $\mathcal{B}_2(\mathfrak{L})$ and $\text{sgn}(\cdot)$ is the signum function. By Proposition 3.14 of [5] the limits

$$|V|^{1/2}(L - \lambda \pm i0)^{-1}|V|^{1/2} = \lim_{\epsilon\to+0} |V|^{1/2}(L - \lambda \pm i\epsilon)^{-1}|V|^{1/2} \tag{2.4}$$

exist in $\mathcal{B}_2(\mathfrak{L})$ for a.e. $\lambda \in \mathbb{R}$. The same holds for the limits

$$|V|^{1/2}(L_0 - \lambda \pm i0)^{-1}|V|^{1/2}.$$

Moreover by Proposition 3.13 of [5] the derivative

$$M_0(\lambda) := \frac{|V|^{1/2}E_0(d\lambda)|V|^{1/2}}{d\lambda} \geq 0 \tag{2.5}$$

exists in $\mathcal{B}_1(\mathfrak{L})$ for a.e. $\lambda \in \mathbb{R}$. We set

$$\mathfrak{Q}_\lambda := \text{clo}\big\{\text{ran}\,(M_0(\lambda))\big\} \subseteq \mathfrak{L}.$$

By $\{Q(\lambda)\}_{\lambda\in\mathbb{R}}$ we denote the family of orthogonal projections from $\mathfrak{L}$ onto $\mathfrak{Q}_\lambda$. One verifies that $\{Q(\lambda)\}_{\lambda\in\mathbb{R}}$ is measurable. Let us consider the standard Hilbert space $L^2(\mathbb{R}, d\lambda, \mathfrak{L})$. On $L^2(\mathbb{R}, d\lambda, \mathfrak{L})$ we introduce the projection Q

$$(Qf)(\lambda) := Q(\lambda)f(\lambda), \quad \lambda \in \mathbb{R}, \quad f \in L^2(\mathbb{R}, d\lambda, \mathfrak{L}),$$

and set $\mathfrak{Q} = \text{ran}\,(Q)$. Further, in $L^2(\mathbb{R}, d\lambda, \mathfrak{L})$ we define the multiplication operator $M_\mathfrak{L}$ by

$$\begin{aligned} (M_\mathfrak{L} f)(\lambda) &:= \lambda f(\lambda), \quad \lambda \in \mathbb{R}, \\ \text{dom}\,(M_\mathfrak{L}) &:= \big\{f \in L^2(\mathbb{R}, d\lambda, \mathfrak{L}) : \lambda f(\lambda) \in L^2(\mathbb{R}, d\lambda, \mathfrak{L})\big\}. \end{aligned}$$

Obviously, the multiplication operator $M_\mathfrak{L}$ and the projection Q commute. We set

$$M_\mathfrak{Q} := M_\mathfrak{L} \upharpoonright \text{dom}\,(M_\mathfrak{L}) \cap \mathfrak{Q}.$$

From Section 4.5 of [5] one gets that the absolutely continuous part L^{ac} of the perturbed operator L and the operator $M_{\mathfrak{Q}}$ are unitarily equivalent. In the following we denote the subspace $\mathfrak{Q}$ by $L^2(\mathbb{R}, d\lambda, \mathfrak{Q}_\lambda)$ which can be regarded as the direct integral of the family of subspaces $\{\mathfrak{Q}_\lambda\}_{\lambda\in\mathbb{R}}$ with respect to the Lebesgue measure $d\lambda$ on $\mathbb{R}$, cf. [5].

Since the scattering operator $S = W_+(L, L_0)^* W_-(L, L_0)$ acts on $\mathfrak{L}^{ac}(L_0)$ and commutes with L_0^{ac} there is a measurable family $\{S(\lambda)\}_{\lambda\in\mathbb{R}}$ of operators

$$S(\lambda) : \mathfrak{Q}_\lambda \longrightarrow \mathfrak{Q}_\lambda$$

such that S is unitarily equivalent to the multiplication operator

$$\begin{aligned} (M_{\mathfrak{Q}}(S)f)(\lambda) &:= S(\lambda)f(\lambda), \\ \operatorname{dom}(M_{\mathfrak{Q}}(S)) &:= L^2(\mathbb{R}, d\lambda, \mathfrak{Q}_\lambda). \end{aligned}$$

The family $\{S(\lambda)\}_{\lambda\in\mathbb{R}}$ is called the *scattering matrix* of the scattering system $\{L, L_0\}$. Since the scattering system $\{L, L_0\}$ is complete the operator $S(\lambda)$ is unitary on $\mathfrak{Q}_\lambda$ for a.e. $\lambda \in \mathbb{R}$.

The following representation theorem of the scattering matrix is a consequence of Corollary 18.9 of [5], see also [5, Section 18.2.2].

Theorem 2.1. *Let L, L_0 and V be self-adjoint operators in $\mathfrak{L}$ as in* (2.1). *Then $\{L, L_0\}$ is a complete scattering system and the corresponding scattering matrix matrix $\{S(\lambda)\}_{\lambda\in\mathbb{R}}$ admits the representation*

$$S(\lambda) = I_{\mathfrak{Q}_\lambda} - 2\pi i\, M_0^{1/2}(\lambda)\left\{C - C|V|^{1/2}(L-\lambda-i0)^{-1}|V|^{1/2}C\right\} M_0^{1/2}(\lambda)$$

for a.e. $\lambda \in \mathbb{R}$.

3. Channel scattering matrices

Let us now assume that the Hilbert space $\mathfrak{L}$ is the orthogonal sum of two subspaces $\mathfrak{H}_1$ and $\mathfrak{H}_2$, $\mathfrak{L} = \mathfrak{H}_1 \oplus \mathfrak{H}_2$, that L_0 is a diagonal block operator matrix of the form

$$L_0 = \begin{pmatrix} H_1 & 0 \\ 0 & H_2 \end{pmatrix} : \begin{matrix} \mathfrak{H}_1 \\ \oplus \\ \mathfrak{H}_2 \end{matrix} \longrightarrow \begin{matrix} \mathfrak{H}_1 \\ \oplus \\ \mathfrak{H}_2 \end{matrix}, \tag{3.1}$$

cf. (1.2), where H_1 and H_2 are self-adjoint operators in $\mathfrak{H}_1$ and $\mathfrak{H}_2$ and that $V \in \mathcal{B}_1(\mathfrak{L})$ is a self-adjoint trace class operator of the form

$$V = \begin{pmatrix} 0 & G \\ G^* & 0 \end{pmatrix} : \begin{matrix} \mathfrak{H}_1 \\ \oplus \\ \mathfrak{H}_2 \end{matrix} \longrightarrow \begin{matrix} \mathfrak{H}_1 \\ \oplus \\ \mathfrak{H}_2 \end{matrix}, \tag{3.2}$$

see (1.4). The operator $G : \mathfrak{H}_2 \longrightarrow \mathfrak{H}_1$ describes the interaction between the channels. Since V is a trace class operator we have

$$G \in \mathcal{B}_1(\mathfrak{H}_2, \mathfrak{H}_1).$$

The perturbed or full Hamiltonian L has the form

$$L := L_0 + V = \begin{pmatrix} H_1 & G \\ G^* & H_2 \end{pmatrix} : \begin{matrix} \mathfrak{H}_1 \\ \oplus \\ \mathfrak{H}_2 \end{matrix} \longrightarrow \begin{matrix} \mathfrak{H}_1 \\ \oplus \\ \mathfrak{H}_2 \end{matrix} . \tag{3.3}$$

The following lemma is known as the Feshbach decomposition in physics, cf. [10, 11]. We use the notation

$$H_1(z) = H_1 + V_1(z) \quad \text{and} \quad H_2(z) = H_2 + V_2(z), \quad z \in \mathbb{C}\backslash\mathbb{R}, \tag{3.4}$$

where

$$V_1(z) = -G(H_2 - z)^{-1}G^* \quad \text{and} \quad V_2(z) = -G^*(H_1 - z)^{-1}G, \tag{3.5}$$

see (1.6), (1.9), (1.5) and (1.8).

Lemma 3.1. *Let L, $H_1(z)$ and $H_2(z)$, $z \in \mathbb{C}\backslash\mathbb{R}$, be given by* (3.3) *and* (3.4), *respectively. Then we have $z \in \operatorname{res}(H_i(z))$, $i = 1,2$, for all $z \in \mathbb{C}\backslash\mathbb{R}$ and*

$$(L - z)^{-1} = \begin{pmatrix} (H_1(z) - z)^{-1} & -(H_1 - z)^{-1}G(H_2(z) - z)^{-1} \\ -(H_2(z) - z)^{-1}G^*(H_1 - z)^{-1} & (H_2(z) - z)^{-1} \end{pmatrix}. \tag{3.6}$$

Proof. From

$$\operatorname{Im}\big((H_1(z) - z)h_1, h_1\big) = \operatorname{Im}\overline{z}\|h_1\|^2 + \operatorname{Im}\overline{z}\|(H_2 - z)^{-1}G^*h_1\|^2,$$

$z \in \mathbb{C}\backslash\mathbb{R}$, $h_1 \in \mathfrak{H}_1$, we conclude that $(H_1(z)-z)^{-1}$ is a bounded everywhere defined operator for all $z \in \mathbb{C}\backslash\mathbb{R}$. Analogously one verifies that $(H_2(z)-z)^{-1}$ is a bounded everywhere defined operator for all $z \in \mathbb{C}\backslash\mathbb{R}$. A straightforward computation shows

$$(L-z)^{-1} = \begin{pmatrix} (H_1-z)^{-1} + (H_1-z)^{-1}G(H_2(z)-z)^{-1}G^*(H_1-z)^{-1} & -(H_1-z)^{-1}G(H_2(z)-z)^{-1} \\ -(H_2(z)-z)^{-1}G^*(H_1-z)^{-1} & (H_2(z)-z)^{-1} \end{pmatrix}$$

for $z \in \mathbb{C}\backslash\mathbb{R}$. From the identity

$$\big(I - G^*(H_1 - z)^{-1}G(H_2 - z)^{-1}\big)^{-1}G^* = G^*\big(I - (H_1 - z)^{-1}G(H_2 - z)^{-1}G^*\big)^{-1}$$

we obtain

$$\begin{aligned} &(H_1-z)^{-1} + (H_1-z)^{-1}G(H_2(z)-z)^{-1}G^*(H_1-z)^{-1} \\ &= (H_1-z)^{-1}\Big\{I + G(H_2-z)^{-1}\big(I - G^*(H_1-z)^{-1}G(H_2-z)^{-1}\big)^{-1}G^*(H_1-z)^{-1}\Big\} \\ &= (H_1-z)^{-1}\big\{I + G(H_2-z)^{-1}G^*(H_1(z)-z)^{-1}\big\} = (H_1(z)-z)^{-1} \end{aligned}$$

for all $z \in \mathbb{C}\backslash\mathbb{R}$, which proves (3.6). □

In the next lemma we calculate the limit $|V|^{1/2}(L - \lambda - i0)^{-1}|V|^{1/2}$, $\lambda \in \mathbb{R}$, cf. (2.4). Here and in the following it is convenient to use the functions

$$\begin{aligned} N_1(z) &:= |G^*|^{1/2}(H_1 - z)^{-1}|G^*|^{1/2}, \\ N_2(z) &:= |G|^{1/2}(H_2 - z)^{-1}|G|^{1/2}, \end{aligned} \qquad z \in \mathbb{C}\backslash\mathbb{R}, \tag{3.7}$$

and

$$\begin{aligned} F_1(z) &:= |G^*|^{1/2}(H_1(z)-z)^{-1}|G^*|^{1/2}, \\ F_2(z) &:= |G|^{1/2}(H_2(z)-z)^{-1}|G|^{1/2}, \end{aligned} \qquad z \in \mathbb{C}\backslash\mathbb{R}. \tag{3.8}$$

Lemma 3.2. *Let $V \in \mathcal{B}_1(\mathfrak{L})$ be given by (3.2) with $G \in \mathcal{B}_1(\mathfrak{H}_2, \mathfrak{H}_1)$ and let U be a partial isometry such that $G = U|G|$. Then the limits*

$$N_i(\lambda) := \lim_{\epsilon \to +0} N_i(\lambda + i\epsilon) \quad \text{and} \quad F_i(\lambda) := \lim_{\epsilon \to +0} F_i(\lambda + i\epsilon), \quad i = 1, 2, \tag{3.9}$$

exist in $\mathcal{B}_2(\mathfrak{H}_i)$ for a.e. $\lambda \in \mathbb{R}$ and the representation

$$|V|^{1/2}(L - \lambda - i0)^{-1}|V|^{1/2} = \begin{pmatrix} F_1(\lambda) & -N_1(\lambda)UF_2(\lambda) \\ -F_2(\lambda)U^*N_1(\lambda) & F_2(\lambda) \end{pmatrix} \tag{3.10}$$

holds for a.e. $\lambda \in \mathbb{R}$.

Proof. By $|G|^{1/2} \in \mathcal{B}_2(\mathfrak{H}_2)$ and $|G^*|^{1/2} \in \mathcal{B}_2(\mathfrak{H}_1)$ the existence of the limits $N_i(\lambda)$ in (3.9) for a.e. $\lambda \in \mathbb{R}$ follows from Proposition 3.13 of [5]. Using the representations $F_1(z) = |G^*|^{1/2}P_1(L-z)^{-1} \restriction_{\mathfrak{H}_1} |G^*|^{1/2}$ and $F_2(z) = |G|^{1/2}P_2(L-z)^{-1} \restriction_{\mathfrak{H}_2} |G|^{1/2}$, $z \in \mathbb{C}\backslash\mathbb{R}$, which follow from (3.6), and taking into account [5, Proposition 3.13] we again obtain the existence of $F_i(\lambda)$, $i = 1, 2$, for a.e. $\lambda \in \mathbb{R}$. It is easy to see that

$$|V|^{1/2} = (V^*V)^{1/4} = \begin{pmatrix} |G^*|^{1/2} & 0 \\ 0 & |G|^{1/2} \end{pmatrix} \tag{3.11}$$

holds. Let U be a partial isometry from $\mathfrak{H}_2$ into $\mathfrak{H}_1$ such that $G = U|G|$. Making use of the factorizations

$$G = |G^*|^{1/2}U|G|^{1/2} \quad \text{and} \quad G^* = |G|^{1/2}U^*|G^*|^{1/2}, \tag{3.12}$$

the block matrix representation of $(L-z)^{-1}$ in Lemma 3.1 and relation (3.11) one verifies (3.10). □

We note that if U is a partial isometry such that $G = U|G|$ holds and C is defined by

$$C := \begin{pmatrix} 0 & U \\ U^* & 0 \end{pmatrix}, \tag{3.13}$$

then the operator V in (3.2) can be written in the form $|V|^{1/2}C|V|^{1/2}$, cf. (2.3). Let $E_1(\cdot)$ and $E_2(\cdot)$ be the spectral measures of H_1 and H_2, respectively. The operator function $M_0(\cdot)$ from (2.5) here admits the representation

$$M_0(\lambda) = \begin{pmatrix} M_1(\lambda) & 0 \\ 0 & M_2(\lambda) \end{pmatrix} \tag{3.14}$$

for a.e $\lambda \in \mathbb{R}$, where the derivatives

$$M_1(\lambda) = \frac{|G^*|^{1/2}E_1(d\lambda)|G^*|^{1/2}}{d\lambda} \quad \text{and} \quad M_2(\lambda) = \frac{|G|^{1/2}E_2(d\lambda)|G|^{1/2}}{d\lambda} \tag{3.15}$$

exist in $\mathcal{B}_1(\mathfrak{H}_1)$ and $\mathcal{B}_1(\mathfrak{H}_2)$ for a.e. $\lambda \in \mathbb{R}$, respectively. Setting

$$\mathfrak{Q}_{j,\lambda} := \text{clo}\{\text{ran}\,(M_j(\lambda))\}, \quad j = 1, 2,$$

and

$$\mathfrak{L}_\lambda := \mathfrak{L}_{1,\lambda} \oplus \mathfrak{L}_{2,\lambda} \tag{3.16}$$

for a.e. $\lambda \in \mathbb{R}$ we obtain the decomposition

$$L^2(\mathbb{R}, d\lambda, \mathfrak{L}_\lambda) = L^2(\mathbb{R}, d\lambda, \mathfrak{L}_{1,\lambda}) \oplus L^2(\mathbb{R}, d\lambda, \mathfrak{L}_{2,\lambda}),$$

cf. Section 2. From (2.2) the conditions

$$\begin{aligned} \mathfrak{H}_1 &= \operatorname{clospan}\{E_1(\Delta)\operatorname{ran}(|G^*|) : \Delta \in \mathfrak{B}(\mathbb{R})\}, \\ \mathfrak{H}_2 &= \operatorname{clospan}\{E_2(\Delta)\operatorname{ran}(|G|) : \Delta \in \mathfrak{B}(\mathbb{R})\} \end{aligned} \tag{3.17}$$

follow. Moreover, the converse is also true, that is, condition (3.17) implies (2.2). Hence, without loss of generality we assume that condition (3.17) is satisfied. Therefore the reduced multiplication operators $M_{\mathfrak{L}_j}$,

$$M_{\mathfrak{L}_j} := M_{\mathfrak{H}_j} \upharpoonright \operatorname{dom}(M_{\mathfrak{H}_j}) \cap L^2(\mathbb{R}, d\lambda, \mathfrak{L}_{j,\lambda}),$$

where

$$\begin{aligned} &(M_{\mathfrak{H}_j} f)(\lambda) := \lambda f(\lambda), \quad \lambda \in \mathbb{R}, \\ &\operatorname{dom}(M_{\mathfrak{H}_j}) := \{f \in L^2(\mathbb{R}, d\lambda, \mathfrak{H}_j) : \lambda f(\lambda) \in L^2(\mathbb{R}, d\lambda, \mathfrak{H}_j)\}. \end{aligned}$$

are unitary equivalent to the absolutely continuous parts H_j^{ac} of the operators H_j, $j = 1, 2$.

With respect to the decomposition (3.16) the scattering matrix $\{S(\lambda)\}_{\lambda \in \mathbb{R}}$ admits the decomposition

$$S(\lambda) = \begin{pmatrix} S_{11}(\lambda) & S_{12}(\lambda) \\ S_{21}(\lambda) & S_{22}(\lambda) \end{pmatrix} : \begin{matrix} \mathfrak{L}_{1,\lambda} \\ \oplus \\ \mathfrak{L}_{2,\lambda} \end{matrix} \longrightarrow \begin{matrix} \mathfrak{L}_{1,\lambda} \\ \oplus \\ \mathfrak{L}_{2,\lambda} \end{matrix} \tag{3.18}$$

for a.e. $\lambda \in \mathbb{R}$. The entries $\{S_{ij}(\lambda)\}_{\lambda \in \mathbb{R}}$, $i, j = 1, 2$, are called *channel scattering matrices.* We note that the multiplication operators induced by the channel scattering matrices are unitary equivalent to the channel scattering operators $S_{ij} = P_i S P_j$, $i, j = 1, 2$, where P_i is the orthogonal projection in $\mathfrak{L}$ onto the subspace $\mathfrak{H}_j$ and S is the scattering operator of the complete scattering system $\{L, L_0\}$.

In the next proposition we give a more explicit description of the channel scattering matrices $S_{ij}(\lambda)$. The proof is an immediate consequence of Theorem 2.1, Lemma 3.2 and relations (3.14), (3.15) and (3.13).

Proposition 3.3. *Let* L_0, V *and* L *be given in accordance with* (3.1), (3.2) *and* (3.3), *respectively. Then the scattering matrix* $\{S(\lambda)\}_{\lambda \in \mathbb{R}}$ *of the complete scattering system* $\{L, L_0\}$ *admits the representation* (3.18) *with entries* $S_{ij}(\lambda)$ *given by*

$$\begin{aligned} S_{11}(\lambda) &= I_{\mathfrak{L}_{1,\lambda}} + 2\pi i M_1(\lambda)^{1/2} U F_2(\lambda) U^* M_1(\lambda)^{1/2}, \\ S_{12}(\lambda) &= -2\pi i M_1(\lambda)^{1/2}\{U + U F_2(\lambda) U^* N_1(\lambda) U\} M_2(\lambda)^{1/2}, \\ S_{21}(\lambda) &= -2\pi i M_2(\lambda)^{1/2}\{U^* + U^* N_1(\lambda) U F_1(\lambda) U^*\} M_1(\lambda)^{1/2}, \\ S_{22}(\lambda) &= I_{\mathfrak{L}_{2,\lambda}} + 2\pi i M_2(\lambda)^{1/2} U^* F_1(\lambda) U M_2(\lambda)^{1/2}. \end{aligned} \tag{3.19}$$

for a.e. $\lambda \in \mathbb{R}$.

4. Dissipative channel scattering

In this section we consider the (dissipative) scattering system $\{H_1(\lambda), H_1\}$ for a.e. $\lambda \in \mathbb{R}$, where

$$H_1(\lambda) = H_1 + V_1(\lambda), \tag{4.1}$$

is defined for a.e. $\lambda \in \mathbb{R}$, and H_1 is the self-adjoint operator in $\mathfrak{H}_1$ from (3.1). The limit $V_1(\lambda) = \lim_{\epsilon\to+0} V_1(\lambda + i\epsilon)$ (see Lemma 4.1) is called the *optical potential* of the channel $\mathfrak{H}_1$. We recall that a linear operator T in a Hilbert space is said to be dissipative if $\operatorname{Im}(Tf, f) \le 0$, $f \in \operatorname{dom}(T)$, and T is called maximal dissipative if T is dissipative and does not admit a proper dissipative extension. In Theorem 4.4 below we establish a connection between the scattering matrices corresponding to the scattering systems $\{H_1(\lambda), H_1\}$ and the channel scattering matrix $S_{11}(\lambda)$ from (3.18) and (3.19).

Lemma 4.1. *Let $V_1(z) = -G(H_2 - z)^{-1}G^*$, $z \in \mathbb{C}\backslash\mathbb{R}$, be defined by (3.5) with $G \in \mathcal{B}_1(\mathfrak{H}_2, \mathfrak{H}_1)$. Then the limit $V_1(\lambda) = \lim_{\epsilon\to+0} V_1(\lambda + i\epsilon)$ exists in $\mathcal{B}_1(\mathfrak{H}_1)$ and $V_1(\lambda)$ is dissipative for a.e. $\lambda \in \mathbb{R}$.*

Proof. Using the factorizations (3.12) of G and G^* we find

$$V_1(z) = -|G^*|^{1/2} U N_2(z) U^* |G^*|^{1/2}, \quad z \in \mathbb{C}\backslash\mathbb{R}, \tag{4.2}$$

where $N_2(z)$ is given by (3.7). According to Lemma 3.2 the limit $\lim_{\epsilon\to+0} N_2(\lambda+i\epsilon)$ exists in $\mathcal{B}_2(\mathfrak{H}_1)$ and since $|G^*|^{1/2} \in \mathcal{B}_2(\mathfrak{H}_1)$ we conclude that the limit

$$V_1(\lambda) = \lim_{\epsilon\to+0} V_1(\lambda + i\epsilon) = \lim_{\epsilon\to+0} -|G^*|^{1/2} U N_2(\lambda + i\epsilon) U^* |G^*|^{1/2}$$

exists in $\mathcal{B}_1(\mathfrak{H}_1)$ for a.e. $\lambda \in \mathbb{R}$. It is not difficult to see that $\operatorname{Im} V_1(z) \le 0$ for $z \in \mathbb{C}^+$ and therefore also the limit $V_1(\lambda)$ is dissipative for a.e. $\lambda \in \mathbb{R}$. □

It follows from Lemma 4.1 that for a.e. $\lambda \in \mathbb{R}$ the operator $H_1(\lambda) = H_1 + V_1(\lambda)$ is maximal dissipative and therefore $\{H_1(\lambda), H_1\}$ is a *dissipative scattering system* in the sense of [19, 20]. By Theorem 4.3 of [20] the corresponding wave operators

$$W_+^D(H_1(\lambda), H_1) = s - \lim_{t\to+\infty} e^{itH_1(\lambda)^*} e^{-itH_1} P^{ac}(H_1)$$

and

$$W_-^D(H_1(\lambda), H_1) = s - \lim_{t\to+\infty} e^{-itH_1(\lambda)} e^{itH_1} P^{ac}(H_1)$$

exist and are complete which yields that $\{H_1(\lambda), H_1\}$ performs a complete dissipative scattering system for a.e. $\lambda \in \mathbb{R}$, see [19] for details. The associated scattering operators are defined by

$$S_D[\lambda] := W_+^D(H_1(\lambda), H_1)^* W_-^D(H_1(\lambda), H_1)$$

and act on the absolutely continuous subspaces $\mathfrak{H}_1^{ac}(H_1)$. Since $S_D[\lambda]$ commutes with H_1 the scattering operator is unitary equivalent to a multiplication operator in the spectral representation $L^2(\mathbb{R}, d\lambda, \mathfrak{Q}_{1,\mu})$ of H_1 induced by a family of contractions $\{S_D[\lambda](\mu)\}_{\mu\in\mathbb{R}}$. The family $\{S_D[\lambda](\mu)\}_{\mu\in\mathbb{R}}$ is called the *scattering matrix* of the complete dissipative scattering system $\{H_1(\lambda), H_1\}$.

Using the fact that every maximal dissipative operator admits a self-adjoint dilation, i.e., there exists a self-adjoint operator in a (in general) larger Hilbert space such that its compressed resolvent coincides with the resolvents of the maximal dissipative operator for all $z \in \mathbb{C}^+$, cf. [7, Section 7], see also [12], one concludes from Proposition 3.14 of [5] that the limit

$$\mathfrak{F}_1[\lambda](\mu) = \lim_{\epsilon \to +0} \mathfrak{F}_1[\lambda](\mu + i\epsilon)$$

exist in $\mathcal{B}_1(\mathfrak{H}_1)$ for a.e. $\mu \in \mathbb{R}$, where

$$\mathfrak{F}_1[\lambda](z) := |G^*|^{1/2}(H_1(\lambda) - z)^{-1}|G^*|^{1/2}, \quad z \in \mathbb{C}_+,$$

is defined for a.e. $\lambda \in \mathbb{R}$.

The next proposition is a direct consequence of Theorem 2.2 of [16], see also [15].

Proposition 4.2. *Let $G \in \mathcal{B}_1(\mathfrak{H}_2, \mathfrak{H}_1)$ and $H_1(\lambda)$ be given by (4.1). Then for a.e. $\lambda \in \mathbb{R}$ the scattering matrix $\{S_D[\lambda](\mu)\}_{\mu \in \mathbb{R}}$ of the complete dissipative scattering system $\{H_1(\lambda), H_1\}$ admits the representation*

$$S_D[\lambda](\mu) = I_{\mathfrak{Q}_{1,\mu}} + 2\pi i M_1(\mu)^{1/2} U \Big\{ N_2(\lambda) + N_2(\lambda) U^* \mathfrak{F}_1[\lambda](\mu) U N_2(\lambda) \Big\} U^* M_1(\mu)^{1/2}$$

for a.e. $(\mu, \lambda) \in \mathbb{R}^2$ with respect to the Lebesgue measure in $\mathbb{R}^2$.

In the next lemma we show that the limit $\mathfrak{F}_1[\lambda](\lambda)$,

$$\mathfrak{F}_1[\lambda](\lambda) = \lim_{\epsilon \to +0} \mathfrak{F}_1[\lambda](\lambda + i\epsilon) = \lim_{\epsilon \to +0} |G^*|^{1/2}(H_1(\lambda) - \lambda - i\epsilon)^{-1}|G^*|^{1/2}, \tag{4.3}$$

exist in $\mathcal{B}_2(\mathfrak{H}_1)$ for a.e. $\lambda \in \mathbb{R}$.

Lemma 4.3. *Let L_0, V and L be given by (3.1), (3.2) and (3.3), respectively, with $G \in \mathcal{B}_1(\mathfrak{H}_2, \mathfrak{H}_1)$. Further, let $F_1(\lambda)$ be as in Lemma 3.2 and let $H_1(\lambda)$ be defined by (4.1). Then the limit $\mathfrak{F}_1[\lambda](\lambda)$ in (4.3) exists in $\mathcal{B}_2(\mathfrak{H}_1)$ for a.e. $\lambda \in \mathbb{R}$ and the relation*

$$\mathfrak{F}_1[\lambda](\lambda) = F_1(\lambda) \tag{4.4}$$

holds for a.e. $\lambda \in \mathbb{R}$.

Proof. We have

$$\begin{aligned} F_1(z) - \mathfrak{F}_1[\lambda](z) &= |G^*|^{1/2}\big((H_1(z) - z)^{-1} - (H_1(\lambda) - z)^{-1}\big)|G^*|^{1/2} \\ &= |G^*|^{1/2}(H_1(z) - z)^{-1}(V_1(\lambda) - V_1(z))(H_1(\lambda) - z)^{-1}|G^*|^{1/2}. \end{aligned} \tag{4.5}$$

From (4.2) we obtain

$$V_1(\lambda) - V_1(z) = |G^*|^{1/2} U \big(N_2(z) - N_2(\lambda)\big) U^* |G^*|^{1/2}$$

and inserting this expression into (4.5) and using the definitions of $F_1(z)$ in (3.8) and $\mathfrak{F}_1[\lambda](z)$ yields

$$F_1(z) - \mathfrak{F}_1[\lambda](z) = F_1(z) U (N_2(z) - N_2(\lambda)) U^* \mathfrak{F}_1[\lambda](z).$$

Hence

$$F_1(z) = \{I_{\mathfrak{H}_1} + F_1(z) U (N_2(z) - N_2(\lambda)) U^*\} \mathfrak{F}_1[\lambda](z)$$

and for $z = \lambda + i\epsilon$, $\epsilon > 0$ sufficiently small, the operator

$$\{I_{\mathfrak{H}_1} + F_1(z)U(N_2(z) - N_2(\lambda))U^*\}$$

is invertible. Therefore we conclude

$$\{I_{\mathfrak{H}_1} + F_1(\lambda + i\epsilon)U(N_2(\lambda + i\epsilon) - N_2(\lambda))U^*\}^{-1} F_1(\lambda + i\epsilon) = \mathfrak{F}_1[\lambda](\lambda + i\epsilon).$$

From this representation we get the existence of $\mathfrak{F}_1[\lambda](\lambda)$ in $\mathcal{B}_2(\mathfrak{H}_1)$ and the equality (4.4) for a.e. $\lambda \in \mathbb{R}$. □

The next theorem is the main result of this section. We show how the channel scattering matrix $S_{11}(\lambda)$ of the scattering system $\{L, L_0\}$ is connected with the scattering matrices $S_D[\lambda](\mu)$ of the dissipative scattering systems $\{H_1(\lambda), H_1\}$.

Theorem 4.4. *Let $\{L, L_0\}$ be the scattering system from Section* 3*, where L_0, V and L are given by* (3.1)*,* (3.2) *and* (3.3)*, respectively, and $G \in \mathcal{B}_1(\mathfrak{H}_2, \mathfrak{H}_1)$. Further, let $\{S_{ij}(\lambda)\}$, $i, j = 1, 2$, be the corresponding scattering matrix from* (3.18) *and let $S_D[\lambda](\mu)$ be the scattering matrices of the dissipative scattering systems $\{H_1(\lambda), H_1\}$. Then the scattering matrix $S_D[\lambda](\lambda)$ exists for a.e $\lambda \in \mathbb{R}$ and satisfies the relation*

$$S_D[\lambda](\lambda) = S_{11}(\lambda)$$

for a.e. $\lambda \in \mathbb{R}$.

Proof. From Proposition 4.2 and Lemma 4.3 we obtain that $S_D[\lambda](\lambda)$ exists for a.e $\lambda \in \mathbb{R}$ and has the form

$$S_D[\lambda](\mu) = I_{\mathfrak{Q}_{1,\mu}} + 2\pi i M_1(\mu)^{1/2} U\Big\{N_2(\lambda) + N_2(\lambda)U^* F_1(\lambda)UN_2(\lambda)\Big\}U^* M_1(\mu)^{1/2} \quad (4.6)$$

A similar calculation as in the proof of Lemma 3.1 shows

$$F_2(z) = N_2(z) + N_2(z)U^* F_1(z)UN_2(z), \quad z \in \mathbb{C}_+.$$

If z tends to $\lambda \in \mathbb{R}$, then we get

$$F_2(\lambda) = N_2(\lambda) + N_2(\lambda)U^* F_1(\lambda)UN_2(\lambda) \quad (4.7)$$

for a.e. $\lambda \in \mathbb{R}$. Inserting (4.7) into (4.6) we obtain

$$S_D[\lambda](\lambda) = I_{\mathfrak{Q}_{1,\lambda}} + 2\pi i M_1(\lambda)^{1/2} UF_2(\lambda)U^* M_1(\lambda)^{1/2}$$

and by Proposition 3.3 this coincides with $S_{11}(\lambda)$ for a.e. $\lambda \in \mathbb{R}$. □

5. Lax-Phillips channel scattering

Similarly to Lemma 4.1 one verifies that $V_2(\lambda) = \lim_{\epsilon\to+0} V_2(\lambda + i\epsilon)$ exists in $\mathcal{B}_1(\mathfrak{H}_2)$ for a.e. $\lambda \in \mathbb{R}$. The limit $V_2(\lambda)$, which is called the optical potential of the channel $\mathfrak{H}_2$, is dissipative for a.e. $\lambda \in \mathbb{R}$. The optical potential defines the maximal dissipative operator

$$H_2(\lambda) := H_2 + V_2(\lambda)$$

for a.e. $\lambda \in \mathbb{R}$. The operator $H_2(\lambda)$ decomposes for a.e. $\lambda \in \mathbb{R}$ into a self-adjoint part and a completely non-self-adjoint part. Let $\Theta_2[\lambda](\xi)$, $\xi \in \mathbb{C}_-$, be the characteristic function, cf. [12], of the completely non-self-adjoint part of $H_2(\lambda)$. We are going to verify $S_{11}(\lambda) = \Theta_2[\lambda](\lambda)^*$ for a.e. $\lambda \in \mathbb{R}$ which shows that for a.e. $\lambda \in \mathbb{R}$ the scattering matrix $S_{11}(\lambda)$ can be regarded as the result of a certain Lax-Phillips scattering theory, cf. [1, 2, 3, 4, 14].

There is an orthogonal decomposition

$$\mathfrak{H}_2 = \mathfrak{H}_{2,\lambda}^{\mathrm{cns}} \oplus \mathfrak{H}_{2,\lambda}^{\mathrm{self}}$$

for a.e. $\lambda \in \mathbb{R}$ such that $\mathfrak{H}_{2,\lambda}^{\mathrm{cns}}$ and $\mathfrak{H}_{2,\lambda}^{\mathrm{self}}$ reduce $H_2(\lambda)$ into a completely non-self-adjoint operator $H_2^{\mathrm{cns}}(\lambda)$ and a self-adjoint operator $H_2^{\mathrm{self}}(\lambda)$,

$$H_2(\lambda) = H_2^{\mathrm{cns}}(\lambda) \oplus H_2^{\mathrm{self}}(\lambda).$$

Taking into account Proposition 3.14 of [5] we get that

$$\Im\mathrm{m}\,(V_2(\lambda)) = -\pi|G|^{1/2}U^*M_1(\lambda)U|G|^{1/2}$$

for a.e. $\lambda \in \mathbb{R}$. Let us introduce the operator

$$\alpha(\lambda) := \sqrt{2\pi M_1(\lambda)}\, U|G|^{1/2}. \tag{5.1}$$

Notice that

$$\mathrm{clo}\{\mathrm{ran}\,(\alpha(\lambda))\} = \mathfrak{Q}_{1,\lambda}$$

for a.e. $\lambda \in \mathbb{R}$. With the completely non-self-adjoint part $H^{\mathrm{cns}}(\lambda)$ one associates the characteristic function $\Theta_2[\lambda](\cdot) : \mathfrak{Q}_{1,\lambda} \longrightarrow \mathfrak{Q}_{1,\lambda}$ defined by

$$\Theta_2[\lambda](\xi) := I_{\mathfrak{Q}_{1,\lambda}} - i\alpha(\lambda)(H_2(\lambda)^* - \xi)^{-1}\alpha(\lambda)^*,$$

$\xi \in \mathbb{C}_-$. The characteristic function is a contraction-valued holomorphic function in $\mathbb{C}_-$. From [12, Section V.2] we get that the boundary values

$$\Theta_2[\lambda](\mu) := s - \lim_{\epsilon \to +0} \Theta_2[\lambda](\mu - i\epsilon)$$

exist for a.e. $\mu \in \mathbb{R}$.

Theorem 5.1. *Let L_0, V and L be given by* (3.1), (3.2) *and* (3.3). *If the condition $G \in \mathcal{B}_1(\mathfrak{H}_2, \mathfrak{H}_1)$ is satisfied, then the limit $\Theta_2[\lambda](\lambda)$,*

$$\Theta_2[\lambda](\lambda) := s - \lim_{\epsilon \to +0} \Theta_2[\lambda](\lambda - i\epsilon)$$

exists for a.e. $\lambda \in \mathbb{R}$ and the relation

$$S_{11}(\lambda) = \Theta_2[\lambda](\lambda)^*$$

holds for a.e. $\lambda \in \mathbb{R}$.

Proof. We set

$$\Theta_2^*[\lambda](\xi) := \Theta_2[\lambda](\overline{\xi})^* = I_{\mathfrak{Q}_{1,\lambda}} + i\alpha(\lambda)(H_2(\lambda) - \xi)^{-1}\alpha(\lambda)^*$$

$\xi \in \mathbb{C}_+$. Using (5.1) we get

$$\Theta_2^*[\lambda](\xi) = I_{\mathfrak{Q}_{1,\lambda}} + 2\pi i\sqrt{M_1(\lambda)}\, U\mathfrak{F}_2[\lambda](\xi)U^*\sqrt{M_1(\lambda)}$$

for a.e. $\lambda \in \mathbb{R}$, where

$$\mathfrak{F}_2[\lambda](\xi) := |G|^{1/2}(H_2(\lambda) - \xi)^{-1}|G|^{1/2}, \quad \xi \in \mathbb{C}_+.$$

Similar to the proof of Lemma 4.3 one verifies that the limit $\mathfrak{F}_2[\lambda](\lambda)$

$$\mathfrak{F}_2[\lambda](\lambda) = \lim_{\epsilon \to +0} \mathfrak{F}_2[\lambda](\lambda + i\epsilon)$$

exist in $\mathcal{B}_2(\mathfrak{H}_2)$ for a.e. $\lambda \in \mathbb{R}$ and satisfies the relation $\mathfrak{F}_2[\lambda](\lambda) = F_2(\lambda)$. Hence the limit $\Theta_2^*[\lambda](\lambda) = s - \lim_{\epsilon \to +0} \Theta_2[\lambda](\lambda - i\epsilon)^*$ exists for a.e. $\lambda \in \mathbb{R}$ and the relation

$$\Theta_2^*[\lambda](\lambda) = I_{\mathfrak{L}_{1,\lambda}} + 2\pi i \sqrt{M_1(\lambda)}\, U \mathfrak{F}_2[\lambda](\lambda) U^* \sqrt{M_1(\lambda)}$$

holds for a.e $\lambda \in \mathbb{R}$. From (3.19) we obtain that $S_{11}(\lambda) = \Theta_2^*[\lambda](\lambda)$ for a.e. $\lambda \in \mathbb{R}$. Since the limit $\Theta_2^*[\lambda](\lambda)$ exists for a.e. $\lambda \in \mathbb{R}$ one concludes that

$$\Theta_2[\lambda](\lambda) := s - \lim_{\epsilon \to +0} \Theta_2[\lambda](\lambda - i\epsilon)$$

exists for a.e. $\lambda \in \mathbb{R}$ and $\Theta_2[\lambda](\lambda)^* = \Theta_2^*[\lambda](\lambda)$ is valid. This completes the proof Theorem 5.1. □

The last theorem admits an interpretation of the scattering matrix $S_{11}(\lambda)$ as the result of a Lax-Phillips scattering. Indeed, let us introduce the minimal self-adjoint dilation $K_2(\lambda)$ of the maximal dissipative operator $H_2(\lambda)$. We set

$$\mathfrak{K}_{2,\lambda} = \mathfrak{D}_{-,\lambda} \oplus \mathfrak{H}_2 \oplus \mathfrak{D}_{+,\lambda},$$

where

$$\mathfrak{D}_{\pm,\lambda} := L^2(\mathbb{R}_\pm, dx, \mathfrak{L}_{1,\lambda}).$$

Further, we define

$$K_2(\lambda) \begin{pmatrix} f_- \\ f \\ f_+ \end{pmatrix} := \begin{pmatrix} -i\frac{d}{dx} f_- \\ \Re\mathrm{e}\,(H_2(\lambda)) f - \frac{1}{2}\alpha(\lambda)^*[f_+(0) + f_-(0)] \\ -i\frac{d}{dx} f_+ \end{pmatrix}$$

for elements of the domain

$$\mathrm{dom}\,(K_2(\lambda)) := \left\{ \begin{pmatrix} f_- \\ f \\ f_+ \end{pmatrix} : \begin{array}{c} f \in \mathrm{dom}\,(H_2(\lambda)) \\ f_\pm \in W^{1,2}(\mathbb{R}_\pm, dx, \mathfrak{L}_{1,\lambda}) \\ f_+(0) - f_-(0) = -i\alpha(\lambda) f \end{array} \right\}.$$

The operator $K_2(\lambda)$ is self-adjoint and is a minimal self-adjoint dilation of the maximal dissipative operator $H_2(\lambda)$, that is,

$$(H_2(\lambda) - z)^{-1} = P_{\mathfrak{H}_2}^{\mathfrak{K}_{2,\lambda}} (K_2(\lambda) - z)^{-1} \restriction \mathfrak{H}_2$$

for $z \in \mathbb{C}_+$ and

$$\mathfrak{K}_{2,\lambda} = \mathrm{clospan}\big\{ E_{K_2(\lambda)}(\Delta)\mathfrak{H}_2 : \Delta \in \mathcal{B}(\mathbb{R}) \big\},$$

where $E_{K_2(\lambda)}(\cdot)$ is the spectral measure of $K_2(\lambda)$. It turns out that $\mathfrak{D}_{\pm,\lambda}$ are incoming and outgoing subspaces with respect to $K_2(\lambda)$, i.e.,

$$e^{-itK_2(\lambda)} \mathfrak{D}_{+,\lambda} \subseteq \mathfrak{D}_{+,\lambda}, \quad t \geq 0,$$

and

$$e^{-itK_2(\lambda)}\mathfrak{D}_{-,\lambda} \subseteq \mathfrak{D}_{-,\lambda}, \quad t \le 0.$$

However, we remark that the completeness condition

$$\mathfrak{K}_{2,\lambda} = \text{clospan}\{e^{-itK_2(\lambda)}\mathfrak{D}_{\pm,\lambda} : t \in \mathbb{R}\} \tag{5.2}$$

is in general not satisfied. Condition (5.2) holds if and only if the maximal dissipative operator $H_2(\lambda)$ is completely non-selfadjoint and H_2 is singular, that means, the absolutely continuous part H_2^{ac} of H_2 is trivial.

On the subspace $\mathfrak{D}_\lambda$,

$$\mathfrak{D}_\lambda = \mathfrak{D}_{-,\lambda} \oplus \mathfrak{D}_{+,\lambda} = L^2(\mathbb{R}, dx, \mathfrak{Q}_{1,\lambda}) \subseteq \mathfrak{K}_{2,\lambda},$$

let us define the operator $K_0(\lambda)$,

$$(K_0(\lambda)g)(x) := -i\frac{d}{dx}g(x), \quad \text{dom}\,(K_0(\lambda)) := W^{1,2}(\mathbb{R}, dx, \mathfrak{Q}_{1,\lambda}).$$

The self-adjoint operator $K_0(\lambda)$ generates the shift group, i.e,

$$(e^{-itK_0(\lambda)}g)(x) = g(x-t), \quad g \in \mathfrak{D}_\lambda.$$

Using the Fourier transform $\mathcal{F} : L^2(\mathbb{R}, dx, \mathfrak{Q}_{1,\lambda}) \longrightarrow L^2(\mathbb{R}, d\mu, \mathfrak{Q}_{1,\lambda})$,

$$(\mathcal{F}f)(\mu) = \frac{1}{\sqrt{2\pi}}\int_{\mathbb{R}} dx\; e^{-i\mu x}f(x),$$

the operator $K_0(\lambda)$ transforms into the multiplication operator on the Hilbert space $L^2(\mathbb{R}, d\mu, \mathfrak{Q}_{1,\lambda})$. Furthermore, one has

$$e^{-itK_2(\lambda)} \restriction \mathfrak{D}_{+,\lambda} = e^{-itK_0(\lambda)} \restriction \mathfrak{D}_{+,\lambda}, \quad t \ge 0,$$

and

$$e^{-itK_2(\lambda)} \restriction \mathfrak{D}_{-,\lambda} = e^{-itK_0(\lambda)} \restriction \mathfrak{D}_{-,\lambda}, \quad t \le 0.$$

The last properties yield the existence of the Lax-Phillips wave operators

$$W_\pm^{LP}[\lambda] := s - \lim_{t\to\pm\infty} e^{itK_2(\lambda)}J_\pm(\lambda)e^{-itK_0(\lambda)},$$

cf. [5, 14] where $J_\pm(\lambda) : \mathfrak{D}_{\pm,\lambda} \longrightarrow \mathfrak{K}_{2,\lambda}$ is the natural embedding operator. The Lax-Phillips scattering operator $S_{LP}(\lambda)$ is defined by

$$S_{LP}[\lambda] := W_+^{LP}[\lambda]^* W_-^{LP}[\lambda],$$

cf. [5, 14]. With respect to the spectral representation $L^2(\mathbb{R}, d\mu, \mathfrak{Q}_{1,\lambda})$ the Lax-Phillips scattering matrix $\{S_{LP}[\lambda](\mu)\}_{\mu\in\mathbb{R}}$ coincides with $\{\Theta_2[\lambda](\mu)^*\}_{\mu\in\mathbb{R}}$, see [1, 2, 3, 4]. Hence the scattering matrix $\{S_{11}(\lambda)\}_{\lambda\in\mathbb{R}}$ can be regarded as the result of a Lax-Phillips scattering for a.e. $\lambda \in \mathbb{R}$.

Acknowledgement

The support of the work by DFG, Grant 1480/2, is gratefully acknowledged. The authors thank Prof. P. Exner for discussion and literature hints.

References

[1] Adamjan, V.M., Arov, D.Z.: *Unitary couplings of semi-unitary operators*, Akad. Nauk Armjan. SSR Dokl. 43, no. 5, 257–263 (1966).

[2] Adamjan, V.M., Arov, D.Z.: *Unitary couplings of semi-unitary operators*, Mat. Issled. 1, vyp. 2, 3–64 (1966).

[3] Adamjan, V.M., Arov, D.Z.: *On scattering operators and contraction semigroups in Hilbert space*, Dokl. Akad. Nauk SSSR 165, 9–12 (1965).

[4] Adamjan, V.M., Arov, D.Z.: *On a class of scattering operators and characteristic operator-functions of contractions*. Dokl. Akad. Nauk SSSR 160, 9–12 (1965).

[5] Baumgärtel, H., Wollenberg, M.: *Mathematical scattering theory*, Akademie-Verlag, Berlin, 1983.

[6] Davies, E.B.: *Two-channel Hamiltonians and the optical model of nuclear scattering*, Ann. Inst. H. Poincaré Sect. A (N.S.) 29, no. 4, 395–413 (1979).

[7] Davies, E.B.: *Quantum theory of open systems*, Academic Press, London-New York, 1976.

[8] Exner, P.: *Open quantum systems and Feynman integrals*, D. Reidel Publishing Co., Dordrecht, 1985.

[9] Exner, P., Úlehla, I.: *On the optical approximation in two-channel systems*, J. Math. Phys. 24, no. 6, 1542–1547 (1983).

[10] Feshbach, H.: *A unified theory of nuclear reactions. II*, Ann. Physics 19, 287–313 (1962).

[11] Feshbach, H.: *Unified theory of nuclear reactions*, Ann. Physics 5, 357–390 (1958).

[12] Foias, C., Sz.-Nagy, B.: *Harmonic analysis of operators on Hilbert space*, North-Holland Publishing Co., Amsterdam-London; American Elsevier Publishing Co., Inc., New York; Akadémiai Kiadó, Budapest 1970.

[13] Kato, T.: *Perturbation theory for linear operators*, Die Grundlehren der mathematischen Wissenschaften, Band 132 Springer-Verlag New York, Inc., New York.

[14] Lax, P.D., Phillips, R.S.: *Scattering theory*, Academic Press, New York-London, 1967.

[15] Neidhardt, H.: *Scattering matrix and spectral shift of the nuclear dissipative scattering theory. II*, J. Operator Theory 19, no. 1, 43–62 (1988).

[16] Neidhardt, H.: *Scattering matrix and spectral shift of the nuclear dissipative scattering theory*, In "Operators in indefinite metric spaces, scattering theory and other topics" (Bucharest, 1985), 237–250, Oper. Theory Adv. Appl., 24, Birkhäuser, Basel, 1987.

[17] Neidhardt, H.: *On the inverse problem of a dissipative scattering theory. I*, In "Advances in invariant subspaces and other results of operator theory" (Timişoara and Herculane, 1984), 223–238, Oper. Theory Adv. Appl., 17, Birkhäuser, Basel, 1986.

[18] Neidhardt, H.: *Eine mathematische Streutheorie für maximal dissipative Operatoren*, Report MATH, 86-3. Akademie der Wissenschaften der DDR, Institut für Mathematik, Berlin, 1986.

[19] Neidhardt, H.: *A nuclear dissipative scattering theory*, J. Operator Theory 14, no. 1, 57–66 (1985).

[20] Neidhardt, H.: *A dissipative scattering theory*, In "Spectral theory of linear operators and related topics" (Timişoara/Herculane, 1983), 197–212, Oper. Theory Adv. Appl., 14, Birkhäuser, Basel, 1984.

Jussi Behrndt
Technische Universität Berlin
Institut für Mathematik
Straße des 17. Juni 136
D–10623 Berlin, Germany
e-mail: behrndt@math.tu-berlin.de

Hagen Neidhardt
Weierstraß-Institut für
Angewandte Analysis und Stochastik
Mohrenstr. 39
D–10117 Berlin, Germany
e-mail: neidhard@wias-berlin.de

Joachim Rehberg
Weierstraß-Institut für
Angewandte Analysis und Stochastik
Mohrenstr. 39
D–10117 Berlin, Germany
e-mail: rehberg@wias-berlin.de

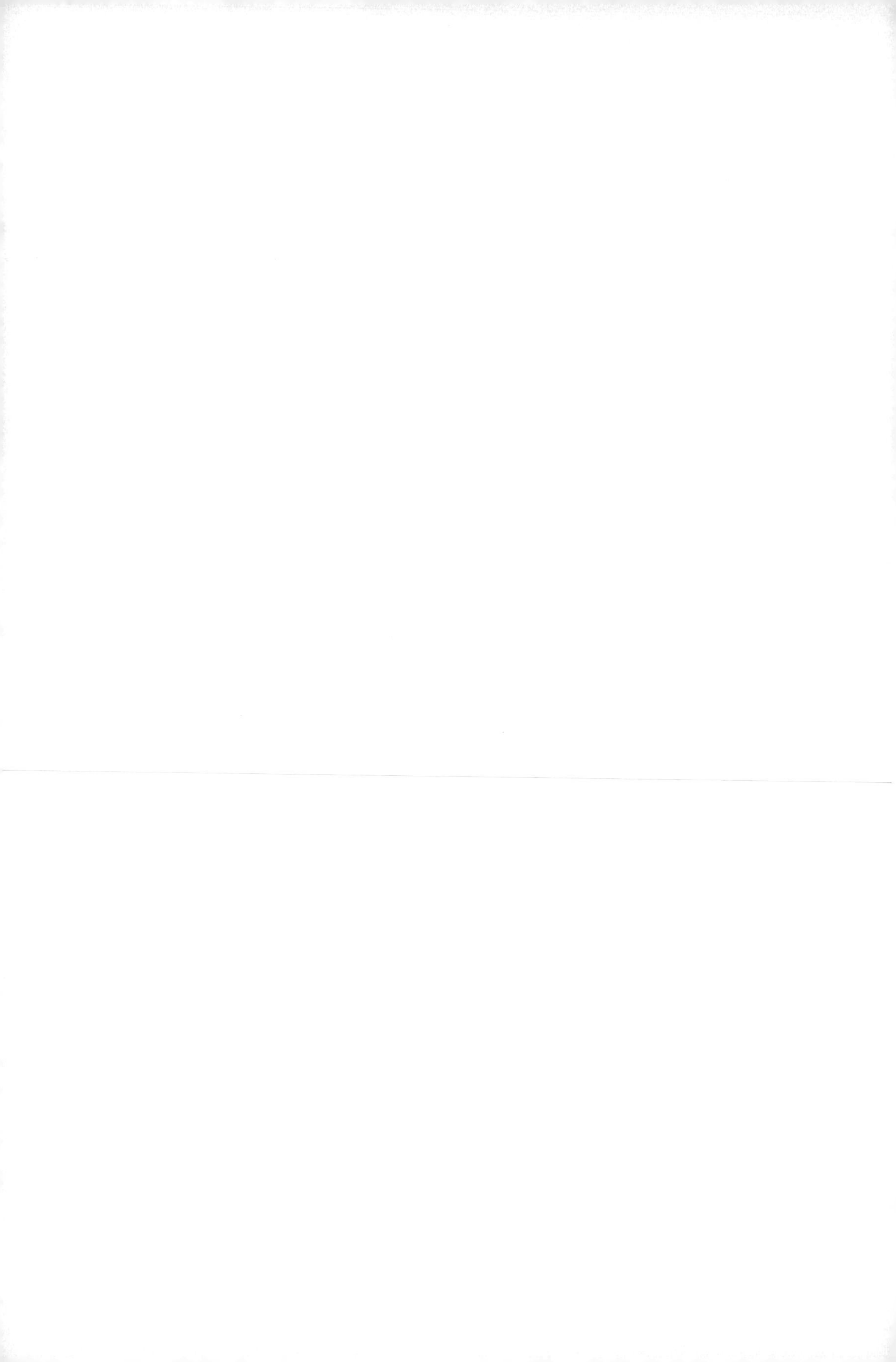

Operator Theory:
Advances and Applications, Vol. 175, 51–88

Asymptotic Expansions of Generalized Nevanlinna Functions and their Spectral Properties

Vladimir Derkach, Seppo Hassi and Henk de Snoo

Abstract. Asymptotic expansions of generalized Nevanlinna functions Q are investigated by means of a factorization model involving a part of the generalized zeros and poles of nonpositive type of the function Q. The main results in this paper arise from the explicit construction of maximal Jordan chains in the root subspace $\mathsf{R}_\infty(S_F)$ of the so-called generalized Friedrichs extension. A classification of maximal Jordan chains is introduced and studied in analytical terms by establishing the connections to the appropriate asymptotic expansions. This approach results in various new analytic characterizations of the spectral properties of selfadjoint relations in Pontryagin spaces and, conversely, translates analytic and asymptotic properties of generalized Nevanlinna functions into the spectral theoretical properties of self-adjoint relations in Pontryagin spaces.

Mathematics Subject Classification (2000). Primary 46C20, 47A06, 47B50; Secondary 47A10, 47A11, 47B25.

Keywords. Generalized Nevanlinna function, asymptotic expansion, Pontryagin space, symmetric operator, selfadjoint extension, operator model, factorization, generalized Friedrichs extension.

1. Introduction

Let $\mathbf{N}_\kappa$ be the class of generalized Nevanlinna functions, i.e., meromorphic functions on $\mathbb{C} \setminus \mathbb{R}$ with $Q(\bar{z}) = \overline{Q(z)}$ and such that the kernel

$$\mathsf{N}_Q(z, \lambda) = \frac{Q(z) - \overline{Q(\lambda)}}{z - \bar{\lambda}}, \quad z, \lambda \in \rho(Q), \quad z \neq \bar{\lambda},$$

The research was supported by the Academy of Finland (project 212150) and the Research Institute for Technology at the University of Vaasa.

has κ negative squares on the domain of holomorphy $\rho(Q)$ of Q, see [20]. If the function $Q \in \mathbf{N}_\kappa$ belongs to the subclass $\mathbf{N}_{\kappa,-2n}$, $n \in \mathbb{N}$, (see [6]) then it admits the following asymptotic expansion

$$Q(z) = \gamma - \sum_{j=1}^{2n+1} \frac{s_{j-1}}{z^j} + o\left(\frac{1}{z^{2n+1}}\right), \quad z\widehat{\rightarrow}\infty, \tag{1.1}$$

where $\gamma, s_j \in \mathbb{R}$ and $z\widehat{\rightarrow}\infty$ means that z tends to ∞ nontangentially ($0 < \varepsilon < \arg z < \pi - \varepsilon$). Asymptotic expansions for $Q \in \mathbf{N}_\kappa$ of the form (1.1) (with $\gamma = 0$) were introduced in [21]. They naturally appear, for instance, in the indefinite moment problem considered in [22]. The expansion (1.1) is equivalent to the following operator representation of the function $Q \in \mathbf{N}_{\kappa,-2n}$:

$$Q(z) = \gamma + [(A-z)^{-1}\omega, \omega], \tag{1.2}$$

where $\omega \in \operatorname{dom} A^n$ and A is a selfadjoint operator in a Pontryagin space $\mathfrak{H}$; see [21, Satz 1.10] and Corollary 3.4 below. The representation (1.2) can be taken to be minimal in the sense that ω is a cyclic vector for A, i.e.,

$$\mathfrak{H} = \overline{\operatorname{span}}\,\{\, (A-z)^{-1}\omega : \, z \in \rho(A) \,\},$$

in which case the negative index $\operatorname{sq}_-(\mathfrak{H})$ of $\mathfrak{H}$ is equal to κ. The representation (1.2) shows that ∞ is a generalized zero of the function $Q(z) - \gamma$, or equivalently, that ∞ is a generalized pole of the function $Q_\infty(z) = -1/(Q(z)-\gamma)$. This means that the underlying symmetric operator S is nondensely defined in $\mathfrak{H}$ with

$$\operatorname{dom} S = \{\, f \in \operatorname{dom} A : \, [f, \omega] = 0 \,\} \tag{1.3}$$

and that

$$S_F = S \widehat{+} (\{0\} \times \operatorname{span}\{\omega\}) \tag{1.4}$$

is a selfadjoint extensions of S in $\mathfrak{H}$ with $\infty \in \sigma_p(S_F)$. Here $\widehat{+}$ stands for the componentwise sum in the Cartesian product $\mathfrak{H} \times \mathfrak{H}$. In other words, the extension S_F is multivalued and, in fact, can be interpreted as the generalized Friedrichs extension of S, see [5] and the references therein. It follows from (1.1) and (1.2) that

$$s_0 = [\omega, \omega] \in \mathbb{R}.$$

If $\kappa > 0$ then it is possible that $s_0 \leq 0$, in which case ∞ is a generalized pole of nonpositive type (GPNT) of the function Q_∞, cf. [23]. More precisely, if ∞ is a GPNT of Q_∞ with multiplicity $\kappa_\infty := \kappa_\infty(Q_\infty)$ (see (2.2) below for the definition), then in (1.1) one automatically has

$$s_0 = \cdots = s_j = 0, \quad \text{for every } j < 2\kappa_\infty - 2.$$

Furthermore, if m is the first nonnegative index in (1.1) such that $s_m \neq 0$ (if it exists), then, equivalently, the function Q_∞ admits an asymptotic expansion of the form

$$Q_\infty(z) = p_{m+1} z^{m+1} + \cdots + p_{2\ell+1} z^{2\ell+1} + o\left(z^{2\ell+1}\right), \quad z\widehat{\rightarrow}\infty, \tag{1.5}$$

where $p_{m+1} = 1/s_m$, $p_i \in \mathbb{R}$, $i = m+1, \ldots, 2\ell+1$, and the integers m, n, and ℓ are connected by $\ell = m - n$ with $m \geq 2\ell$; see Theorem 5.4 below for further details. It turns out that (1.5) holds for some $\ell \leq 0$ if and only if ∞ is a regular critical point of S_F, or equivalently, if and only if the corresponding root subspace

$$\mathsf{R}_\infty(S_F) = \{\, h \in \mathfrak{H} : \{0, h\} \in S_F^k \text{ for some } k \in \mathbb{N}\,\}$$

of the generalized Friedrichs extensions S_F in (1.4) is nondegenerate. In this case the GPNT ∞ of Q_∞ as well as the corresponding root subspace $\mathsf{R}_\infty(S_F)$ are shortly called *regular*. On the other hand, if ∞ is a singular critical point of S_F, then in (1.5) $\ell > 0$ and, moreover, the minimal integer ℓ such that the expansion (1.5) exists coincides with the dimension κ_∞^0 of the *isotropic subspace* of the root subspace $\mathsf{R}_\infty(S_F)$, see Theorem 5.6. In this case the GPNT ∞ of Q_∞ and the corresponding root subspace $\mathsf{R}_\infty(S_F)$ are shortly called *singular* with *the index of singularity* κ_∞^0.

The above-mentioned results reflect the close connections between the asymptotic expansions (1.1), (1.5), and the root subspace $\mathsf{R}_\infty(S_F)$ of S_F. The given assertions are examples of the results in the present paper which have been derived by means of the factorization model of the function Q_∞ recently constructed by the authors in [9]. This model is based on the following "proper" factorization of the function $Q_\infty \in \mathbf{N}_\kappa$:

$$Q_\infty(z) = q(z)q^\sharp(z)Q_0(z), \tag{1.6}$$

where q is a (monic) polynomial, $q^\sharp(z) = \overline{q(\bar{z})}$, and $Q_0 \in \mathbf{N}_{\kappa'}$ such that

$$\kappa_\infty(Q_0) = 0 \quad \text{and} \quad \kappa' = \kappa - \deg q,$$

see Lemma 4.3 below. Such a factorization for Q_∞ is in general not unique, but the factorization model based on such a factorization carries the complete information about the root subspace $\mathsf{R}_\infty(S_F)$ of S_F.

A major part of the results presented in this paper is associated with the structure of the root subspace $\mathsf{R}_\infty(S_F)$ of S_F in a model space and the various connections to the asymptotic expansions (1.1) and (1.5). By using the factorization model based on a proper factorization (1.6) of Q_∞ maximal Jordan chains in $\mathsf{R}_\infty(S_F)$ are constructed in explicit terms. Their construction leads to *three different types of maximal Jordan chains* in $\mathsf{R}_\infty(S_F)$. Each of these three types of maximal Jordan chains admits its own characteristic features, reflecting various properties of the root subspace $\mathsf{R}_\infty(S_F)$. The construction shows explicitly, for instance, when the root subspace $\mathsf{R}_\infty(S_F)$ is regular and when it is singular. The length of the maximal Jordan chain as well as the *signature of the root subspace* $\mathsf{R}_\infty(S_F)$ can be easily read off from their construction. In the case that the root subspace $\mathsf{R}_\infty(S_F)$ is regular, the three types of maximal Jordan chain can be characterized by their length. The first type of maximal Jordan chain is of length $2k+1$, where $k = \deg q = \kappa_\infty(Q_\infty)$, and the second and third type of maximal Jordan chains are of length $2k$ and $2k-1$, respectively. The classification of these maximal Jordan chains remains the same in the case when the root subspace $\mathsf{R}_\infty(S_F)$ is singular. In that case the index of singularity κ_∞^0 as introduced above enters to the formulas, while the difference $\kappa_-(\mathsf{R}_\infty(S_F)) - \kappa_+(\mathsf{R}_\infty(S_F))$ of the negative and the positive index of $\mathsf{R}_\infty(S_F)$ remains unaltered, see Theorem 4.12. All of these facts can be translated into the analytical properties of the functions Q_∞ and $Q = \gamma - 1/Q_\infty$ via the asymptotic expansions (1.1) and (1.5), and conversely.

The classification of maximal Jordan chains in $\mathsf{R}_\infty(S_F)$ motivates an analogous *classification of generalized zeros and poles of nonpositive type* of the function $Q \in \mathbf{N}_\kappa$, which turns out to be connected with the characterization of the multiplicities of GZNT and GPNT of the function Q due to H. Langer in [24]; see Subsection 3.2 for the definitions of generalized zeros and poles of types (T1)–(T3). This induces a classification for the asymptotic expansions for the functions Q and Q_∞; see Theorems 5.3 and 5.4. Some further characterizations of the three different types of generalized zeros and poles are obtained by means of the factorized integral representations of the functions Q and Q_∞, which are based on their canonical factorizations, see [11]; for definitions, see Subsection 2.1, cf. also [5]. In particular, Theorem 6.1 and Theorem 6.3 extend some earlier results by the authors in [6] (where $\kappa = 1$) and in [8], from the regular case to the singular case in an explicit manner involving the index of singularity κ_∞^0, which is characterized in Theorem 5.6 below.

The construction of the maximal Jordan chains in $\mathsf{R}_\infty(S_F)$ using the factorization model for Q_∞ in (1.6) is carried out in Section 4. The most careful treatment of the model is required in the construction of maximal Jordan chains which are of the third type (T3). The reason is that the factorization of Q_∞ does not produce a minimal model for the function Q_∞ directly. In the minimal factorization model the maximal Jordan chains of type (T3) are roughly speaking the shortest ones, cf. (4.21), (4.24), (4.28); see also Theorem 5.3 and Theorem 6.3. The results in Lemma 4.11 and part (iii) of Theorem 4.12 characterize maximal Jordan chains of type (T3). In this case the underlying symmetric relation $S(Q)$ is multivalued (before the auxiliary part of the space is factored out). This statement is true more generally: for an arbitrary $\mathbf{N}_\kappa$-function Q the occurrence of generalized zeros and poles of type (T3) in $\mathbb{R} \cup \{\infty\}$ is an indication that point spectrum $\sigma_p(S(Q))$ of $S(Q)$ is nonempty, see Lemma 6.4 below, which by part (i) of Theorem 4.6 is equivalent to $S(Q)$ being not simple. In fact, the existence of maximal Jordan chains of type (T3) or, equivalently, the existence of GZNT and GPNT of type (T3) can be used to give criteria for minimality of various factorization models for $\mathbf{N}_\kappa$-functions, see Propositions 6.6 and 6.7 below.

The topics considered in this paper have connections to some other recent studies involving asymptotic expansions of $\mathbf{N}_\kappa$-functions, see in particular [6], [8], [9], [12], [13], [15], and their canonical factorization, see, e.g., [3], [5], [7], [11], [14]. For instance, in [13] the authors investigate the subclass of $\mathbf{N}_\kappa$-functions with $\kappa = \kappa_\infty(Q)$ and extend some results, e.g., from [6], [8]. General operator models based on the canonical factorization of $\mathbf{N}_\kappa$-functions have been introduced in [3]; for another model not using the canonical factorization of Q, see [18]. The construction of a minimal canonical factorization model by using reproducing kernel Pontryagin space methods has been recently worked out in [12], cf. also [3, Theorem 4.1]. Some of the results in the present paper can be naturally augmented by the results which can be found from [15], where characteristic properties of the generalized zeros and poles of $\mathbf{N}_\kappa$-functions have been studied with the aid of their operator representations.

The present paper forms a continuation of the paper [9], where the details concerning the construction of the announced factorization model can be found. Some basic definitions and concepts which will be used throughout the paper are given in Section 2. In Section 3 some additions concerning the subclasses $\mathbf{N}_{\kappa,-\ell}$ as introduced in [6] are given, including a proof for [6, Proposition 6.2] as announced in that paper, cf. Theorem 3.3 below; see also Theorem 5.4 for an extension of these results. Asymptotic expansions are introduced in Section 3 and a classification of generalized zeros and poles is given. In Section 4 the main ingredients concerning the factorization model are given and the construction of maximal Jordan chains in $\mathsf{R}_\infty(S_F)$ is carried out. The connection between the properties of the root subspace $\mathsf{R}_\infty(S_F)$ and the asymptotic expansions of the form (1.1) and (1.5) is investigated in Section 5. Finally, in Section 6 the classification of GZNT and GPNT is connected with factorized integral representations of the functions Q and $Q_\infty(z)$. In this section also the generalized zeros and poles of nonpositive type of $\mathbf{N}_\kappa$-functions which belong to $\mathbb{R}$ are briefly treated and some consequences as announced above are established.

2. Preliminaries

2.1. Canonical factorization of $Q \in \mathbf{N}_\kappa$

The notions of generalized poles and generalized zeros of nonpositive type were introduced in [23]. The following definitions are based on [24]. A point $\alpha \in \mathbb{R}$ is called a *generalized pole* of nonpositive type (GPNT) of the function $Q \in \mathbf{N}_\kappa$ with multiplicity κ_α $(= \kappa_\alpha(Q))$ if

$$-\infty < \lim_{z \widehat{\to} \alpha} (z-\alpha)^{2\kappa_\alpha+1} Q(z) \le 0, \quad 0 < \lim_{z \widehat{\to} \alpha} (z-\alpha)^{2\kappa_\alpha-1} Q(z) \le \infty. \tag{2.1}$$

Similarly, the point ∞ is called a generalized pole of nonpositive type (GPNT) of Q with multiplicity κ_∞ $(= \kappa_\infty(Q))$ if

$$0 \le \lim_{z \widehat{\to} \infty} \frac{Q(z)}{z^{2\kappa_\infty+1}} < \infty, \quad -\infty \le \lim_{z \widehat{\to} \infty} \frac{Q(z)}{z^{2\kappa_\infty-1}} < 0. \tag{2.2}$$

A point $\beta \in \mathbb{R}$ is called a *generalized zero* of nonpositive type (GZNT) of the function $Q \in \mathbf{N}_\kappa$ if β is a generalized pole of nonpositive type of the function $-1/Q$. The multiplicity π_β $(= \pi_\beta(Q))$ of the GZNT β of Q can be characterized by the inequalities:

$$0 < \lim_{z \widehat{\to} \beta} \frac{Q(z)}{(z-\beta)^{2\pi_\beta+1}} \le \infty, \quad -\infty < \lim_{z \widehat{\to} \beta} \frac{Q(z)}{(z-\beta)^{2\pi_\beta-1}} \le 0. \tag{2.3}$$

Similarly, the point ∞ is called a generalized zero of nonpositive type (GZNT) of Q with multiplicity π_∞ $(=\pi_\infty(Q))$ if

$$-\infty \leq \lim_{z\widehat{\to}\infty} z^{2\pi_\infty+1}Q(z) < 0, \quad 0 \leq \lim_{z\widehat{\to}\infty} z^{2\pi_\infty-1}Q(z) < \infty. \tag{2.4}$$

It was shown in [23] that for $Q\in\mathbf{N}_\kappa$ the total number (counting multiplicities) of poles (zeros) in $\mathbb{C}_+$ and generalized poles (zeros) of nonpositive type in $\mathbb{R}\cup\{\infty\}$ is equal to κ. Let $\alpha_1,\dots,\alpha_l$ $(\beta_1,\dots,\beta_m)$ be all the generalized poles (zeros) of nonpositive type in $\mathbb{R}$ and the poles (zeros) in $\mathbb{C}_+$ with multiplicities $\kappa_1,\dots,\kappa_l$ $(\pi_1,\dots,\pi_m)$. Then the function Q admits a canonical factorization of the form

$$Q(z)=r(z)r^\sharp(z)Q_{00}(z), \quad Q_{00}\in\mathbf{N}_0, \quad r=\frac{\widetilde{p}}{\widetilde{q}}, \tag{2.5}$$

where $\widetilde{p}(z)=\prod_{j=1}^m(z-\beta_j)^{\pi_j}$ and $\widetilde{q}(z)=\prod_{j=1}^l(z-\alpha_j)^{\kappa_j}$ are relatively prime polynomials of degree $\kappa-\pi_\infty(Q)$ and $\kappa-\kappa_\infty(Q)$, respectively; see [11], [5]. It follows from (2.5) that the function Q admits the (factorized) integral representation

$$Q(z)=r(z)r^\sharp(z)\left(a+bz+\int_{\mathbb{R}}\left(\frac{1}{t-z}-\frac{t}{1+t^2}\right)d\rho(t)\right), \quad r=\frac{\widetilde{p}}{\widetilde{q}}, \tag{2.6}$$

where $a\in\mathbb{R}$, $b\geq 0$, and $\rho(t)$ is a nondecreasing function satisfying the integrability condition

$$\int_{\mathbb{R}}\frac{d\rho(t)}{t^2+1}<\infty. \tag{2.7}$$

2.2. The subclasses $\mathbf{N}_{\kappa,1}$ and $\mathbf{N}_{\kappa,0}$

A function $Q\in\mathbf{N}_\kappa$ is said to belong to the subclass $\mathbf{N}_{\kappa,1}$, if

$$\lim_{z\widehat{\to}\infty}\frac{Q(z)}{z}=0 \text{ and } \int_\eta^\infty \frac{|\operatorname{Im} Q(iy)|}{y}\,dy<\infty,$$

with $\eta>0$ large enough. Similarly $Q\in\mathbf{N}_\kappa$ is said to belong to the subclass $\mathbf{N}_{\kappa,0}$, if

$$\lim_{z\widehat{\to}\infty}\frac{Q(z)}{z}=0 \text{ and } \limsup_{z\widehat{\to}\infty}|z\operatorname{Im} Q(z)|<\infty,$$

see [5]. In the following theorems the subclasses $\mathbf{N}_{\kappa,1}$ and $\mathbf{N}_{\kappa,0}$ are characterized both in terms of the integral representation (2.6) and in terms of operator representations of the form (1.2). Let E_t be a spectral function of a selfadjoint operator A in a Pontryagin space $\mathfrak{H}$, see [1]. Denote by $\mathfrak{H}_\ell:=\mathfrak{H}_\ell(A)$, $\ell\in\mathbb{N}$, the set of all elements $h\in\mathfrak{H}$ such that $\int_\Delta |t|^\ell d[E_th,h]<\infty$ for some neighborhood Δ of $\pm\infty$. Moreover, let $\mathfrak{H}_{-\ell}(A)$, $\ell\in\mathbb{N}$, be the corresponding dual spaces. Here, for instance, $\mathfrak{H}_{-1}(A)$ can be identified as the set of all generalized elements obtained by completing $\mathfrak{H}$ with respect to the inner product $\int_\Delta(1+|t|)^{-1}d[E_th,h]<\infty$ with some neighborhood Δ of $\pm\infty$. The operator A admits a natural continuation $\widetilde{A}$ from $\mathfrak{H}$ into $\mathfrak{H}_{-1}$, see [5] for further details. The classes $\mathbf{N}_{\kappa,1}$ and $\mathbf{N}_{\kappa,0}$ are characterized in the following two theorems, see [5].

Theorem 2.1. ([5]) *For $Q \in \mathbf{N}_\kappa$ the following statements are equivalent:*

(i) *Q belongs to $\mathbf{N}_{\kappa,1}$;*

(ii) *$Q(z) = \gamma + [(\widetilde{A} - z)^{-1}\omega, \omega]$, $z \in \rho(A)$, for some selfadjoint operator A in a Pontryagin space $\mathfrak{H}$, a cyclic vector $\omega \in \mathfrak{H}_{-1}$, and $\gamma \in \mathbb{R}$;*

(iii) *Q has the integral representation* (2.6) *with $\deg \widetilde{q} - \deg \widetilde{p} = \pi_\infty(Q) > 0$, or with $\deg \widetilde{p} = \deg \widetilde{q}$ $(\pi_\infty(Q) = 0)$, $b = 0$, and*

$$\int_{\mathbb{R}} (1 + |t|)^{-1} d\rho(t) < \infty. \tag{2.8}$$

Theorem 2.2. ([5]) *For $Q \in \mathbf{N}_\kappa$ the following statements are equivalent:*

(i) *Q belongs to $\mathbf{N}_{\kappa,0}$;*

(ii) *$Q(z) = \gamma + O(1/z)$, $z \widehat{\rightarrow} \infty$;*

(iii) *$Q(z) = \gamma + [(A - z)^{-1}\omega, \omega]$, $z \in \rho(A)$, for some selfadjoint operator A in a Pontryagin space $\mathfrak{H}$, a cyclic vector $\omega \in \mathfrak{H}$, and $\gamma \in \mathbb{R}$;*

(iv) *Q has the integral representation* (2.6) *with $\deg \widetilde{q} - \deg \widetilde{p} = \pi_\infty(Q) > 0$, or with $\deg \widetilde{p} = \deg \widetilde{q}$ $(\pi_\infty(Q) = 0)$, $b = 0$, and*

$$\int_{\mathbb{R}} d\rho(t) < \infty. \tag{2.9}$$

Remark 2.3. If $Q \in \mathbf{N}_{\kappa,0}$, then the operator representation of Q in part (iii) of Theorem 2.2 implies that

$$\lim_{z \widehat{\rightarrow} \infty} -z(Q(z) - \gamma) = [\omega, \omega].$$

Hence, the statement (ii) in Theorem 2.2 can be strengthened in the sense that for every function $Q \in \mathbf{N}_{\kappa,0}$ there are real numbers γ and s_0, such that

$$Q(z) = \gamma - \frac{s_0}{z} + o\left(\frac{1}{z}\right), \quad z \widehat{\rightarrow} \infty. \tag{2.10}$$

3. Asymptotic expansions of generalized Nevanlinna functions

Asymptotic expansions of generalized Nevanlinna functions (as in (2.10)) can be used for studying operator and spectral theoretical properties of selfadjoint extensions of symmetric operators in Pontryagin and Hilbert spaces, see [20], [16]. In this section a subdivision of the class $\mathbf{N}_\kappa$ of generalized Nevanlinna functions is given along the lines of [16], [6]. Moreover, a classification for generalized zeros of nonpositive type is introduced and interpreted via asymptotic expansions.

3.1. The subclasses $\mathbf{N}_{\kappa,-\ell}$ of generalized Nevanlinna functions

Definition 3.1. A function $Q \in \mathbf{N}_\kappa$ is said to belong to the subclass $\mathbf{N}_{\kappa,-2n}$, $n \in \mathbb{N}$, if there are real numbers γ and $s_0, \dots, s_{2n-1}$ such that the function

$$\widetilde{Q}(z) = z^{2n} \left(Q(z) - \gamma + \sum_{j=1}^{2n} \frac{s_{j-1}}{z^j} \right) \tag{3.1}$$

is $O(1/z)$ as $z\widehat{\to}\infty$. Moreover, $Q \in \mathbf{N}_\kappa$ is said to belong to the subclass $\mathbf{N}_{\kappa,-2n+1}$ if the function $\widetilde{Q}$ in (3.1) belongs to $\mathbf{N}_{\kappa',1}$ for some $\kappa' \in \mathbb{N}$.

The next lemma clarifies the above definition of the subclasses $\mathbf{N}_{\kappa,-\ell}$, $\ell \in \mathbb{N}$.

Lemma 3.2. *If the function Q belongs to the subclass $\mathbf{N}_{\kappa,-2n}$ ($\mathbf{N}_{\kappa,-2n+1}$) for some $n \in \mathbb{N}$, then the function $\widetilde{Q}$ in (3.1) belongs to the subclass $\mathbf{N}_{\kappa',0}$ (resp. $\mathbf{N}_{\kappa',1}$) with $\kappa' \leq \kappa$. Moreover, the following inclusions are satisfied*

$$\cdots \subset \mathbf{N}_{\kappa,-2n-1} \subset \mathbf{N}_{\kappa,-2n} \subset \mathbf{N}_{\kappa,-2n+1} \subset \cdots \subset \mathbf{N}_{\kappa,0} \subset \mathbf{N}_{\kappa,1}. \tag{3.2}$$

Proof. Rewrite the expression for the function $\widetilde{Q}$ in (3.1) in the form $\widetilde{Q}(z) = z^{2n}\widehat{Q}(z)$. Then $\widehat{Q}(z)$ as a sum of two generalized Nevanlinna functions is also a generalized Nevanlinna function, and therefore, in view of (2.5), $\widetilde{Q}$ is a generalized Nevanlinna function, too. Next it is shown that the inequality $\kappa(\widetilde{Q}) \leq \kappa(Q)$ is satisfied. First observe that the condition $\widetilde{Q}(z) = O(1)$ and hence, in particular, the condition $\widetilde{Q}(z) = O(1/z)$ as $z\widehat{\to}\infty$ implies that

$$\kappa_\infty(\widetilde{Q}) = 0, \tag{3.3}$$

cf. (2.2), (2.4). Clearly, $\kappa_\alpha(\widetilde{Q}) = \kappa_\alpha(Q)$ for every $\alpha \neq 0, \infty$, while for $\alpha = 0$ one derives from (2.1) the estimate

$$\kappa_0(\widetilde{Q}) \leq \kappa_0(Q). \tag{3.4}$$

Therefore, one can conclude from (3.3) and (3.4) that $\kappa(\widetilde{Q}) \leq \kappa(Q)$. Now by Theorem 2.2 the condition $\widetilde{Q}(z) = O(1/z)$, $z\widehat{\to}\infty$, is equivalent to $\widetilde{Q} \in \mathbf{N}_{\kappa',0}$ with $\kappa' \leq \kappa$, which proves the first statement for the subclasses $\mathbf{N}_{\kappa,-2n}$. If $Q \in \mathbf{N}_{\kappa,-2n+1}$, then $\widetilde{Q}(z) = O(1)$ and since $\kappa(\widetilde{Q}) \leq \kappa(Q)$, one actually has $\widetilde{Q} \in \mathbf{N}_{\kappa',1}$ for $\kappa' \leq \kappa$.

Since $\mathbf{N}_{\kappa,0} \subset \mathbf{N}_{\kappa,1}$ the inclusions $\mathbf{N}_{\kappa,-2n} \subset \mathbf{N}_{\kappa,-2n+1}$, $n \in \mathbb{N}$, follow from the first part of the lemma. Now let $Q \in \mathbf{N}_{\kappa,-2n-1}$. Then by definition

$$z^2\widetilde{Q}(z) + zs_{2n} + s_{2n+1} \in \mathbf{N}_{\kappa',1}, \tag{3.5}$$

where $\widetilde{Q}$ is as in (3.1) and $\kappa' \leq \kappa$. It is clear from (3.5) (see Theorem 2.1) that $\widetilde{Q}(z) = O(1/z)$ as $z\widehat{\to}\infty$. Hence, $Q \in \mathbf{N}_{\kappa,-2n}$ and this proves the remaining inclusions in (3.2). □

The subclasses $\mathbf{N}_{\kappa,-\ell}$, $\ell \in \mathbb{N}$, are now characterized by means of the operator and the integral representation of Q in (1.2) and (2.6), respectively.

Theorem 3.3. *For $Q \in \mathbf{N}_\kappa$ the following statements are equivalent:*

(i) *$Q \in \mathbf{N}_{\kappa,-\ell}$, $\ell \in \mathbb{N}$;*
(ii) *$Q(z) = \gamma + [(A - z)^{-1}\omega, \omega]$, $z \in \rho(A)$, for some selfadjoint operator A in a Pontryagin space $\mathfrak{H}$, a cyclic vector $\omega \in \mathfrak{H}_\ell$, and $\gamma \in \mathbb{R}$;*

(iii) Q *has an integral representation* (2.6) *with* $\pi_\infty(Q) = \deg \widetilde{q} - \deg \widetilde{p} \geq 0$ *(and* $b = 0$ *if* $\pi_\infty(Q) = 0$*), such that*

$$\int_{\mathbb{R}} (1 + |t|)^{\ell - 2\pi_\infty} d\rho(t) < \infty. \tag{3.6}$$

Proof. (i) $\Rightarrow$ (iii) Let $Q \in \mathbf{N}_{\kappa,-\ell}$, where ℓ is either $2n$ or $2n-1$, $n \in \mathbb{N}$. In view of (3.2) and Theorems 2.1, 2.2 one has $\pi_\infty(Q) = \deg \widetilde{q} - \deg \widetilde{p} \geq 0$, and if $\pi_\infty(Q) = 0$ then $b = 0$ and (2.8) or (2.9) is satisfied. By Lemma 3.2 the function $\widetilde{Q}$ in (3.1) belongs to $\mathbf{N}_{\kappa',2n-\ell}$ with $\kappa' \leq \kappa$. Hence, $\widetilde{Q}$ admits the factorization

$$\widetilde{Q} = \widetilde{r}(z)\widetilde{r}^\sharp(z) \left(\widetilde{a} + \widetilde{b}z + \int_{\mathbb{R}} \left(\frac{1}{t-z} - \frac{t}{1+t^2} \right) \right) d\widetilde{\rho}(t), \quad \widetilde{r} = \frac{\widetilde{p}_2}{\widetilde{q}_2}, \tag{3.7}$$

where $\widetilde{p}_2$ and $\widetilde{q}_2$ are the polynomials associated to $\widetilde{Q}$, cf. (2.5). Moreover, the inequality $\pi_\infty(\widetilde{Q}) = \deg \widetilde{q}_2 - \deg \widetilde{p}_2 \geq 0$ holds by Theorems 2.1 and 2.2. On the other hand, it follows from (2.6) and (3.1) that $\widetilde{Q}$ admits also the representation

$$\widetilde{Q}(z) = z^{2n} r(z) r^\sharp(z) \left(a + bz + \int_{\mathbb{R}} \left(\frac{1}{t-z} - \frac{t}{1+t^2} \right) \right) d\rho(t) + p_1(z), \tag{3.8}$$

where p_1 is a polynomial with $\deg p_1 \leq 2n$. An application of the generalized Stieltjes inversion formula (see [19]) shows that the measures $d\widetilde{\rho}(t)$ in (3.7) and $d\rho(t)$ in (3.8) are connected by

$$|\widetilde{r}(t)|^2 d\widetilde{\rho}(t) = t^{2n} |r(t)|^2 \, d\rho(t). \tag{3.9}$$

Therefore, if $\widetilde{Q} \in \mathbf{N}_{\kappa',1} \setminus \mathbf{N}_{\kappa',0}$ so that $\ell = 2n-1$, then $\deg \widetilde{p}_2 = \deg \widetilde{q}_2$ and $d\widetilde{\rho}(t)$ satisfies the condition (2.8) in Theorem 2.1. The condition (3.6) follows now from (3.9). If $\widetilde{Q} \in \mathbf{N}_{\kappa',0}$ so that $\ell = 2n$, then either $\deg \widetilde{p}_2 = \deg \widetilde{q}_2$ in which case $d\widetilde{\rho}(t)$ satisfies the condition (2.9) in Theorem 2.2, or $\pi_\infty(\widetilde{Q}) = \deg \widetilde{q}_2 - \deg \widetilde{p}_2 > 0$ in which case $d\widetilde{\rho}(t)$ satisfies the condition (2.7). In both cases

$$\int_{|t|>M} |\widetilde{r}(t)|^2 d\widetilde{\rho}(t) < \infty \qquad \text{for } M > 0 \text{ large enough.}$$

Hence, again the condition (3.6) follows from (3.9).

(ii) $\Leftrightarrow$ (iii) Let E_t be the spectral function of a selfadjoint operator A in the minimal representation (1.2) of Q. It follows from (1.2), (2.6), and the generalized Stieltjes inversion formula that

$$d[E_t\omega, \omega] = |r(t)|^2 \, d\rho(t), \quad t \in \Delta,$$

in some neighborhood Δ of $\pm\infty$. This implies that

$$\int_{\mathbb{R}} (1 + |t|)^{\ell - 2\pi_\infty} \, d\rho(t) < \infty \;\text{ if and only if }\; \int_{\Delta} (1 + |t|)^{\ell} \, d[E_t\omega, \omega] < \infty,$$

i.e., $\omega \in \mathfrak{H}_\ell$, which proves the equivalence of (ii) and (iii).

(ii) $\Rightarrow$ (i) First consider the case $\ell = 2n$. Then $\omega \in \mathfrak{H}_\ell$ means that $\omega \in \operatorname{dom} A^n$. Define the function $\widetilde{Q}$ in (3.1) by setting

$$s_j = [A^j\omega, \omega], \quad s_{n+j} = [A^j\omega, A^n\omega], \quad j = 0, \ldots, n. \tag{3.10}$$

Then a straightforward calculation shows that $\widetilde{Q}$ admits the operator representation

$$\widetilde{Q}(z) = [(A - z)^{-1}\omega', \omega'], \quad \omega' = A^n\omega \in \mathfrak{H}. \tag{3.11}$$

Therefore, $\widetilde{Q}(z) = O(1/z)$ and $Q \in \mathbf{N}_{\kappa,-\ell}$.

Now let $\ell = 2n - 1$. Then $\omega \in \mathfrak{H}_\ell$ means that $\omega' := A^n\omega \in \mathfrak{H}_{-1}$. Hence $s_{2n-1} := [\ddot{A}\omega', \omega']$ is well defined. Moreover, by defining $s_0, \ldots, s_{2n-2}$ as in (3.10) it follows that the function $\widetilde{Q}$ in (3.1) admits the operator representation

$$\widetilde{Q}(z) = [(\widetilde{A} - z)^{-1}\omega', \omega'], \quad \omega' = A^n\omega \in \mathfrak{H}_{-1}. \tag{3.12}$$

Hence, by Theorem 2.1 $\widetilde{Q} \in \mathbf{N}_{\kappa',1}$ for some $\kappa' \in \mathbb{N}$ and thus $Q \in \mathbf{N}_{\kappa,-\ell}$. This completes the proof. □

In the case of even indices $\ell = 2n$ the equivalence of (i) and (ii) in Theorem 3.3 coincides with the following result of M.G. Kreĭn and H. Langer, see [21, Satz 1.10].

Corollary 3.4. ([21]) *The function $Q \in \mathbf{N}_\kappa$ admits an operator representation $Q(z) = \gamma + [(A - z)^{-1}\omega, \omega]$ with $\gamma \in \mathbb{R}$ and $\omega \in \mathfrak{H}_{2n}$ $(= \operatorname{dom} A^n)$ if and only if there are real numbers γ and $s_0, \ldots, s_{2n}$, such that*

$$Q(z) = \gamma - \sum_{j=1}^{2n+1} \frac{s_{j-1}}{z^j} + o\left(\frac{1}{z^{2n+1}}\right), \quad z \widehat{\to} \infty. \tag{3.13}$$

In this case the numbers $s_0, \ldots, s_{2n}$ are given by (3.10).

Proof. The proof of Theorem 3.3 shows that the condition $\omega \in \operatorname{dom} A^n$ is equivalent to the operator representation (3.11) of the function $\widetilde{Q}(z)$ in (3.1). Now by applying (2.10) in Remark 2.3 to the function $\widetilde{Q}(z)$ in (3.11) and taking into account (3.1) the equivalence to the expansion (3.13) follows. □

The criterion of M.G. Kreĭn and H. Langer formulated in Corollary 3.4 does not hold in the case of an odd index $\ell = 2n - 1$. However, it is clear that if $\omega \in \mathfrak{H}_{2n-1}$ then the analog of the expansion (3.13) exists.

Corollary 3.5. *If the function $Q \in \mathbf{N}_\kappa$ admits an operator representation $Q(z) = \gamma + [(A - z)^{-1}\omega, \omega]$ with $\gamma \in \mathbb{R}$ and $\omega \in \mathfrak{H}_{2n-1}$ $(= \operatorname{dom} |A|^{n-1/2})$ then there are real numbers γ and $s_0, \ldots, s_{2n}$, such that*

$$Q(z) = \gamma - \sum_{j=1}^{2n} \frac{s_{j-1}}{z^j} + o\left(\frac{1}{z^{2n}}\right), \quad z \widehat{\to} \infty. \tag{3.14}$$

Proof. Since $\omega \in \mathfrak{H}_{2n-1}$ the operator representation (3.12) in the proof of Theorem 3.3 shows that $\widetilde{Q}(z) = o(1)$. The expansion (3.14) for the function Q follows now from the definition of $\widetilde{Q}$ in (3.1). □

It is emphasized that the existence of the expansion (3.14) does not imply that $\omega \in \mathfrak{H}_{2n-1}$ or, equivalently, that $\omega' := A^n\omega$ belongs to $\mathfrak{H}_{-1}$. In this case $[\widetilde{A}\omega', \omega']$ need not be defined and hence it cannot coincide with the coefficient s_{2n-1} in (3.14).

3.2. A classification of generalized zeros of nonpositive type

For what follows it will be useful to give a classification for generalized zeros and poles of nonpositive type of a function $Q \in \mathbf{N}_\kappa$.

Let ∞ be a GZNT of Q with multiplicity $\pi_\infty > 0$. It follows from (2.4) that precisely one of the following three cases can occur:

$$\text{(T1)} \qquad -s_{2\pi_\infty} := \lim_{z \widehat{\to} \infty} z^{2\pi_\infty+1}Q(z) < 0, \quad \lim_{z \widehat{\to} \infty} z^{2\pi_\infty-1}Q(z) = 0;$$

$$\text{(T2)} \qquad \lim_{z \widehat{\to} \infty} z^{2\pi_\infty+1}Q(z) = \infty, \quad \lim_{z \widehat{\to} \infty} z^{2\pi_\infty-1}Q(z) = 0;$$

$$\text{(T3)} \qquad \lim_{z \widehat{\to} \infty} z^{2\pi_\infty+1}Q(z) = \infty, \quad -s_{2\pi_\infty-2} := \lim_{z \widehat{\to} \infty} z^{2\pi_\infty-1}Q(z) > 0.$$

In these cases ∞ is said to be a generalized zero of type (T1), (T2), or (T3), respectively; the shorter notations GZNT1, GZNT2, and GZNT3 are used accordingly. The corresponding classification for a finite generalized zero $\beta \in \mathbb{R}$ of Q is defined analogously:

$$\text{(T1)} \qquad \lim_{z \widehat{\to} \beta} \frac{Q(z)}{(z-\beta)^{2\pi_\beta+1}} > 0, \quad \lim_{z \widehat{\to} \beta} \frac{Q(z)}{(z-\beta)^{2\pi_\beta-1}} = 0;$$

$$\text{(T2)} \qquad \lim_{z \widehat{\to} \beta} \frac{Q(z)}{(z-\beta)^{2\pi_\beta+1}} = \infty, \quad \lim_{z \widehat{\to} \beta} \frac{Q(z)}{(z-\beta)^{2\pi_\beta-1}} = 0;$$

$$\text{(T3)} \qquad \lim_{z \widehat{\to} \beta} \frac{Q(z)}{(z-\beta)^{2\pi_\beta+1}} = \infty, \quad \lim_{z \widehat{\to} \beta} \frac{Q(z)}{(z-\beta)^{2\pi_\beta-1}} < 0.$$

A generalized pole of nonpositive type $\beta \in \mathbb{R} \cup \{\infty\}$ of Q is said to be of type (T1), (T2), or (T3), if β is a generalized zero of nonpositive type of the function $-1/Q$ which is of type (T1), (T2), or (T3), respectively.

To give some immediate implications of the above classification consider the generalized zero ∞ of Q. If it is of the first type, then it follows from (T1) that $Q \in \mathbf{N}_{\kappa,-2\pi_\infty}$. Moreover, Q has the following asymptotic expansion:

$$Q(z) = -\frac{s_{2\pi_\infty}}{z^{2\pi_\infty+1}} + o\left(\frac{1}{z^{2\pi_\infty+1}}\right), \quad z \widehat{\to} \infty, \quad s_{2\pi_\infty} > 0. \tag{3.15}$$

If the generalized zero ∞ of Q is of type (T3), then $Q \in \mathbf{N}_{\kappa,-2\pi_\infty-2}$ and Q has the following asymptotic expansion

$$Q(z) = -\frac{s_{2\pi_\infty-2}}{z^{2\pi_\infty-1}} + o\left(\frac{1}{z^{2\pi_\infty-1}}\right), \quad z \widehat{\to} \infty, \quad s_{2\pi_\infty-2} < 0. \tag{3.16}$$

In the case that the generalized zero ∞ is of type (T2) there are two possibilities: either Q belongs to $\mathbf{N}_{\kappa,-2\pi_\infty}$, in which case both of the moments $s_{2\pi_\infty-1}$ and $s_{2\pi_\infty}$ are finite and Q has the asymptotic expansion

$$Q(z) = -\frac{s_{2\pi_\infty-1}}{z^{2\pi_\infty}} - \frac{s_{2\pi_\infty}}{z^{2\pi_\infty+1}} + o\left(\frac{1}{z^{2\pi_\infty+1}}\right), \quad z \widehat{\to} \infty, \quad s_{2\pi_\infty-1} \neq 0, \tag{3.17}$$

or Q belongs to $\mathbf{N}_{\kappa,-2(\pi_\infty-1)} \setminus \mathbf{N}_{\kappa,-2\pi_\infty}$ and it has the asymptotic expansion

$$Q(z) = -\frac{s_{2\pi_\infty-1}}{z^{2\pi_\infty}} + o\left(\frac{1}{z^{2\pi_\infty}}\right), \quad z\widehat{\to}\infty, \tag{3.18}$$

or

$$Q(z) = o\left(\frac{1}{z^{2\pi_\infty-1}}\right), \quad z\widehat{\to}\infty. \tag{3.19}$$

Observe, that the expansions (3.17) and (3.18) are also special cases of the expansion (3.19). Hence, if ∞ is a generalized zero of type (T2), then Q has an expansion of the form (3.16), but now with $s_{2\pi_\infty-2} = 0$; however, Q does not have an expansion of the form (3.15). Similar observations remain true for generalized zeros $\beta \in \mathbb{R}$ and poles $\alpha \in \mathbb{R} \cup \{\infty\}$. For instance, to get the analogous expansions for a generalized zero $\beta \in \mathbb{R}$ apply the transform $-Q(1/(z-\beta))$ to the expansions in (3.15)–(3.19); cf. also [15]. The role of the above classification for generalized zeros and poles of nonpositive type will be described in detail in Sections 4–6.

4. An operator model for the generalized Friedrichs extension

4.1. Boundary triplets and Weyl functions

The construction of the model uses the notion of a boundary triplet in a Pontryagin space setting. Let $\mathfrak{H}$ be a Pontryagin space with negative index κ, let S be a closed symmetric relation in $\mathfrak{H}$ with defect numbers (n, n), and let S^* be the adjoint of S. A triplet $\Pi = \{\mathbb{C}^n, \Gamma_0, \Gamma_1\}$ is said to be a *boundary triplet* for S^*, if the following two conditions are satisfied:

(i) the mapping $\Gamma : \widehat{f} \to \{\Gamma_0\widehat{f}, \Gamma_1\widehat{f}\}$ from S^* to $\mathbb{C}^n \oplus \mathbb{C}^n$ is surjective;
(ii) the abstract Green's identity

$$[f', g] - [f, g'] = (\Gamma_1\widehat{f}, \Gamma_0\widehat{g}) - (\Gamma_0\widehat{f}, \Gamma_1\widehat{g}) \tag{4.1}$$

holds for all $\widehat{f} = \{f, f'\}$, $\widehat{g} = \{g, g'\} \in S^*$,

see, e.g., [2], [10]. It is easily seen that $A_0 = \ker \Gamma_0$ and $A_1 = \ker \Gamma_1$ are selfadjoint extensions of S. Associated to every boundary triplet there is the Weyl function Q defined by

$$Q(z)\Gamma_0\widehat{f}_z = \Gamma_1\widehat{f}_z, \quad z \in \rho(A_0),$$

where $\widehat{f}_z := \{f_z, zf_z\} \in \widehat{\mathfrak{N}}_z$, and $\mathfrak{N}_z = \ker(S^* - z)$ denotes the defect subspace of S at $z \in \mathbb{C}$. It follows from (4.1) that the Weyl function Q is also a Q-function of the pair $\{S, A_0\}$ in the sense of Kreĭn and Langer, see [20]. If S is *simple*, so that

$$\mathfrak{H} = \overline{\operatorname{span}}\,\{\mathfrak{N}_z : z \in \rho(A_0)\},$$

then the Weyl function Q belongs to the class $\mathbf{N}_\kappa$, otherwise $Q \in \mathbf{N}_{\kappa'}$ with $\kappa' \le \kappa$. Moreover, if S is simple and H is a selfadjoint extension of S in $\mathfrak{H}$, then the point spectrum of H is also simple, that is, every eigenspace of H is one-dimensional, and if $\alpha \in \mathbb{R} \cup \{\infty\}$, then the root subspace at α is at most $2\kappa + 1$-dimensional.

In the case where S is given by (1.3) one can define a boundary triplet for S^* as follows.

Proposition 4.1. (cf. [5]) *Let A be a selfadjoint operator in a Pontryagin space $\mathfrak{H}$ and let the restriction S of A be defined by (1.3) with $\omega \in \mathfrak{H}$. Then the adjoint S^* of S in $\mathfrak{H}$ is of the form*

$$S^* = \{\{f, Af + c\omega\} : f \in \operatorname{dom} A,\, c \in \mathbb{C}\}$$

and a boundary triplet $\Pi^\infty = \{\mathbb{C}, \Gamma_0^\infty, \Gamma_1^\infty\}$ for S^* is determined by

$$\Gamma_0^\infty \widehat{f} = [f, \omega], \quad \Gamma_1^\infty \widehat{f} = c, \quad \widehat{f} = \{f, Af + c\omega\} \in S^*.$$

The corresponding Weyl function Q_∞ is given by

$$Q_\infty(z) = -\frac{1}{[(A-z)^{-1}\omega, \omega]}, \qquad z \in \rho(A). \tag{4.2}$$

4.2. The model operator $S(Q_\infty)$ corresponding to a proper factorization

Operator models for generalized Nevanlinna functions whose only generalized pole of nonpositive type is at ∞ have been constructed in [6] and [14]. Such functions admit a canonical factorization of the form

$$Q_\infty(z) = q(z)q^\sharp(z)Q_0(z), \tag{4.3}$$

where $Q_0 \in \mathbf{N}_0$, $q(z) = z^k + q_{k-1}z^{k-1} + \cdots + q_0$ is a polynomial, and $q^\sharp(z) = \overline{q(\bar{z})}$. In general, models which are based on the canonical factorization of $Q \in \mathbf{N}_\kappa$ are not necessarily minimal, i.e., the underlying model operator $S(Q_\infty)$ need not be simple and it can even be a symmetric relation (multivalued operator). However, with the canonical factorization the nonsimple part of $S(Q_\infty)$ can be easily identified and factored out to produce a simple symmetric operator from $S(Q_\infty)$, cf. [3]. The model constructed for $S(Q_\infty)$ in [6] uses an orthogonal coupling of two symmetric operators. In [9] this model was adapted to the case where the function Q_0 is allowed to be a generalized Nevanlinna function, too. In this case the situation becomes more involved and, in general, one cannot represent $S(Q_\infty)$ as an orthogonal sum of a simple symmetric operator and a selfadjoint relation. However, such a simple orthogonal decomposition for $S(Q_\infty)$ can still be obtained if the factorization (4.3) of Q_∞ is proper. This concept is defined as follows.

Definition 4.2. ([9]) The factorization $Q_\infty(z) = q(z)q^\sharp(z)Q_0(z)$ is said to be *proper* if q is a divisor of degree $\kappa_\infty(Q_\infty) > 0$ of the polynomial $\widetilde{q}$ in the canonical factorization of the function Q_∞, cf. (2.5).

Clearly, proper factorizations of $Q_\infty \in \mathbf{N}_\kappa$ always exist, but they are not unique if $\widetilde{q}$ has more than one zero and $\kappa_\infty(Q_\infty) < \kappa$. Proper factorizations $Q_\infty = qq^\sharp Q_0$ can be characterized also without using the canonical factorization of Q_∞.

Lemma 4.3. *Let $Q_\infty \in \mathbf{N}_\kappa$ have a factorization of the form*

$$Q_\infty(z) = q(z)q^\sharp(z)Q_0(z), \quad \deg q = k \geq 1, \tag{4.4}$$

where $q(z)$ is a monic polynomial, and let $\alpha \in \sigma(q)$ be a zero of q with multiplicity k_α. Then the following statements are equivalent:

(i) *the factorization (4.4) of Q_∞ is proper;*

(ii) *the multiplicities $\kappa_\infty(Q_\infty)$ and $\pi_\alpha(Q_\infty)$ satisfy the following relations:*

$$\kappa_\infty(Q_\infty) = \deg q \text{ and } \pi_\alpha(Q_\infty) \geq k_\alpha \text{ for all } \alpha \in \sigma(q); \tag{4.5}$$

(iii) *$\kappa_\infty(Q_0)$ and $\kappa(Q_\infty) = \kappa$ satisfy the following identities*

$$\kappa_\infty(Q_0) = 0 \text{ and } \kappa(Q_\infty) = \deg q + \kappa(Q_0). \tag{4.6}$$

Proof. (i) $\Leftrightarrow$ (ii) In a proper factorization (4.4) $\kappa_\infty(Q_\infty) = \deg q$ and clearly the inequalities in (4.5) just mean that q divides the polynomial $\widetilde{q}$ in the canonical factorization of Q_∞.

(i) $\Rightarrow$ (iii) If the factorization (4.4) is proper, then in the canonical factorization of the function Q_0 the numerator $\widetilde{q}_0$ and denominator $\widetilde{p}_0\,(= \widetilde{p})$ of the corresponding rational factor r_0 are of the same degree $\kappa(Q_0)$, and this implies (4.6).

(iii) $\Rightarrow$ (i) It follows from the second equality in (4.6) that q and the polynomial $\widetilde{p}_0$ in the canonical factorization of the function Q_0 are relatively prime and, therefore, q is a factor of the polynomial $\widetilde{q}$ in the canonical factorization of Q_∞. Moreover, $\pi_\infty(Q_0) = 0$. Now the assumption $\kappa_\infty(Q_0) = 0$ implies that $\kappa_\infty(Q_\infty) = \deg q$. □

The construction of factorization models is now briefly described. Let q be a polynomial as in (4.4) of degree $k = \deg q$. Define the $k \times k$ matrices $\mathcal{B}_q$ and $\mathcal{C}_q$ by

$$\mathcal{B}_q = \begin{pmatrix} q_1 & \dots & q_{k-1} & 1 \\ \vdots & \cdot^{\cdot^{\cdot}} & 1 & 0 \\ q_{k-1} & \cdot^{\cdot^{\cdot}} & \cdot^{\cdot^{\cdot}} & \vdots \\ 1 & 0 & \dots & 0 \end{pmatrix}, \quad \mathcal{C}_q = \begin{pmatrix} 0 & 1 & \dots & 0 \\ \vdots & \ddots & \ddots & 0 \\ 0 & 0 & \dots & 1 \\ -q_0 & -q_1 & \dots & -q_{k-1} \end{pmatrix},$$

so that $\sigma(\mathcal{C}_q) = \sigma(q)$. Moreover, let $\mathfrak{H}_q$ be a $2k$-dimensional Pontryagin space defined by

$$(\mathbb{C}^k \oplus \mathbb{C}^k, \langle \mathcal{B}\cdot, \cdot \rangle), \quad \mathcal{B} = \begin{pmatrix} 0 & B_q \\ B_{q^\sharp} & 0 \end{pmatrix}.$$

A general factorization model for functions Q_∞ of the form (4.4) was constructed in [9] and can be applied, in particular, for proper factorizations of Q_∞.

Theorem 4.4. (cf. [9]) *Let $Q_\infty \in \mathbf{N}_\kappa$ be a generalized Nevanlinna function and let*

$$Q_\infty(\lambda) = q(\lambda) q^\sharp(\lambda) Q_0(\lambda), \tag{4.7}$$

be a proper factorization of Q_∞, where q is a monic polynomial of degree $k = \deg q \geq 1$. Let S_0 be a closed symmetric relation in a Pontryagin space $\mathfrak{H}_0$ with the boundary triplet $\Pi^0 = \{\mathbb{C}, \Gamma_0^0, \Gamma_1^0\}$ whose Weyl function is Q_0.

Then:

(i) *the function* Q_0 *in* (4.7) *belongs to the class* $\mathbf{N}_{\kappa-k}$*;*

(ii) *the linear relation*

$$S(Q_\infty) = \left\{ \left\{ \begin{pmatrix} f_0 \\ f \\ \widetilde{f} \end{pmatrix}, \begin{pmatrix} f_0' \\ C_{q^\sharp} f \\ C_q \widetilde{f} + \Gamma_0^0 \widehat{f}_0 e_k \end{pmatrix} \right\} : \begin{array}{c} \widehat{f}_0 = \{f_0, f_0'\} \in S_0^*, \\ f_1 = \Gamma_1^0 \widehat{f}_0, \\ \widetilde{f}_1 = 0 \end{array} \right\} \tag{4.8}$$

is closed and symmetric in $\mathfrak{H} := \mathfrak{H}_0 \oplus \mathfrak{H}_q$ *and has defect numbers* (1, 1)*;*

(iii) *the adjoint* $S(Q_\infty)^*$ *of* $S(Q_\infty)$ *is given by*

$$S(Q_\infty)^* = \left\{ \left\{ \begin{pmatrix} f_0 \\ f \\ \widetilde{f} \end{pmatrix}, \begin{pmatrix} f_0' \\ C_{q^\sharp} f + \widetilde{\varphi} e_k \\ C_q \widetilde{f} + \Gamma_0^0 \widehat{f}_0 e_k \end{pmatrix} \right\} : \begin{array}{c} \widehat{f}_0 = \{f_0, f_0'\} \in S_0^*, \\ f_1 = \Gamma_1^0 \widehat{f}_0, \\ \widetilde{\varphi} \in \mathbb{C} \end{array} \right\};$$

(iv) *a boundary triplet* $\Pi = \{\mathbb{C}, \Gamma_0, \Gamma_1\}$ *for* $S(Q_\infty)^*$ *is determined by*

$$\Gamma_0(\widehat{f}_0 \oplus \widehat{F}) = \widetilde{f}_1, \quad \Gamma_1(\widehat{f}_0 \oplus \widehat{F}) = \widetilde{\varphi}, \quad \widehat{f}_0 \oplus \widehat{F} \in S(Q_\infty)^*; \tag{4.9}$$

(v) *the corresponding Weyl function coincides with* Q_∞*.*

Proof. Since the factorization (4.7) is proper the statement (i) is immediate from the equality (4.6) in Lemma 4.3. All the other statements are contained in [9]. □

In fact, the statement (iv) in Theorem 4.4 can be obtained directly also from Proposition 4.1, since $S(Q_\infty)$ in (4.8) is a restriction of the selfadjoint relation

$$A(Q_\infty) = \left\{ \left\{ \begin{pmatrix} f_0 \\ f \\ \widetilde{f} \end{pmatrix}, \begin{pmatrix} f_0' \\ C_{q^\sharp} f \\ C_q \widetilde{f} + \Gamma_0^0 \widehat{f}_0 e_k \end{pmatrix} \right\} : \begin{array}{c} \widehat{f}_0 = \{f_0, f_0'\} \in S_0^*, \\ f_1 = \Gamma_1^0 \widehat{f}_0 \end{array} \right\} \tag{4.10}$$

to the subspace $\mathfrak{H} \ominus \omega_0$, where $\omega_0 = \operatorname{col}(0, e_k, 0)$; compare (1.3). The generalized Friedrichs extension of $S(Q_\infty)$ is given by

$$S_F(Q_\infty) = \left\{ \left\{ \begin{pmatrix} f_0 \\ f \\ \widetilde{f} \end{pmatrix}, \begin{pmatrix} f_0' \\ C_{q^\sharp} f + \widetilde{\varphi} e_k \\ C_q \widetilde{f} + \Gamma_0^0 \widehat{f}_0 e_k \end{pmatrix} \right\} : \begin{array}{c} \widehat{f}_0 = \{f_0, f_0'\} \in S_0^*, \\ f_1 = \Gamma_1^0 \widehat{f}_0, \\ \widetilde{f}_1 = 0, \ \widetilde{\varphi} \in \mathbb{C} \end{array} \right\}. \tag{4.11}$$

According to (4.2) in Proposition 4.1 the Weyl function $Q_\infty(z)$ corresponding to the boundary triplet (4.9) is of the form

$$Q_\infty(z) = -\frac{1}{[(A(Q_\infty) - z)^{-1}\omega_0, \omega_0]}.$$

Thus the function $Q = -1/Q_\infty$ has the representation

$$Q(z) = [(A(Q_\infty) - z)^{-1}\omega, \omega], \tag{4.12}$$

which is, however, not necessarily minimal, since $\operatorname{mul} S$ and $\ker(S - \alpha)$, $\alpha \in \sigma(q)$, can be nontrivial. The following lemma describes these subspaces.

Lemma 4.5. ([9]) *Under the assumptions of Theorem* 4.4 *let* S_0 *be a simple closed symmetric operator in the Pontryagin space* $\mathfrak{H}_0$ *and let* $A_i^0 = \ker \Gamma_i^0 (\supset S_0)$, $i = 0, 1$. *Then:*

(i) $\operatorname{mul} S(Q_\infty)$ *is nontrivial if and only if* $\operatorname{mul} A_1^0$ *is nontrivial and in this case*

$$\operatorname{mul} S(Q_\infty) = \{ (g, 0, \Gamma_0^0 \widehat{g} e_k)^\top : \widehat{g} = \{0, g\} \in A_1^0 \}; \tag{4.13}$$

(ii) *if* $\operatorname{mul} S(Q_\infty)$ *is nontrivial, then it is spanned by a positive vector;*
(iii) *if* $\operatorname{mul} A_0^0$ *is nontrivial, then it is spanned by a positive vector;*
(iv) $\sigma_p(S(Q_\infty)) = \sigma_p(A_0^0) \cap \sigma(q^\sharp)$ *and for* $\alpha \in \sigma_p(A_0^0) \cap \sigma(q^\sharp)$ *one has*

$$\ker (S(Q_\infty) - \alpha) = \{ (g_0, \Gamma_1^0 \widehat{g}_0 \Lambda|_{\lambda=\alpha}, 0)^\top : g_0 \in \ker (A_0^0 - \alpha) \}, \tag{4.14}$$

where $\Lambda = (1, \lambda, \dots, \lambda^{k-1})$, $\lambda \in \mathbb{C}$;
(v) *if* $\ker (S(Q_\infty) - \alpha)$ *or, equivalently,* $\ker (A_0^0 - \alpha)$ *is nontrivial, then it is spanned by a positive vector.*

It follows from (ii) and (iv) that the linear relation $S(Q_\infty)$ can be decomposed into a direct sum of an operator S' with an empty point spectrum and a selfadjoint part in a Hilbert space which is the sum of $\operatorname{mul} S(Q_\infty)$ and $\ker (S(Q_\infty) - \alpha)$, $\alpha \in \sigma_p(A_0^0) \cap \sigma(q^\sharp)$. The next theorem shows that the reduced operator S' is simple.

Theorem 4.6. *Let the assumptions of Theorem* 4.4 *and Lemma* 4.5 *be satisfied and let* $S(Q_\infty)$, $A(Q_\infty)$, *and* $S_F(Q_\infty)$ *be given by* (4.8), (4.10), *and* (4.11), *respectively. Then:*

(i) $S(Q_\infty)$ *is simple if and only if* $\sigma_p(S(Q_\infty)) = \emptyset$. *In this case the linear relations* $S = S(Q_\infty)$, $A = A(Q_\infty)$, *and* $S_F = S_F(Q_\infty)$ *satisfy the equalities* (1.3) *and* (1.4) *with* $\omega = \omega_0$ *and the operator representation* (4.12) *of* $Q = -1/Q_\infty$ *is minimal.*
(ii) *If* $S(Q_\infty)$ *is not simple, then the subspace*

$$\mathfrak{H}'' = \operatorname{span} \{ \operatorname{mul} S(Q_\infty), \ker (S(Q_\infty) - \alpha) : \alpha \in \sigma_p(A_0^0) \cap \sigma(q) \} \tag{4.15}$$

is positive and reducing for $S(Q_\infty)$. *The simple part of* $S(Q_\infty)$ *coincides with the restriction* S' *of* $S(Q_\infty)$ *to* $\mathfrak{H}' := \mathfrak{H} \ominus \mathfrak{H}''$. *The compressions* S', A', *and* S'_F *of* $S(Q_\infty)$, $A(Q_\infty)$, *and* $S_F(Q_\infty)$ *to the subspace* $\mathfrak{H}'$ *satisfy the equalities* (1.3) *and* (1.4), *with* $\omega \in \mathfrak{H}'$ *given by*

$$\omega = \begin{cases} \omega_0, & \text{if } k > 1, \\ (g, -1/\overline{\Gamma_0 \widehat{g}}, \Gamma_0 \widehat{g})^\top, & \text{if } k = 1, \end{cases}$$

and the function $Q = -1/Q_\infty$ *admits the minimal representation*

$$Q(\lambda) = -1/Q_\infty(\lambda) = [(A' - \lambda)^{-1} \omega, \omega].$$

4.3. The root subspace of the generalized Friedrichs extension at ∞

Let the assumptions of Theorem 4.6 be satisfied and let the linear relations $S(Q_\infty)$, $A(Q_\infty)$, $S_F(Q_\infty)$ and S', A', S'_F be as defined in that theorem. Since S' is a simple symmetric operator in the Pontryagin space $\mathfrak{H}'$ its selfadjoint extension S'_F has a simple point spectrum. In particular, the multivalued part $\operatorname{mul} S'_F$ of S'_F is at most one-dimensional. The corresponding root subspace

$$\mathsf{R}_\infty(S'_F) = \operatorname{span}\{\, g \in \mathfrak{H}' : \{0, g\} \in {S'_F}^k, \ \text{ for some } k \in \mathbb{N}\,\}$$

is at most $2\kappa+1$-dimensional. The root subspace $\mathsf{R}_\infty(S'_F)$ is spanned by the vectors ω_j which form a maximal Jordan chain in $\mathsf{R}_\infty(S'_F)$:

$$\mathsf{R}_\infty(S'_F) = \operatorname{span}\{\, \omega_j \in \mathfrak{H}' : \{\omega_{j-1}, \omega_j\} \in S'_F, \quad j = 0, \dots, \nu - 1\,\},$$

where $\nu = \dim \mathsf{R}_\infty(S'_F)$ and $\omega_{-1} = 0$.

The main properties of the root subspace $\mathsf{R}_\infty(S'_F)$ of the selfadjoint extension S'_F of S' are given in Lemmas 4.7–4.11 below. The proofs of these lemmas are constructive, since they are based on the factorization model for $S(Q_\infty)$ given in Theorem 4.4, cf. also [9]. As a consequence a detailed description for the structure of the corresponding Jordan chains in the factorization model is obtained. It turns out that three different types of maximal Jordan chains can appear in this model. In each of these cases maximal Jordan chains in $\mathsf{R}_\infty(S_F)$ can be regular or singular and the signature of $\mathsf{R}_\infty(S_F)$ has its own specific nature in each case.

The first lemma concerns the dimension and nondegeneracy of the root subspace $\mathsf{R}_\infty(S'_F)$; further equivalent conditions and descriptions (without a specific model space) for arbitrary generalized poles and zeros of $Q \in \mathbf{N}_\kappa$ with direct (nonconstructive) proofs can be found from [15].

Lemma 4.7. *Let $Q_\infty \in \mathbf{N}_\kappa$, let $\kappa_\infty(Q_\infty) = \deg q (= k) > 0$, and let S', S'_F, and $\omega \in \mathfrak{H}$, $[\omega, \omega] \le 0$, be as in Theorem 4.6. Then:*

(i) *$\dim \mathsf{R}_\infty(S'_F) \ge \deg q$;*

(ii) *$\dim \mathsf{R}_\infty(S'_F)$ is equal to ν if and only if $\omega \in \operatorname{dom} {S'_F}^{\nu-1} \setminus \operatorname{dom} {S'_F}^{\nu}$;*

(iii) *$\mathsf{R}_\infty(S'_F)$ is a regular subspace of dimension ν if and only if*

$$\omega \in \operatorname{dom} {S'_F}^{\nu-1} \setminus \operatorname{dom} {S'_F}^{\nu} \ \text{ and } [\omega, \omega_{\nu-1}] \ne 0 \text{ for } \{\omega, \omega_{\nu-1}\} \in {S'_F}^{\nu-1};$$

(iv) *$\mathsf{R}_\infty(S'_F)$ is a singular subspace of dimension ν if and only if*

$$\omega \in \operatorname{dom} {S'_F}^{\nu-1} \setminus \operatorname{dom} {S'_F}^{\nu} \ \text{ and } [\omega, \omega_{\nu-1}] = 0 \text{ for } \{\omega, \omega_{\nu-1}\} \in {S'_F}^{\nu-1}. \tag{4.16}$$

Proof. (i) Consider a proper factorization (4.7) for Q_∞ and the model operator $S(Q_\infty)$ constructed in Theorem 4.4 and denote by $A_0 = \ker \Gamma_0$ the selfadjoint extension $S_F(Q_\infty)$ in (4.11) of $S(Q_\infty)$ in (4.8). Clearly, the vectors

$$w_0 = \begin{pmatrix} 0 \\ e_k \\ 0 \end{pmatrix}, \dots, w_{k-2} = \begin{pmatrix} 0 \\ e_2 \\ 0 \end{pmatrix}, \ w_{k-1} = \begin{pmatrix} 0 \\ e_1 \\ 0 \end{pmatrix} \tag{4.17}$$

form a Jordan chain in $\mathsf{R}_\infty(A_0)$, that is, $w_0 \in \operatorname{mul} A_0$ and $\{w_{j-1}, w_j\} \in A_0$ for $j = 1, \dots, k-1$. If $S(Q_\infty)$ is simple then $A_0 = S'_F$ and this proves (i). In the case when $S(Q_\infty)$ is not simple, but $\operatorname{mul} S(Q_\infty)$ is trivial the vectors $w_0, \dots, w_{k-1}$ still belong to $\mathsf{R}_\infty(S'_F)$ since for all $\alpha \in \sigma_p(A_0^0) \cap \sigma(q^\sharp)$ these vectors are orthogonal to the eigenspaces $\ker(S(Q_\infty) - \alpha)$ which were described in (4.14).

Assume that $\operatorname{mul} S(Q_\infty)$ is not trivial. Again it follows from Lemma 4.5 that the vectors $w_0, \dots, w_{k-1}$ are orthogonal to $\ker(S(Q_\infty) - \alpha)$ for all $\alpha \in \sigma_p(A_0^0) \cap \sigma(q^\sharp)$. Moreover, by using (4.13) in Lemma 4.5 it is seen that the vectors $w_0, \dots, w_{k-2}$ are orthogonal to $\operatorname{mul} S(Q_\infty)$. Since $\operatorname{mul} S(Q_\infty)[\perp]\ker(S(Q_\infty) - \alpha)$ for all $\alpha \in \sigma_p(A_0^0) \cap \sigma(q^\sharp)$, it is easy to check that the projections $w'_j = P' w_j$ of w_j, $j = 1, \dots, k-1$, to the subspace $\mathfrak{H}' = \mathfrak{H} \ominus \mathfrak{H}''$, where $\mathfrak{H}''$ is given by (4.15), take the form

$$w'_0 = w_0, \dots, w'_{k-2} = w_{k-2}, w'_{k-1} = \begin{pmatrix} -(\overline{\Gamma_0^0 \widehat{g}})g \\ e_1 \\ -|\Gamma_0^0 \widehat{g}|^2 e_k \end{pmatrix}, \tag{4.18}$$

where $g \in \operatorname{mul} A_1^0$, $[g, g] = 1$, $\widehat{g} = \{0, g\} \in A_1^0$, and $\Gamma_0^0 \widehat{g} \neq 0$. Therefore, the sequence $\{w'_0, \dots, w'_{k-1}\}$ forms a Jordan chain in $\mathsf{R}_\infty(S'_F)$ and hence again $\dim \mathsf{R}_\infty(S'_F) \geq k$.

(ii) This statement holds true since S'_F has a simple spectrum.

(iii)&(iv) Let $\omega_0, \dots, \omega_{\nu-1}$ be a maximal chain in $\mathsf{R}_\infty(S'_F)$. Since $\{0, \omega_0\} \in S'_F$ one obtains immediately $[\omega_0, \omega_j] = 0$ for all $j < \nu - 1$. Therefore, the subspace $\mathsf{R}_\infty(S'_F)$ is regular if and only if $[\omega_0, \omega_{\nu-1}] \neq 0$. This proves (iii) and (iv). □

The explicit construction of a maximal Jordan chain $\{\omega_0, \dots, \omega_{\nu-1}\}$ spanning $\mathsf{R}_\infty(S'_F)$ in the factorization model will be carried out by a continuation of the chains (4.17) and (4.18), which correspond to the cases $\operatorname{mul} S(Q_\infty) = \{0\}$ and $\operatorname{mul} S(Q_\infty) \neq \{0\}$, respectively. Observe, that by Lemma 4.5 $\operatorname{mul} S(Q_\infty)$ is nontrivial if and only if $\operatorname{mul} A_1^0$ is nontrivial. Moreover, since S_0 is simple at most one of the selfadjoint extensions A_0^0 or A_1^0 of S_0 can be multivalued.

Case I: $\operatorname{mul} S(Q_\infty) = \{0\}$. The proof of Lemma 4.7 shows that $\{\omega_0, \dots, \omega_{\nu-1}\}$ can be constructed as a continuation of the chain (4.17) if $\nu > k$ with $\omega_j = w_j$, $j = 0, \dots, k-1$. The formula (4.11) implies that the vector ω_k should be of the form $\omega_k = (g_0, f, \widetilde{f})^\top$, where

$$\widehat{g}_0 = \{0, g_0\} \in S_0^*, \quad \Gamma_1^0 \widehat{g}_0 = 1, \quad f = C_{q^\sharp} e_1 + \widetilde{\varphi} e_k, \quad \widetilde{f} = \Gamma_0^0 \widehat{g}_0 e_k.$$

By choosing $\widetilde{\varphi} e_k = -C_{q^\sharp} e_1$ one obtains

$$\omega_k = \begin{pmatrix} g_0 \\ 0 \\ \Gamma_0^0 \widehat{g}_0 e_k \end{pmatrix}, \quad \widehat{g}_0 = \{0, g_0\} \in S_0^*, \quad \Gamma_1^0 \widehat{g}_0 = 1. \tag{4.19}$$

Here $g_0 \neq 0$ if and only if $\omega_k \neq 0$. Therefore, one can continue the chain (4.17) if and only if $\operatorname{mul} S_0^*$ is nontrivial. This proves the following

Lemma 4.8. *If* $\operatorname{mul} S_0^* = \{0\}$, *then* $\mathsf{R}_\infty(S'_F)$ *is isotropic and*

$$\dim \mathsf{R}_\infty(S'_F) = \kappa_0(\mathsf{R}_\infty(S'_F)) = k.$$

Now assume that $\operatorname{mul} S_0^*$ is nontrivial, so that one can continue the chain (4.17) by a vector $\omega_k \neq 0$ of the form (4.19). Here two further different cases can occur: either $\operatorname{mul} A_0^0$ is nontrivial, in which case $\operatorname{mul} A_0^0 = \operatorname{mul} S_0^*$, or $\operatorname{mul} A_0^0 = \{0\}$. These two cases give rise to two different types of maximal Jordan chains in $\mathsf{R}_\infty(S_F')$.

Lemma 4.9. *Let* $\operatorname{mul} A_0^0$ *be nontrivial. Then* $\mathsf{R}_\infty(S_F')$ *is regular if and only if* $\nu = 2k+1$. *If* $\mathsf{R}_\infty(S_F')$ *is singular then* $k+1 \leq \nu \leq 2k$ *and*

$$\kappa_0(\mathsf{R}_\infty(S_F')) = 2k+1-\nu, \quad \kappa_-(\mathsf{R}_\infty(S_F')) = \nu - k - 1, \quad \kappa_+(\mathsf{R}_\infty(S_F')) = \nu - k. \tag{4.20}$$

Proof. Since $\operatorname{mul} A_0^0 \neq \{0\}$ one has $\Gamma_0^0 \widehat{g}_0 = 0$. It follows from (4.11) and (4.19) that the chain $\{\omega_k, \ldots, \omega_{\nu-1}\}$ can be taken to be of the form

$$\omega_k = \begin{pmatrix} g_0 \\ 0 \\ 0 \end{pmatrix}, \ \omega_{k+1} = \begin{pmatrix} g_1 \\ 0 \\ \widetilde{f}_{(1)} \end{pmatrix}, \ldots, \omega_{\nu-1} = \begin{pmatrix} g_{\nu-k-1} \\ 0 \\ \widetilde{f}_{(\nu-k-1)} \end{pmatrix}, \tag{4.21}$$

with $\{g_{j-1}, g_j\} \in A_1^0$ and

$$\widetilde{f}_{(j)} = a\widetilde{e}_{k-j+1} + \sum_{i=0}^{j-2} c_{j,i}\widetilde{e}_{k-i}, \quad c_{j,i} \in \mathbb{C}, \quad j = 1, \ldots, \nu-k-1,$$

where

$$a := \Gamma_0^0\{g_0, g_1\} = [\omega_{k-1}, \omega_{k+1}] = [\omega_k, \omega_k] = [g_0, g_0] > 0,$$

since the vector $g_0 \in \operatorname{mul} A_0^0$ is positive in view of Lemma 4.5. There are two reasons for the chain to break: either $g_{\nu-k-1} \notin \operatorname{dom} A_1^0$, or one has $(\widetilde{f}_{(\nu-k-1)})_1 \neq 0$ in which case $\nu = 2k+1$. Since $[\omega_{\nu-1}, \omega_0] = (\widetilde{f}_{\nu-k-1})_1$, the subspace $\mathsf{R}_\infty(A_0)$ is regular if and only if $\nu = 2k+1$.

Observe, that the maximal chain $\{\omega_0, \ldots, \omega_{\nu-1}\}$ in $\mathsf{R}_\infty(S_F)$ is also a maximal chain in $\mathsf{R}_\infty(S_F')$, because the vectors $\omega_j \in \mathsf{R}_\infty(S_F)$, $j = 0, \ldots, \nu-1$, are orthogonal to the eigenspaces $\ker(S(Q_\infty) - \alpha)$ for $\alpha \in \sigma_p(A_0^0) \cap \sigma(q)$.

If $\mathsf{R}_\infty(S_F')$ is singular then $k+1 \leq \nu \leq 2k$. The isotropic subspace of $\mathsf{R}_\infty(S_F')$ is spanned by the vectors ω_j, $j = 0, \ldots, 2k-\nu$, and

$$[\omega_{2k-\nu+1}, \omega_{\nu-1}] = [\omega_k, \omega_k] = [g_0, g_0] = a > 0.$$

Therefore, the Gram matrix of the chain $\omega_0, \ldots, \omega_{\nu-1}$ is a Hankel matrix of the form

$$G = \begin{pmatrix} 0 & 0 \\ 0 & G_0 \end{pmatrix} \text{ with } G_0 = \begin{pmatrix} 0 & & a \\ & \cdot^{\cdot^{\cdot}} & \\ a & & * \end{pmatrix}, \tag{4.22}$$

where the left upper corner of G is the matrix $0_{2k+1-\nu}$. The other two equalities in (4.20) are implied by the structure of G_0 in (4.22), since $a > 0$. □

Lemma 4.10. *Let* $\operatorname{mul} A_0^0 = \operatorname{mul} A_1^0 = \{0\}$ *and let* $\operatorname{mul} S_0^* \neq \{0\}$. *Then* $\mathsf{R}_\infty(S_F')$ *is regular if and only if* $\nu = 2k$. *If* $\mathsf{R}_\infty(S_F')$ *is singular then* $k+1 \leq \nu \leq 2k-1$ *and*

$$\kappa_0(\mathsf{R}_\infty(S_F')) = 2k-\nu, \quad \kappa_-(\mathsf{R}_\infty(S_F')) = \kappa_+(\mathsf{R}_\infty(S_F')) = \nu - k. \tag{4.23}$$

Proof. By assumptions $\Gamma_0^0 \widehat{g}_0 \neq 0$. In this case the chain $\{\omega_k, \dots, \omega_{\nu-1}\}$ takes the form

$$\omega_k = \begin{pmatrix} g_0 \\ 0 \\ a\widetilde{e}_k \end{pmatrix}, \ \omega_{k+1} = \begin{pmatrix} g_1 \\ 0 \\ \widetilde{f}_{(1)} \end{pmatrix}, \dots, \omega_{\nu-1} = \begin{pmatrix} g_{\nu-k-1} \\ 0 \\ \widetilde{f}_{(\nu-k-1)} \end{pmatrix}, \tag{4.24}$$

where $a := \Gamma_0^0 \widehat{g}_0 \neq 0$, $\{g_{j-1}, g_j\} \in A_1^0$, and $\widetilde{f}_{(j)}$ are given by

$$\widetilde{f}_{(j)} = a\widetilde{e}_{k-j} + \sum_{i=0}^{j-1} c_{j,i}\widetilde{e}_{k-i}, \quad c_{j,i} \in \mathbb{C}, \quad j = 1, \dots, \nu-k-1. \tag{4.25}$$

As in Lemma 4.9 it is seen from (4.24) that the subspace $\mathsf{R}_\infty(S_F')$ is regular if and only if $\nu = 2k$.

If $\mathsf{R}_\infty(S_F')$ is singular then $k+1 \leq \nu \leq 2k-1$. The isotropic subspace of $\mathsf{R}_\infty(S_F')$ is spanned by the vectors ω_j, $j = 0, \dots, 2k-1-\nu$, and

$$[\omega_{2k-\nu}, \omega_{\nu-1}] = [\omega_k, \omega_{k-1}] = a \neq 0.$$

Therefore, the Gram matrix of the chain $\omega_0, \dots, \omega_{\nu-1}$ takes the form (4.22), where the left upper corner of G is the matrix $0_{2k-\nu}$. The equalities in (4.23) are implied by the structure of G. □

Case II: $\operatorname{mul} S(Q_\infty) \neq \{0\}$. The proof of Lemma 4.7 shows that a maximal Jordan chain $\{\omega_0, \dots, \omega_{\nu-1}\}$ in $\mathsf{R}_\infty(S_F')$ can now be constructed as a continuation of the chain (4.18) with $\omega_j = w_j'$, $j = 0, \dots, k-1$.

Lemma 4.11. *Let* $\operatorname{mul} S(Q_\infty) \neq \{0\}$. *Then* $\mathsf{R}_\infty(S_F')$ *is regular if and only if* $\nu = 2k-1$. *If* $\mathsf{R}_\infty(S_F')$ *is singular then* $1 < k \leq \nu \leq 2k-2$ *and*

$$\kappa_0(\mathsf{R}_\infty(S_F')) = 2k-1-\nu, \quad \kappa_-(\mathsf{R}_\infty(S_F')) = \nu - k + 1, \quad \kappa_+(\mathsf{R}_\infty(S_F')) = \nu - k. \tag{4.26}$$

Proof. If $k = 1$ then $\omega_0 = (-(\overline{\Gamma_0^0 \widehat{g}})g, 1, -|\Gamma_0^0 \widehat{g}|^2)^\top$ and it follows from (4.11) that the chain (4.18) cannot be continued, in which case $\nu = k = 1$ and $[\omega_0, \omega_0] = -|\Gamma_0^0 \widehat{g}|^2 < 0$.

Now let $k > 1$ and consider the continuation of the chain (4.18). According to (4.11) the condition $\{\omega_{k-1}, \omega_k\} \in A_0$ for some ω_k means that for some vector g_1 one has

$$\widehat{h} := \{-(\overline{\Gamma_0^0 \widehat{g}})g, g_1\} \in S_0^*, \quad \Gamma_1^0 \widehat{h} = 1, \tag{4.27}$$

where $\widehat{g} = \{0, g\} \in A_1^0$, $[g, g] = 1$, $\Gamma_0^0 \widehat{g} \neq 0$. Observe, that here the conditions $\Gamma_1^0 \widehat{h} = 1$ and $[g, g] = 1$ are equivalent, since by Green's identity (4.1)

$$-\Gamma_0^0 \widehat{g}[g, g] = \Gamma_1^0 \widehat{g} \overline{\Gamma_0^0 \widehat{h}} - \Gamma_0^0 \widehat{g} \overline{\Gamma_1^0 \widehat{h}} = -\Gamma_0^0 \widehat{g} \overline{\Gamma_1^0 \widehat{h}}.$$

Observe also, that $g \notin \operatorname{dom} A_1^0$, since $g \in \operatorname{mul} A_1^0$ and the vector g is positive. Moreover, it follows from (4.11) that the continuation of the chain (4.18) can be taken to be of the form

$$\omega_k = \begin{pmatrix} g_1 \\ 0 \\ \widetilde{f}_{(1)} \end{pmatrix}, \dots, \omega_{\nu-1} = \begin{pmatrix} g_{\nu-k} \\ 0 \\ \widetilde{f}_{(\nu-k)} \end{pmatrix}, \tag{4.28}$$

where $\{g_{j-1}, g_j\} \in A_1^0$ for $j = 2, \dots, \nu - k$ and $\widetilde{f}_{(j)}$, $j = 1, \dots, \nu - k$, are given by (4.25), where

$$a := -|\Gamma_0^0 \widehat{g}|^2 = [\omega_{k-1}, \omega_{k-1}] < 0.$$

By (4.13) the vector ω_j is orthogonal to $\operatorname{mul} S(Q_\infty)$ for all $j = k, \dots, \nu - 2$. In addition, one can select $g_{\nu-k}$ in (4.28) such that $[g_{\nu-k}, g] = 0$ and then also $\omega_{\nu-1}$ is orthogonal to $\operatorname{mul} S(Q_\infty)$. Moreover, the vectors ω_j, $j = k, \dots, \nu - 1$, are orthogonal to the eigenspaces $\ker (S(Q_\infty) - \alpha)$, $\alpha \in \sigma_p(A_0^0) \cap \sigma(q)$. Hence (4.18) together with (4.28) forms a maximal Jordan chain in $\mathsf{R}_\infty(S'_F)$. Moreover, the subspace $\mathsf{R}_\infty(S'_F)$ is regular if and only if $\nu = 2k - 1$.

If $\mathsf{R}_\infty(S'_F)$ is singular then $k > 1$ and $k \le \nu \le 2k-2$, see (4.27). The isotropic subspace of $\mathsf{R}_\infty(S'_F)$ is spanned by the vectors ω_j, $j = 0, \dots, 2k - \nu - 2$, and

$$[\omega_{2k-\nu-1}, \omega_{\nu-1}] = [\omega_{k-1}, \omega_{k-1}] = a < 0.$$

Therefore, the Gram matrix of the chain $\omega_0, \dots, \omega_{\nu-1}$ is a Hankel matrix of the form (4.22), where the left upper corner of G is the matrix $0_{2k-1-\nu}$ and $a < 0$. This proves the equalities (4.26). □

The above considerations show that three different types of maximal Jordan chains in $\mathsf{R}_\infty(S'_F)$ can occur, cf. (4.21), (4.24), and (4.28). The longest Jordan chains appear in the first case, where $\operatorname{mul} A_0^0 \neq \{0\}$, see (4.21), while the shortest Jordan chains appear in the third case, where $\operatorname{mul} A_1^0 \neq \{0\}$, or equivalently, $\operatorname{mul} S(Q_\infty) \neq \{0\}$, see (4.28). The main properties associated with each of these maximal Jordan chains in Lemmas 4.8–4.11 are collected in the next theorem.

Theorem 4.12. *Let $Q_\infty \in \mathbf{N}_\kappa$, $\kappa_\infty(Q_\infty) = k > 0$, $\nu = \dim(\mathsf{R}_\infty(S'_F))$, and let S', S'_F be as in Theorem 4.6. Then one of the following three cases occurs:*

(i) *If $\operatorname{mul} A_0^0 \neq \{0\}$, then $k + 1 \le \nu \le 2k + 1$, $\kappa_-(\mathsf{R}_\infty(S'_F)) = \kappa_+(\mathsf{R}_\infty(S'_F)) - 1$, and $\kappa_0(\mathsf{R}_\infty(S'_F)) = 2k + 1 - \nu$. Moreover, $\mathsf{R}_\infty(S'_F)$ is singular if and only if $\nu \le 2k$.*

(ii) *If $\operatorname{mul} A_0^0 = \operatorname{mul} A_1^0 = \{0\}$, then $k \le \nu \le 2k$, $\kappa_-(\mathsf{R}_\infty(S'_F)) = \kappa_+(\mathsf{R}_\infty(S'_F))$, and $\kappa_0(\mathsf{R}_\infty(S'_F)) = 2k - \nu$. Moreover, $\mathsf{R}_\infty(S'_F)$ is singular if and only if $\nu \le 2k - 1$.*

(iii) *If $\operatorname{mul} A_1^0 \neq \{0\}$, then $k \le \nu \le 2k - 1$, $\kappa_-(\mathsf{R}_\infty(S'_F)) = \kappa_+(\mathsf{R}_\infty(S'_F)) + 1$, and $\kappa_0(\mathsf{R}_\infty(S'_F)) = 2k - 1 - \nu$. Moreover, $\mathsf{R}_\infty(S'_F)$ is singular if and only if $\nu \le 2k - 2$ and $k > 1$.*

Proof. It is enough to prove the statement (ii). For this one combines the results in Lemma 4.8 and Lemma 4.10. Indeed, if $\operatorname{mul} S_0^* = \{0\}$ then by Lemma 4.8 one has $\nu = \dim \mathsf{R}_\infty(S'_F) = k \le 2k - 1$ and $\kappa_0(\mathsf{R}_\infty(S'_F)) = k = 2k - \nu$. □

5. Analytic characterizations of the root subspace at ∞

Let $Q_\infty \in \mathbf{N}_\kappa$ have a minimal operator representation (1.2) with $\gamma = 0$, where A is a selfadjoint operator in $\mathfrak{H}$, and let S and S_F be defined by (1.3) and (1.4), respectively. In Section 4 one such minimal operator representation was constructed in Theorem 4.6 by using the factorization (4.2) of Q_∞ in Theorem 4.4. By unitary equivalence general statements concerning S, A, and S_F in a minimal representation of the function Q_∞ can be obtained by considering the corresponding objects S', A', and S'_F in the model of Theorem 4.6. In this section the results concerning the root subspace of the generalized Friedrichs extension $S_{F'}$ in Section 4 are connected with the asymptotic expansions in Section 3. In particular, the connection between the classification of generalized poles and zeros introduced in Subsection 3.2 and the three different types of maximal Jordan chains in Section 4 is explained.

5.1. The root subspace of the generalized Friedrichs extension at ∞ and operator representations

In order to establish the connection between the Jordan chains in the root subspace $\mathsf{R}_\infty(S_F)$ of the generalized Friedrichs extension S_F and the asymptotic expansions in Section 3 the following lemma will be useful.

Lemma 5.1. *Let A be a selfadjoint operator in a Pontryagin $\mathfrak{H}$, let $\omega \in \mathfrak{H}$, and let S and S_F be defined by (1.3) and (1.4). Then:*

(i) $\operatorname{dom} S^n = \{ f \in \operatorname{dom} A^n : [A^j f, \omega] = 0$ *for all* $j < n\}$, $n \in \mathbb{N}$;

(ii) $\omega \in \operatorname{dom} S^n$ *if and only if* $\omega \in \operatorname{dom} S_F^n$, $n \in \mathbb{N}$;

(iii) *if* $\omega_0, \omega_1, \dots, \omega_n$ *with* $\omega_0 = \omega$ *is a Jordan chain in* $\mathsf{R}_\infty(S_F)$ *such that*

$$[\omega_i, \omega_j] = 0 \quad \textit{for all} \quad i + j < l \le 2n, \quad l \in \mathbb{N}, \tag{5.1}$$

then

$$[S^i\omega, S^j\omega] = [\omega_i, \omega_j] \quad \textit{for all} \quad i + j \le l \le 2n. \tag{5.2}$$

Proof. (i) If $f \in \operatorname{dom} S^n$, then the definition of S in (1.3) shows that $f \in \operatorname{dom} A^n$ and $[A^j f, \omega] = 0$ for all $j < n$, since $A^j f \in \operatorname{dom} S$ for all $j < n$ and ω is orthogonal to $\operatorname{dom} S$. Conversely, assume that $f \in \operatorname{dom} A^n$ and $[A^j f, \omega] = 0$ for all $j < n$. Then the definition (1.3) shows that $A^j f \in \operatorname{dom} S$ for all $j < n$. Hence $f \in \operatorname{dom} S^n$.

(ii) Note that the condition $\omega \in \operatorname{dom} S_F^n$ means that there is a chain of vectors $\omega_0, \omega_1, \dots, \omega_n$, $\omega_0 = \omega$, such that $\{\omega_{j-1}, \omega_j\} \in S_F$ for $j \le n$. According to (1.4) this is equivalent to

$$\omega_j = S\omega_{j-1} + c_j\omega \quad \text{for some } c_j \in \mathbb{C}, \quad j = 1, \dots, n. \tag{5.3}$$

Hence, if $\omega \in \operatorname{dom} S_F^n$, then $\omega_j \in \operatorname{dom} S$ and $\omega_{j-1} \in \operatorname{dom} S^2$ for all $j < n$, and this leads to $\omega \in \operatorname{dom} S^n$. The reverse implication is also clear from (5.3).

(iii) It follows from (5.1) and (5.3) with $l \geq 1$ that

$$[\omega_0, \omega_1] = [\omega_0, S\omega_0].$$

Now, if $[S^i\omega_0, S^j\omega_0] = [\omega_i, \omega_j] = 0$ for all $i+j < m \leq l$, then one obtains from (5.3) that

$$\begin{aligned}[\omega_{i+1}, \omega_j] &= [S^{i+1}\omega_0 + \sum_{\alpha=0}^{i} c_{i+1-\alpha} S^\alpha \omega_0, S^j\omega_0 + \sum_{\beta=0}^{j-1} c_{j-\beta} S^\beta \omega_0] \\ &= [S^{i+1}\omega_0, S^j\omega_0],\end{aligned}$$

which completes the proof. □

The statements (iii) and (iv) in Lemma 4.7 can now be reformulated in terms of the moments $s_j = [A^j\omega, \omega]$ of the operator A, cf. also [15].

Proposition 5.2. *Let S, S_F, $\omega \in \mathfrak{H}$, and $[\omega, \omega] \leq 0$, be as in (1.3) and (1.4), in a minimal representation of $Q_\infty \in \mathbf{N}_\kappa$ with $\kappa_\infty(Q_\infty) = k > 0$. Then:*

(i) $\mathsf{R}_\infty(S_F)$ *is a regular subspace of dimension ν if and only if*

$$\omega \in \operatorname{dom} A^{\nu-1}, \ \textit{and } s_{\nu-1} \neq 0, \ s_j = 0 \textit{ for all } j < \nu - 1; \tag{5.4}$$

(ii) $\mathsf{R}_\infty(S_F)$ *is a singular subspace of dimension ν if and only if*

$$\omega \in \operatorname{dom} A^{\nu-1} \setminus \operatorname{dom} A^\nu, \ \textit{and } s_j = 0 \textit{ for all } j \leq \nu - 1. \tag{5.5}$$

Proof. (i) Assume that $\mathsf{R}_\infty(S_F)$ is a regular subspace of dimension ν. Then by Lemma 4.7 $\omega \in \operatorname{dom} S_F^{\nu-1} \setminus \operatorname{dom} S_F^\nu$ and by Lemma 5.1 $\omega \in \operatorname{dom} S^{\nu-1} \subset \operatorname{dom} A^{\nu-1}$. This implies that $s_j = [A^j\omega, \omega] = [S^j\omega, \omega] = 0$ for $j < \nu - 1$. Moreover, if $\{\omega, \omega_{\nu-1}\} \in S_F^{\nu-1}$ then

$$s_{\nu-1} = [A^{\nu-1}\omega, \omega] = [S^{\nu-1}\omega, \omega] = [\omega_{\nu-1}, \omega] \neq 0,$$

so that (5.4) follows.

Conversely, if (5.4) holds, then $\omega \in \operatorname{dom} S_F^{\nu-1}$ by Lemma 5.1. Moreover, for $\{\omega, \omega_{\nu-1}\} \in S_F^{\nu-1}$ one has

$$[\omega_{\nu-1}, \omega] = [A^{\nu-1}\omega, \omega] = s_{\nu-1} \neq 0,$$

so that $\omega \notin \operatorname{dom} S_F^\nu$. Hence, $\mathsf{R}_\infty(S_F)$ is a regular subspace of dimension ν.

(ii) Assume that $\mathsf{R}_\infty(S_F)$ is a singular subspace of dimension ν. Then it follows from Lemma 4.7 and Lemma 5.1 that $\omega \in \operatorname{dom} S^{\nu-1} \subset \operatorname{dom} A^{\nu-1}$ and $[A^j\omega, \omega] = [S^j\omega, \omega] = 0$ for all $j \leq \nu - 1$. If $\omega \in \operatorname{dom} A^\nu$, then by Lemma 5.1 $\omega \in \operatorname{dom} S_F^\nu$, a contradiction to (4.16). Thus, $\omega \notin \operatorname{dom} A^\nu$ and (5.5) follows.

Conversely, assume that (5.5) holds. Then one has $\omega \in \operatorname{dom} S_F^{\nu-1} \setminus \operatorname{dom} S_F^\nu$ by Lemma 5.1. Moreover, by (5.2) one has

$$[\omega_{\nu-1}, \omega] = [A^{\nu-1}\omega, \omega] = s_{\nu-1} = 0.$$

Hence, (4.16) holds and $\mathsf{R}_\infty(S_F)$ is a singular subspace of dimension ν. □

5.2. Asymptotic expansions and the classification for the generalized zero (pole) ∞ in the model space

The classification of generalized poles of Q_∞ or, equivalently, of the generalized zeros of $Q = -1/Q_\infty$ in Subsection 3.2 is now connected with the maximal Jordan chains constructed in Section 4. For this purpose observe that the assumptions of Theorem 4.12 can be expressed in terms of the Weyl function $Q_0(\lambda)$ of S_0 in the following equivalent form:

$$\operatorname{mul} A_0^0 = \{0\} \quad \Leftrightarrow \quad \lim_{z \widehat{\to} \infty} \frac{Q_0(z)}{z} = 0, \tag{5.6}$$

$$\operatorname{mul} A_1^0 = \{0\} \quad \Leftrightarrow \quad \lim_{z \widehat{\to} \infty} zQ_0(z) = \infty. \tag{5.7}$$

Now one can reformulate Theorem 4.12 in the form which makes clear the connection with the classification of generalized zeros and poles of nonpositive type introduced in Section 3.

Theorem 5.3. *Let $Q \in \mathbf{N}_\kappa$ have a minimal representation (1.2) and let S and S_F be defined by (1.3) and (1.4). Let ∞ be a generalized zero of negative type of Q with multiplicity $\pi_\infty(Q) = k > 0$ and let the root subspace $\mathsf{R}_\infty(S_F)$ be of dimension ν. Then $\mathsf{R}_\infty(S_F)$ is regular if and only if Q has an asymptotic expansion of the form*

$$Q(z) = -\frac{s_{\nu-1}}{z^\nu} - \cdots - \frac{s_{2\nu-2}}{z^{2\nu-1}} + o\left(\frac{1}{z^{2\nu-1}}\right), \quad z \widehat{\to} \infty, \quad s_{\nu-1} \neq 0. \tag{5.8}$$

Moreover, precisely one of the following three cases occurs:

(i) *If ∞ is a GZNT1, then $k+1 \le \nu \le 2k+1$ and Q has the asymptotic expansion*

$$Q(z) = -\frac{s_{2k}}{z^{2k+1}} - \cdots - \frac{s_{2\nu-2}}{z^{2\nu-1}} + o\left(\frac{1}{z^{2\nu-1}}\right), \quad z \widehat{\to} \infty, \tag{5.9}$$

where $s_{2k} > 0$. In this case $\kappa_-(\mathsf{R}_\infty(S_F)) = \kappa_+(\mathsf{R}_\infty(S_F)) - 1$, $\kappa_0(\mathsf{R}_\infty(S_F)) = 2k + 1 - \nu$, and $\mathsf{R}_\infty(S_F)$ is singular if and only if $\nu \le 2k$.

(ii) *If ∞ is a GZNT2, then $k \le \nu \le 2k$ and for $\nu \ge k+1$ Q has the asymptotic expansion*

$$Q(z) = -\frac{s_{2k-1}}{z^{2k}} - \cdots - \frac{s_{2\nu-2}}{z^{2\nu-1}} + o\left(\frac{1}{z^{2\nu-1}}\right), \quad z \widehat{\to} \infty, \tag{5.10}$$

where $s_{2k-1} \neq 0$, and for $\nu = k$ Q has the asymptotic expansion

$$Q(z) = o\left(\frac{1}{z^{2\nu-1}}\right), \quad z \widehat{\to} \infty. \tag{5.11}$$

In this case $\kappa_-(\mathsf{R}_\infty(S_F)) = \kappa_+(\mathsf{R}_\infty(S_F))$, $\kappa_0(\mathsf{R}_\infty(S_F)) = 2k-\nu$, and $\mathsf{R}_\infty(S_F)$ is singular if and only if $\nu \le 2k-1$.

(iii) *If ∞ is a GZNT3, then $k \le \nu \le 2k-1$ and Q has the asymptotic expansion*

$$Q(z) = -\frac{s_{2k-2}}{z^{2k-1}} - \cdots - \frac{s_{2\nu-2}}{z^{2\nu-1}} + o\left(\frac{1}{z^{2\nu-1}}\right), \quad z \widehat{\to} \infty, \tag{5.12}$$

where $s_{2k-2} < 0$. *In this case* $\kappa_-(\mathsf{R}_\infty(S_F)) = \kappa_+(\mathsf{R}_\infty(S_F))+1$, $\kappa_0(\mathsf{R}_\infty(S_F)) = 2k-1-\nu$, *and* $\mathsf{R}_\infty(S_F)$ *is singular if and only if* $\nu \le 2k-2$ *and* $k>1$.

Proof. Since the root subspace $\mathsf{R}_\infty(S_F)$ is of dimension ν it follows from Proposition 5.2 that $\omega \in \operatorname{dom} A^{\nu-1}$. By Theorem 3.3 this means that $Q \in \mathbf{N}_{\kappa,-2(\nu-1)}$. Now Corollary 3.4 and (5.4) show that in the regular case the asymptotic expansion is of the form (5.8).

If the root subspace $\mathsf{R}_\infty(S_F)$ is singular then Proposition 5.2 and Theorem 3.3 yield $\omega \in \operatorname{dom} A^{\nu-1} \setminus \operatorname{dom} A^{\nu}$ and $Q \in \mathbf{N}_{\kappa,-2(\nu-1)} \setminus \mathbf{N}_{\kappa,-2\nu}$. Now consider the classification given in Subsection 3.1.

(i) Assume that ∞ is a GZNT1 of Q. Then by (T1) the following limit exists:

$$\lim_{z \widehat{\to} \infty} z^{2k+1} Q(z) < 0.$$

It follows from the factorization (4.7) that

$$\lim_{z \widehat{\to} \infty} \frac{Q_0(z)}{z} = \lim_{z \widehat{\to} \infty} \frac{Q_\infty(z)}{z^{2k+1}} = -\lim_{z \widehat{\to} \infty} \frac{1}{z^{2k+1} Q(z)} > 0.$$

According to (5.6) this means that $\operatorname{mul} A_0^0 \neq \{0\}$. The asymptotic expansion (5.9) is implied by (3.15) and Corollary 3.4. The remaining statements are obtained from part (i) of Theorem 4.12.

(ii) Assume that ∞ is a GZNT2 of Q. Then by (T2) one has

$$\lim_{z \widehat{\to} \infty} z^{2k+1} Q(z) = \infty, \qquad \lim_{z \widehat{\to} \infty} z^{2k-1} Q(z) = 0.$$

It follows from the factorization (4.7) that

$$\lim_{z \widehat{\to} \infty} \frac{Q_0(z)}{z} = -\lim_{z \widehat{\to} \infty} \frac{1}{z^{2k+1} Q(z)} = 0, \quad \lim_{z \widehat{\to} \infty} z Q_0(z) = -\lim_{z \widehat{\to} \infty} \frac{1}{z^{2k-1} Q(z)} = \infty.$$

According to (5.6) and (5.7) this means that $\operatorname{mul} A_0^0 = \operatorname{mul} A_1^0 = \{0\}$. The asymptotic expansion (5.10) is implied by (3.17), (3.18), and Corollary 3.4, while the expansion (5.11) is obtained from (3.18), (3.19). The remaining statements are obtained from part (ii) of Theorem 4.12.

(iii) Finally, assume that ∞ is a GZNT3 of Q. Then by (T3) the following limit exists:

$$\lim_{z \widehat{\to} \infty} z^{2k-1} Q(z) > 0.$$

It follows from (4.7) that

$$\lim_{z \widehat{\to} \infty} z Q_0(z) = -\lim_{z \widehat{\to} \infty} \frac{1}{z^{2k-1} Q(z)} < 0,$$

and in view of (5.7) this means that $\operatorname{mul} A_1^0 \neq \{0\}$. The asymptotic expansion (5.12) is implied by (3.16) and Corollary 3.4, and the remaining statements are obtained from part (iii) of Theorem 4.12. □

The characterizations in Theorem 5.3 can be translated also for the generalized poles of nonpositive type of the function Q_∞ by means of the following theorem. In fact, this result is an extension of [6, Theorem 5.2] and can be seen to augment also the result stated in Theorem 3.3.

Theorem 5.4. *Let $Q \in \mathbf{N}_\kappa$, $Q \neq 0$, with $\lim_{z\widehat{\to}\infty} Q(z) = 0$ and let $Q_\infty = -1/Q$. Then the following statements are equivalent:*

(i) *$Q \in \mathbf{N}_{\kappa,-2n} \setminus \mathbf{N}_{\kappa,-2n-2}$ and $m \geq 0$ is the maximal integer such that $s_j = 0$ for all $j \leq m-1 (\leq 2n)$;*
(ii) *$Q(z) = [(A-z)^{-1}\omega, \omega]$, $z \in \rho(A)$, for some selfadjoint operator A in a Pontryagin space $\mathfrak{H}$ and a cyclic vector $\omega \in \operatorname{dom} A^n \setminus \operatorname{dom} A^{n+1}$ satisfying $[A^j\omega, A^i\omega] = 0$ for all $i+j \leq m-1 (\leq 2n)$, $i,j \leq n$;*
(iii) *Q has an asymptotic expansion of the form*

$$Q(z) = -\frac{s_m}{z^{m+1}} - \cdots - \frac{s_{2n}}{z^{2n+1}} + o\left(\frac{1}{z^{2n+1}}\right), \quad z\widehat{\to}\infty, \tag{5.13}$$

where $m (\leq 2n+1)$ and n are maximal nonnegative integers, such that (5.13) holds;
(iv) *$Q_\infty = -1/Q$ has an asymptotic expansion of the form*

$$Q_\infty(z) = p_{m+1} z^{m+1} + \cdots + p_{2\ell+1} z^{2\ell+1} + o\left(z^{2\ell+1}\right), \quad z\widehat{\to}\infty, \tag{5.14}$$

where $p_{m+1} \neq 0$ if $m \geq 2\ell$ and $\ell \in \mathbb{Z}$ (with $2\ell \leq m+1$) is minimal such that (5.14) holds.

In this case the integers $m, n \geq 0$ and ℓ are connected by $\ell = m-n$. Moreover, in (5.13) $s_m \neq 0$ if and only if $p_{m+1} \neq 0$ in (5.14), in which case $p_{m+1} = 1/s_m$ and $m \leq 2n$ or, equivalently, $2\ell \leq m$.

Proof. (i) $\Leftrightarrow$ (ii) This equivalence follows from Theorem 3.3 and the formulas (3.10) for the moments s_j, $j \leq 2n$.

(i) $\Leftrightarrow$ (iii) If (i) holds then by Corollary 3.4 Q has an asymptotic expansion of the form (5.13) and maximality of n in this expansion follows from the assumption $Q \notin \mathbf{N}_{\kappa,-2n+2}$. The converse statement is also clear.

(iii) $\Leftrightarrow$ (iv) Assume that Q satisfies (5.13). If $m \leq 2n$ then $s_m \neq 0$. Otherwise $m = 2n+1$ and the expansion (5.13) reduces to

$$Q(z) = o\left(\frac{1}{z^{2n+1}}\right), \quad z\widehat{\to}\infty. \tag{5.15}$$

In the case that $m \leq 2n$, $s_m \neq 0$ and the expansion (5.13) can be rewritten in the form

$$z^{m+1} Q(z) = -s_m - \cdots - \frac{s_{2n}}{z^{2n-m}} + o\left(\frac{1}{z^{2n-m}}\right), \quad z\widehat{\to}\infty. \tag{5.16}$$

Since $Q_\infty = -1/Q$, by inverting the expansion (5.16) one concludes that the expansion (5.13) for Q with $s_m \neq 0$ is equivalent for Q_∞ to admit an expansion of the form

$$Q_\infty(z) = p_{m+1}z^{m+1} + \cdots + p_{2(m-n)+1}z^{2(m-n)+1} + o\left(z^{2(m-n)+1}\right), \quad z\widehat{\to}\infty,$$

where $p_{m+1} = 1/s_m \neq 0$. This means that Q_∞ has an asymptotic expansion of the form (5.14), where the integer $\ell = m - n$, $2\ell < m + 1$, is minimal if n is maximal, and conversely. Hence the equivalence of (iii) and (iv) is shown in the case that $m \leq 2n$.

Next consider the case that $m = 2n + 1$. Then Q satisfies (5.15) and this is an expansion of the form (5.13) with the maximal integers $n \geq 0$ and $m = 2n + 1$. Since $Q_\infty = -1/Q$, one concludes that

$$Q_\infty(z) = o(z^{2n+3}), \quad z\widehat{\to}\infty,$$

so that Q_∞ has an expansion of the form (5.14) with $\ell := n+1 > 0$ and $m+1 = 2\ell$. Moreover, here $\ell = n+1$ is the minimal integer, such that Q_∞ admits an expansion of the form (5.14). Observe, that since $\ell = n + 1$ is minimal, the maximal m in (5.13) is equal to $m = 2n + 1$, so that the equality $\ell = m - n$ holds also in this case. In particular, (5.13) and (5.14) are still equivalent if $m = 2n + 1$, in which case one can take $m + 1 = 2\ell$. □

To establish the classification of the asymptotic expansions for the function Q_∞ along the lines of Theorem 5.3 it is enough to consider expansions of the form (5.14) with $\ell = m - n \geq 0$ (so that in (5.13) $n \leq m$) and then apply the last statement of Theorem 5.3. In this case (5.14) takes the form

$$Q_\infty(z) = P(z) + o\left(z^{2\ell+1}\right), \quad z\widehat{\to}\infty,$$

where

$$P(z) := p_{m+1}z^{m+1} + \cdots + p_{2\ell+1}z^{2\ell+1}, \quad \ell \geq 0,$$

is a real polynomial of degree $\deg P = m + 1$, whose leading coefficient is given by $p_{m+1} = 1/s_m$ if $\deg P > 2\ell$. If $m + 1 = 2\ell$, then one can take $P = 0$.

5.3. Asymptotic expansions and the index of singularity

As another consequence of Theorem 5.4 some characterizations for the index of singularity of the generalized pole (zero) of nonpositive type of Q_∞ (of $Q = -1/Q_\infty$) at ∞ are given. The motivation for this notion is given in the end of this section.

Definition 5.5. Let ∞ be a generalized pole of Q_∞ (zero of Q) of nonpositive type of order $\nu\,(= \dim \mathsf{R}_\infty(S_F))$ and let $H_\nu = (s_{i+j})_{i,j=0}^{\nu-1}$ be the $\nu \times \nu$ Hankel matrix which is determined by the finite moments s_j, $0 \leq j \leq 2\nu - 2$, of Q. Then $\kappa_\infty^0 = \dim(\ker H_\nu)$ is called the *index of singularity* of Q_∞ at ∞.

Theorem 5.6. *Let n and m be maximal nonnegative integers with $n \leq m$, such that $Q \in \mathbf{N}_{\kappa,-2n}$ and $s_j = 0$ for all $j \leq m - 1 (\leq 2n)$, and let S, A, S_F, ω be from the minimal operator representation* (1.2) *of Q.*

Then the following assertions are equivalent:

(i) ∞ *is a generalized zero of nonpositive type of* Q *with the index of singularity equal to* κ_∞^0*;*

(ii) $\kappa_\infty^0 = m - n(\geq 0)$*;*

(iii) $Q_\infty = -1/Q$ *admits an asymptotic expansion of the form*

$$Q_\infty(z) = P_{n+1}(z) + o(z^{2\kappa_\infty^0+1}), \quad z\widehat{\to}\infty, \tag{5.17}$$

where P_{n+1} *is a polynomial of degree* $n+1 \geq 2\kappa_\infty^0$ *and* $\kappa_\infty^0 \geq 0$ *is the minimal integer such that* (5.17) *holds;*

(iv) κ_∞^0 *is equal to the dimension of the isotropic subspace of* $\mathsf{R}_\infty(S_F)$.

Proof. (i) $\Leftrightarrow$ (ii) It follows from Theorem 5.3 and the maximality of n and m, $n \leq m$, that $n = \nu - 1$, where $\nu = \dim \mathsf{R}_\infty(S_F)$, cf. [15, Theorem 5.2]. Since clearly $\dim(\ker H_\nu) = m - n(\geq 0)$ the equivalence of (i) and (ii) is shown.

(ii) $\Leftrightarrow$ (iii) This follows immediately from Theorem 5.4 with $\ell = m - n = \kappa_\infty^0 \geq 0$. Here minimality of $\kappa_\infty^0 \geq 0$ in (5.17) is equivalent to the maximality of $n\,(= m - \kappa_\infty^0)$ in (5.13).

(ii) $\Leftrightarrow$ (iv) To prove this equivalence the indices m and n are calculated in each of the cases (i)–(iii) in Theorem 5.3. In the case (i) one obtains from (5.9) $m = 2k$, $n = \nu - 1$. Hence

$$\kappa_\infty^0 = m - n = 2k - \nu + 1 = \kappa_0(\mathsf{R}_\infty(S_F)).$$

Similarly in the case (ii) for both expansions (5.10) and (5.11) one has $m = 2k - 1$, $n = \nu - 1$. Thus

$$\kappa_\infty^0 = m - n = 2k - \nu = \kappa_0(\mathsf{R}_\infty(S_F)).$$

Finally in the case (iii) $m = 2k - 2$, $n = \nu - 1$, and

$$\kappa_\infty^0 = m - n = 2k - 1 - \nu = \kappa_0(\mathsf{R}_\infty(S_F)).$$

This completes the proof. □

The equivalence of (i) and (iv) in Theorem 5.6 is also a direct consequence of [15, Corollary 4.4]. From Theorem 5.6 one obtains immediately the following characterization for the regularity of a critical point.

Corollary 5.7. ([6, Theorem 4.1], [17, Proposition 1.6]) *The root subspace* $\mathsf{R}_\infty(S_F)$ *is nondegenerate* (*equivalently* ∞ *is a regular critical point of* S_F) *if and only if* Q_∞ *admits the representation*

$$Q_\infty(z) = P(z) + Q_0(z), \quad \text{where } Q_0(z) = o(z), \quad z\widehat{\to}\infty.$$

Proof. This is immediate from the equivalence of (iii) and (iv) in Theorem 5.6. □

The index κ_∞^0 measures the degree of singularity of the singular critical point ∞ of S_F. The characterization of κ_∞^0 via the asymptotic expansion (5.17) in Theorem 5.6 is particularly appealing: it extends the result stated in Corollary 5.7 to the case of *singular critical points* in an explicit manner.

6. Spectral characterizations via the underlying Weyl functions

In this section the structure of the underlying root subspace corresponding to the three different types of maximal Jordan chains constructed in Section 4 is studied by means of the factorized integral representations of the underlying Weyl functions. First detailed results are presented for the point ∞. Then the case of finite generalized zeros and poles of nonpositive type is treated briefly. Furthermore, it is shown how the classification of all generalized zeros and poles of $Q \in \mathbf{N}_\kappa$ belonging to $\mathbb{R} \cup \{\infty\}$ can be applied in establishing analytic criteria for the minimality of (not necessarily canonical) factorization models of $\mathbf{N}_\kappa$-functions.

6.1. The classification of generalized zeros and poles at ∞

The canonical factorization of generalized Nevanlinna functions in (1.6) implies that $Q(z) = -1/Q_\infty(z)$ has the following integral representation:

$$Q(z) = -1/Q_\infty(z) = \frac{\widetilde{p}(z)\widetilde{p}^\sharp(z)}{\widetilde{q}(z)\widetilde{q}^\sharp(z)} \left(a + bz + \int_\mathbb{R} \left(\frac{1}{t-z} - \frac{t}{t^2+1} \right) d\rho(t) \right), \tag{6.1}$$

where $\widetilde{p}$ and $\widetilde{q}$ are as in (2.5), $a \in \mathbb{R}$, $b \geq 0$, and $\rho(t)$ satisfies

$$\int_\mathbb{R} \frac{d\rho(t)}{t^2+1} < \infty. \tag{6.2}$$

In the case that $\int_\mathbb{R} d\rho(t) < \infty$, denote $a_0 = a - \int_\mathbb{R} t/(t^2+1)\, d\rho(t) \in \mathbb{R}$.

In the next theorem the classification of the Jordan chains is characterized via the spectral properties of the function Q using the factorized integral representation (6.1).

Theorem 6.1. *Let $Q_\infty \in \mathbf{N}_\kappa$, let $k = \kappa_\infty(Q_\infty) > 0$, and let $Q(z) = -1/Q_\infty(z)$ have the factorized integral representation* (6.1). *Then:*

(i) *the GZNT ∞ of Q is regular and of type* (T1) *if and only if*

$$b = a_0 = 0, \quad \int_\mathbb{R} d\rho(t) > 0, \text{ and } \int_\mathbb{R} (1+|t|)^{2k} d\rho(t) < \infty; \tag{6.3}$$

(ii) *the GZNT ∞ of Q is regular and of type* (T2) *if and only if*

$$b = 0,\ a_0 \neq 0, \text{ and } \int_\mathbb{R} (1+|t|)^{2(k-1)} d\rho(t) < \infty; \tag{6.4}$$

(iii) *the GZNT ∞ of Q is regular and of type* (T3) *if and only if*

$$b > 0 \text{ and } \int_\mathbb{R} (1+|t|)^{2(k-2)} d\rho(t) < \infty; \tag{6.5}$$

(iv) *the GZNT ∞ of Q is singular and of type* (T1) *with the index of singularity $\kappa_\infty^0 (> 0)$ if and only if $b = a_0 = 0$ and*

$$\int_\mathbb{R} (1+|t|)^{2(k-\kappa_\infty^0)} d\rho(t) < \infty, \quad \int_\mathbb{R} (1+|t|)^{2(k-\kappa_\infty^0+1)} d\rho(t) = \infty; \tag{6.6}$$

(v) *the GZNT* ∞ *of* Q *is singular of type* (T2) *with the index of singularity* $(0<)\kappa_\infty^0<k$ *if and only if* $b=0$, $a_0\neq 0$, *and*

$$\int_{\mathbb{R}}(1+|t|)^{2(k-1-\kappa_\infty^0)}d\rho(t)<\infty,\quad \int_{\mathbb{R}}(1+|t|)^{2(k-\kappa_\infty^0)}d\rho(t)=\infty, \tag{6.7}$$

and with the index of singularity $\kappa_\infty^0=k(>0)$ *if and only if* $b=0$ *and*

$$\int_{\mathbb{R}}d\rho(t)=\infty; \tag{6.8}$$

(vi) *the GZNT* ∞ *of* Q *is singular and of type* (T3) *with the index of singularity* $\kappa_\infty^0(>0)$ *if and only if* $b>0$ *and*

$$\int_{\mathbb{R}}(1+|t|)^{2(k-2-\kappa_\infty^0)}d\rho(t)<\infty,\quad \int_{\mathbb{R}}(1+|t|)^{2(k-1-\kappa_\infty^0)}d\rho(t)=\infty. \tag{6.9}$$

Proof. By Proposition 5.2 the root subspace $\mathsf{R}_\infty(S_F)$ is regular (singular) of dimension ν if and only if (5.4) holds (respectively (5.5) holds). According to Theorem 3.3 the condition $\omega\in\operatorname{dom}A^{\nu-1}$ is equivalent to the condition

$$\int_{\mathbb{R}}(1+|t|)^{2(\nu-k-1)}d\rho(t)<\infty. \tag{6.10}$$

This leads to the integrability conditions in (6.3)–(6.5) in the regular case and to the integrability conditions (6.7)–(6.9) in the singular case, see Theorem 4.12.

It follows from the expansions (5.9)–(5.12) of Q in Theorem 5.3 and the factorized integral representation of Q in (6.1) that

$$s_j=0 \text{ for } j<2k-2,\quad s_{2k-2}(Q)=-b, \tag{6.11}$$

and moreover, that

$$\text{if } \int_{\mathbb{R}}d\rho(t)<\infty \text{ and } b=0,\quad \text{then } s_{2k-1}=-a_0, \tag{6.12}$$

and

$$\text{if } \int_{\mathbb{R}}d\rho(t)<\infty \text{ and } b=a_0=0,\quad \text{then } s_{2k}=\int_{\mathbb{R}}d\rho(t). \tag{6.13}$$

All the statements of the theorem can now be obtained from Theorem 5.3 by combining Proposition 5.2 with (6.10)–(6.13). □

Observe that the regular cases (i)–(iii) are obtained from the singular cases (iv)–(vi) by taking $\kappa_\infty^0=0$ and excluding the second condition in (6.6), (6.7), and (6.9), respectively. All of the conditions in Theorem 6.1 are based on the canonical factorization of $Q=rr^\sharp Q_{00}$ in (6.1) involving the ordinary Nevanlinna function $Q_{00}\in\mathbf{N}_0$ in (2.5). In the factorization model of Theorem 4.4 the factor Q_0 belongs to the class $\mathbf{N}_{\kappa-k}$, where $k=\kappa_\infty(Q_\infty)$. The classification of Jordan chains was described in Theorem 4.12 by means of the selfadjoint extensions A_0^0 and A_1^0 of S_0 in the Pontryagin space $\mathfrak{H}_0$ whose negative index is equal to $\kappa(Q_0)=\kappa(Q_\infty)-\kappa_\infty(Q_\infty)$. Analogous descriptions remain true also for the canonical factorization of Q_∞.

Proposition 6.2. *Let $Q = rr^\sharp Q_{00}$ be the canonical factorization of $Q \in \mathbf{N}_\kappa$ in (6.1) with $k = \pi_\infty(Q) > 0$, let S_{00} be a simple symmetric operator in a Hilbert space $\mathfrak{H}_{00}$ with a boundary triplet $\Pi_{00} = \{\mathbb{C}, \Gamma_0^{00}, \Gamma_1^{00}\}$ whose Weyl function is equal to Q_{00}, and let $A_0^{00} = \ker \Gamma_0^{00}$ and $A_1^{00} = \ker \Gamma_1^{00}$ be the corresponding selfadjoint extensions of S_{00} in $\mathfrak{H}_{00}$. Then:*

(i) *the GZNT ∞ of Q is of type* (T1) *if and only if* $\operatorname{mul} A_1^{00} \neq \{0\}$;
(ii) *the GZNT ∞ of Q is of type* (T2) *if and only if* $\operatorname{mul} A_0^{00} = \operatorname{mul} A_1^{00} = \{0\}$;
(iii) *the GZNT ∞ of Q is of type* (T3) *if and only if* $\operatorname{mul} A_0^{00} \neq \{0\}$.

Proof. The identities (6.11) and (6.12) concern the function Q_{00}. The conditions which describe GZNT ∞ of type (T1) are $b = a_0 = 0$ and $\int_{\mathbb{R}} d\rho(t) < \infty$. These conditions are equivalent to $-\lim_{z \widehat{\to} \infty} zQ_{00}(z) < \infty$, which holds if and only if

$$-\lim_{z \widehat{\to} \infty} \frac{1}{zQ_{00}(z)} > 0 \quad \Leftrightarrow \quad \operatorname{mul} A_1^{00} \neq \{0\},$$

where the last equivalence follows from the simplicity of the operator S_{00} in $\mathfrak{H}_{00}$. Similarly, for type (T2) one has the conditions $b = 0$ and $a_0 \neq 0$ if $\int_{\mathbb{R}} d\rho(t) < \infty$, or the conditions $b = 0$ and $\int_{\mathbb{R}} d\rho(t) = \infty$. These conditions are equivalent to

$$\lim_{z \widehat{\to} \infty} \frac{Q_{00}(z)}{z} = 0, \quad \lim_{z \widehat{\to} \infty} zQ_{00} = \infty \Leftrightarrow \quad \operatorname{mul} A_0^{00} = \operatorname{mul} A_1^{00} = \{0\}.$$

Finally, for type (T3) one has the condition $b > 0$ which is equivalent to

$$\lim_{z \widehat{\to} \infty} \frac{Q_{00}(z)}{z} > 0 \quad \Leftrightarrow \quad \operatorname{mul} A_0^{00} \neq \{0\},$$

which completes the proof. □

The spectral theoretic characterization in Theorem 6.1 was based on the canonical factorization of the function $Q = -1/Q_\infty$. Since ∞ is a GPNT of the function Q_∞ it is natural to translate this result for the canonical factorization of Q_∞:

$$Q_\infty(z) = \frac{\widetilde{q}(z)\widetilde{q}^\sharp(z)}{\widetilde{p}(z)\widetilde{p}^\sharp(z)} \left(a_\infty + b_\infty z + \int_{\mathbb{R}} \left(\frac{1}{t-z} - \frac{t}{t^2+1} \right) d\sigma_\infty(t) \right), \tag{6.14}$$

where $\widetilde{p}$, $\widetilde{q}$ are as in (6.1), $a_\infty \in \mathbb{R}$, $b_\infty \geq 0$, and $\sigma_\infty(t)$ satisfies the analog of (6.2). In the case that $\int_{\mathbb{R}} d\sigma_\infty(t) < \infty$, denote $\gamma_\infty = a_\infty - \int_{\mathbb{R}} t/(t^2+1)\, d\sigma_\infty(t) \in \mathbb{R}$.

Theorem 6.3. *Let $Q_\infty \in \mathbf{N}_\kappa$ with $k = \kappa_\infty(Q_\infty) > 0$ have the factorized integral representation* (6.14). *Then:*

(i) *the GPNT ∞ of Q_∞ is regular and of type* (T1) *if and only if $b_\infty > 0$ and*

$$\int_{\mathbb{R}} (1+|t|)^{2(k-1)} d\sigma_\infty(t) < \infty; \tag{6.15}$$

(ii) *the GPNT ∞ of Q_∞ is regular and of type* (T2) *if and only if $b_\infty = 0$, $\gamma_\infty \neq 0$, and the integrability condition* (6.15) *is satisfied;*
(iii) *the GPNT ∞ of Q_∞ is regular and of type* (T3) *if and only if $b_\infty = \gamma_\infty = 0$, $\int_{\mathbb{R}} d\rho(t) > 0$, and the integrability condition* (6.15) *is satisfied;*

(iv) *the GPNT* ∞ *of* Q_∞ *is singular and of type* (T1) *with the index of singularity* $\kappa_\infty^0(>0)$ *if and only if* $b_\infty > 0$ *and*

$$\int_{\mathbb{R}} (1+|t|)^{2(k-1-\kappa_\infty^0)} d\sigma_\infty(t) < \infty, \quad \int_{\mathbb{R}} (1+|t|)^{2(k-\kappa_\infty^0)} d\sigma_\infty(t) = \infty; \tag{6.16}$$

(v) *the GPNT* ∞ *of* Q_∞ *is singular and of type* (T2) *with the index of singularity* $(0<)\kappa_\infty^0 < k$ *if and only if* $b_\infty = 0$, $\gamma_\infty \neq 0$, *and the integrability conditions* (6.16) *are satisfied, and with the index of singularity* $\kappa_\infty^0 = k(>0)$ *if and only if* $b_\infty = 0$ *and* $\int_{\mathbb{R}} d\sigma_\infty(t) = \infty$;

(vi) *the GPNT* ∞ *of* Q_∞ *is singular and of type* (T3) *with the index of singularity* $\kappa_\infty^0(>0)$ *if and only if* $b_\infty = \gamma_\infty = 0$, *and the integrability conditions* (6.16) *are satisfied.*

Proof. In each case the conditions concerning the parameters b_∞, a_∞, and γ_∞ are immediate from Proposition 6.2. It remains to establish the integrability conditions for $\sigma_\infty(t)$.

First observe that the integrability conditions for $\rho(t)$ in Theorem 6.1 concern the function Q_{00} in (2.5), while the integrability conditions for $\sigma_\infty(t)$ concern the function $-1/Q_{00}$. Now if the GPNT ∞ is of type (T1) then due to the conditions $b = a_0 = 0$ the measure $d\rho(t)$ has two more finite moments than the measure $d\sigma_\infty(t)$. Therefore, the integrability conditions in (6.3) and (6.6) are equivalent to those in (6.15) and (6.16), respectively; cf. [16, Theorem 4.2]. Similarly, due to $b_\infty = \gamma_\infty = 0$, the integrability conditions in (6.5) and (6.9) are equivalent to those in (6.15) and (6.16), respectively. Moreover, by [16, Theorem 4.2] the integrability conditions in (6.4) and (6.7) are equivalent to those in (6.15) and (6.16), respectively, while the conditions $b = 0$ and (6.8) are clearly equivalent to the conditions $b_\infty = 0$ and $\int_{\mathbb{R}} d\sigma_\infty(t) = \infty$. □

In the case that $\kappa(Q_\infty) = \kappa_\infty(Q_\infty) = 1$ the result in Theorem 6.3 simplifies to [8, Theorem 4.1]. Observe also that if the function $\sigma_\infty(t)$ has a compact support in $\mathbb{R}$, then (6.16) shows that ∞ cannot be a singular critical point of S_F. Moreover, if $\int_{\mathbb{R}} d\sigma_\infty(t) = 0$ and $\kappa_\infty(Q_\infty) > 0$, then only the cases (i) and (ii) in Theorem 6.3 can occur.

6.2. The classification of generalized zeros and poles in $\mathbb{R}$

The classification of generalized zeros and poles of nonpositive type has been studied in detail at the point $z = \infty$. Similar results hold true also for finite generalized zeros and poles of nonpositive type of $Q \in \mathbf{N}_\kappa$ which belong to $\mathbb{R}$. Here some main characterizations for the classification of finite generalized zeros $\beta \in \mathbb{R}$ and finite generalized poles $\alpha \in \mathbb{R}$ of Q are presented.

First the classification of zeros and poles is characterized by means of the canonical factorization of $Q \in \mathbf{N}_\kappa$.

Lemma 6.4. *Let* $Q \in \mathbf{N}_\kappa$, $\kappa > 0$, *be factorized as in* (6.1). *Then the types* (T1)–(T3) *of a generalized zero* $\beta \in \mathbb{R}$ *of* Q *are characterized as follows:*

(i) *the GZNT* $\beta \in \mathbb{R}$ *of* Q *is of type* (T1) *if and only if* $\beta \in \sigma_p(A_1^{00})$;
(ii) *the GZNT* $\beta \in \mathbb{R}$ *of* Q *is of type* (T2) *if and only if* $\beta \notin \sigma_p(A_0^{00}) \cup \sigma_p(A_1^{00})$;
(iii) *the GZNT* $\beta \in \mathbb{R}$ *of* Q *is of type* (T3) *if and only if* $\beta \in \sigma_p(A_0^{00})$.

Moreover, the types (T1)–(T3) *of a generalized pole* $\alpha \in \mathbb{R}$ *of* Q *are characterized as follows:*

(iv) *the GPNT* $\alpha \in \mathbb{R}$ *of* Q *is of type* (T1) *if and only if* $\alpha \in \sigma_p(A_0^{00})$;
(v) *the GPNT* $\alpha \in \mathbb{R}$ *of* Q *is of type* (T2) *if and only if* $\alpha \notin \sigma_p(A_0^{00}) \cup \sigma_p(A_1^{00})$;
(vi) *the GPNT* $\alpha \in \mathbb{R}$ *of* Q *is of type* (T3) *if and only if* $\alpha \in \sigma_p(A_1^{00})$.

Proof. (i)–(iii) In view of the canonical factorization of Q in (6.1) one obtains the following representation for the limits in (2.3):

$$\lim_{z \hat{\to} \beta} \frac{Q(z)}{(z-\beta)^{2\pi_\beta - 1}} = \lim_{z \hat{\to} \beta} (z-\beta) Q_{00}(z) = -(\rho(\beta+) - \rho(\beta-)) \, (\leq 0) \tag{6.17}$$

and if $\int_{\mathbb{R}} \frac{d\rho(t)}{(t-\beta)^2} < \infty$ and $\lim_{z \hat{\to} \beta} Q_{00}(z) = 0$, then

$$\lim_{z \hat{\to} \beta} \frac{Q(z)}{(z-\beta)^{2\pi_\beta + 1}} = \lim_{z \hat{\to} \beta} \frac{Q_{00}(z)}{(z-\beta)} = b + \int_{\mathbb{R}} \frac{d\rho(t)}{(t-\beta)^2} \, (> 0), \tag{6.18}$$

and otherwise the limit in (6.18) is not finite. Observe that the limit in (6.17) is negative if and only if $\beta \in \sigma_p(A_0^{00})$. The limit in (6.18) is finite if and only if for the function $-1/Q_{00}$ the limit in (6.17) is negative, which is equivalent to $\beta \in \sigma_p(A_1^{00})$. The statements (i)–(iii) are now obvious from the defining properties of the classifications given in Subsection 3.2.

(iv)–(vi) Apply the characterizations of generalized zeros in the first part of the lemma to the function $-1/Q$. □

Next some characteristic properties of the underlying root subspaces associated with the classification of generalized zeros $\beta \in \mathbb{R}$ (generalized poles $\alpha \in \mathbb{R}$) are established in a minimal representation of Q (of $-1/Q$, respectively).

Let $S(Q)$ be a simple symmetric operator in a Pontryagin space $\mathfrak{H}$, let $\{\Gamma_0, \Gamma_1, \mathcal{H}\}$ be a boundary triplet for S^* such that the corresponding Weyl function is the given $\mathbf{N}_\kappa$-function Q, and let $A(Q) = \ker \Gamma_0$ and $A(-1/Q) = \ker \Gamma_1$. Moreover, let $\mathsf{R}_\alpha(A(Q))$ and $\mathsf{R}_\beta(A(-1/Q))$ be the root subspaces of the selfadjoint extension $A(Q)$ and $A(-1/Q)$ of $S(Q)$ associated with the generalized pole α and the generalized zero $\beta \in \mathbb{R}$ of Q, respectively. The following result is analogous to Theorem 4.12. For simplicity, the classification of a GPNT $\alpha \in \mathbb{R}$ and a GZNT $\beta \in \mathbb{R}$ of Q is characterized here by using the signature of the corresponding root subspace. Here the following notations will be used:

$$\nu_\alpha := \dim \mathsf{R}_\alpha(A(Q)), \quad \nu_\beta := \dim \mathsf{R}_\beta(A(-1/Q)),$$
$$\kappa_\pm^\alpha := \kappa_\pm(\mathsf{R}_\alpha(A(Q))), \quad \kappa_\pm^\beta := \kappa_\pm(\mathsf{R}_\beta(A(-1/Q))),$$
$$\kappa_0^\alpha := \kappa_0(\mathsf{R}_\alpha(A(Q))), \quad \kappa_0^\beta := \kappa_\pm(\mathsf{R}_\beta(A(-1/Q))).$$

Proposition 6.5. *With the notations given above the following assertions hold for a GPNT* $\alpha \in \mathbb{R}$ *and a GZNT* $\beta \in \mathbb{R}$ *of* $Q \in \mathbf{N}_\kappa$*:*

(i) α *is of type* (T1) *if and only if* $\kappa_-^\alpha = \kappa_+^\alpha - 1$, *in this case* $\kappa_\alpha + 1 \le \nu_\alpha \le 2\kappa_\alpha + 1$ *and* $\kappa_0^\alpha = 2\kappa_\alpha + 1 - \nu_\alpha$;
(ii) α *is of type* (T2) *if and only if* $\kappa_-^\alpha = \kappa_+^\alpha$, *in this case* $\kappa_\alpha \le \nu_\alpha \le 2\kappa_\alpha$ *and* $\kappa_0^\alpha = 2\kappa_\alpha - \nu_\alpha$;
(iii) α *is of type* (T3) *if and only if* $\kappa_-^\alpha = \kappa_+^\alpha + 1$, *in this case* $\kappa_\alpha \le \nu_\alpha \le 2\kappa_\alpha - 1$ *and* $\kappa_0^\alpha = 2\kappa_\alpha - 1 - \nu_\alpha$;
(iv) β *is of type* (T1) *if and only if* $\kappa_-^\beta = \kappa_+^\beta - 1$, *in this case* $\pi_\beta + 1 \le \nu_\beta \le 2\pi_\beta + 1$ *and* $\nu_0^\beta = 2\pi_\beta + 1 - \nu_\beta$;
(v) β *is of type* (T2) *if and only if* $\kappa_-^\beta = \kappa_+^\beta$, *in this case* $\pi_\beta \le \nu_\beta \le 2\pi_\beta$ *and* $\nu_0^\beta = 2\pi_\beta - \nu_\beta$;
(vi) β *is of type* (T3) *if and only if* $\kappa_-^\beta = \kappa_+^\beta + 1$, *in this case* $\pi_\beta \le \nu_\beta \le 2\pi_\beta - 1$ *and* $\nu_0^\beta = 2\pi_\beta - 1 - \nu_\beta$.

Proof. The statements (i)–(iii) are obtained from Theorem 4.12 by considering the transform $Q_\infty(z) := -Q(\alpha + 1/z)$. Namely, the root subspace $\mathsf{R}_\alpha(A(Q))$ associated with the generalized pole α of Q coincides with the root subspace $\mathsf{R}_\infty(S_F')$ associated with the generalized pole ∞ of the function Q_∞.

The statements (iv)–(vi) follow by applying the results in (i)–(iii) to the function $-1/Q$. □

The classification for a GZNT $\beta \in \mathbb{R}$ and a GPNT α of Q can be characterized also via asymptotic expansions of the function Q in a neighborhood of these points. The defining properties of the types (T1)–(T3) are reflected in these expansions in a similar manner as in Theorems 5.3 and 5.4 above. Such expansions have been studied in [15] and, for instance, the analog of Theorem 5.3 for a GZNT $\beta \in \mathbb{R}$ of Q can be easily derived from [15, Theorem 5.2] and for a GPNT α of Q form [15, Theorem 5.4]. Moreover, these expansions can be characterized also via the canonical factorization of Q along the lines of Theorems 6.1 and 6.3 by using Lemma 6.4; see also [15, Section 6]. A detailed formulation of these results is left for the reader.

6.3. Analytic criteria for the minimality of (localized) factorization models

The classification of generalized zeros and poles of nonpositive type of $Q \in \mathbf{N}_\kappa$ can be used to give analytic criteria for general (localized) factorizations models based on (proper) factorizations of Q of the form

$$Q(z) = \widehat{r}(z)\widehat{r}^\#(z)\widehat{Q}_0, \quad \widehat{r} = \frac{\widehat{p}}{\widehat{q}}, \tag{6.19}$$

where $\widehat{p}$ and $\widehat{q}$ are divisors of the polynomials $\widetilde{p}$ and $\widetilde{q}$ in the canonical factorization of Q in (6.1), with multiplicity of a zero equal to its original multiplicity. Such factorization models can be used, for instance, in studying *local spectral properties* of the function $Q \in \mathbf{N}_\kappa$, along the lines carried out in the previous sections of the present paper at ∞ with the aid of the model in Theorem 4.6 for proper factorizations of Q at ∞.

The basic observation here is that according to Lemma 4.11 the occurrence of a maximal Jordan chain of type (T3) means that $\operatorname{mul} S(Q_\infty) \neq \{0\}$. Therefore, the existence of such a Jordan chain is connected with the non-minimality of the factorization model of Q. The next result is formulated for the canonical factorization model as constructed in [3, Theorem 3.3]; by unitary equivalence the result holds for all other canonical factorization models, too; cf. [12].

Proposition 6.6. *The canonical factorization model constructed for $Q \in \mathbf{N}_\kappa$ in* [3, Theorem 3.3] *is simple if and only if all generalized zeros $\beta \in \mathbb{R} \cup \{\infty\}$ and generalized poles $\alpha \in \mathbb{R} \cup \{\infty\}$ of Q of nonpositive type are either of type* (T1) *or of type* (T2).

Proof. By Lemma 6.4 a real GZNT β (a real GPNT α) of Q is not of type (T3) if and only if $\beta \notin \sigma_p(A_0^{00})$ (respectively $\alpha \notin \sigma_p(A_1^{00})$). Moreover, by Proposition 6.2 the GZNT $\beta = \infty$ (the GPNT $\alpha = \infty$) is not of type (T3) if and only if $\operatorname{mul} A_0^{00} = \{0\}$ (respectively, $\operatorname{mul} A_1^{00} = \{0\}$). Now, according to [3, Theorem 4.1], these conditions characterize the simplicity of the corresponding factorization model for Q. □

These observations lead to the following analytic characterization for the simplicity of the factorization model in Theorem 4.4 ([9, Theorem 4.2]) which is based on a proper factorization of Q_∞ at ∞ (see Definition 4.2).

Proposition 6.7. *Let $Q_\infty \in \mathbf{N}_\kappa$ with $k = \kappa_\infty(Q_\infty) > 0$ and let $Q_\infty = qq^\sharp Q_0$ be a proper factorization of Q_∞ with some monic polynomial q, $\deg q = k$. Moreover, let the symmetric operator S_0 be simple in the Pontryagin space $\mathfrak{H}_0$ (see Lemma 4.5). Then the factorization model for Q_∞ in Theorem 4.4 is minimal if and only if the following two conditions are satisfied:*

(1) *$\alpha = \infty$ is not a GPNT of Q_∞ of type* (T3);
(2) *all the real zeros of q as GZNT of Q_∞ are either of the type* (T1) *or* (T2).

Proof. By Theorem 4.6 the symmetric relation $S(Q_\infty)$ is simple if and only if it has an empty point spectrum, or equivalently, the subspace $\mathfrak{H}''$ in (4.15) is trivial. In view of Lemma 4.5 the last condition is equivalent to

$$\operatorname{mul} A_1^0 = \{0\} \quad \text{and} \quad \sigma_p(A_0^0) \cap \sigma(q) = \emptyset. \tag{6.20}$$

By Theorem 4.12 the first condition in (6.20) is equivalent to the property formulated in part (1). The condition $\alpha \in \sigma_p(A_0^0) \cap \sigma(q)$ means that $q(\alpha) = 0$ and

$$\lim_{z \widehat{\to} \alpha} (z - \alpha) Q_0(z) < 0, \tag{6.21}$$

since by Lemma 4.5 $\ker (A_0^0 - \alpha)$ is spanned by a positive vector, cf. [9, Lemma 2.3]. Hence, if the multiplicity of α as a root of q is κ_α, then

$$\lim_{z \widehat{\to} \alpha} \frac{Q_\infty(z)}{(z - \alpha)^{2\kappa_\alpha - 1}} < 0, \tag{6.22}$$

and thus $\pi_\alpha(Q_\infty) = \kappa_\alpha$ and α is of type (T3). Conversely, (6.22) implies (6.21) with $\kappa_\alpha = \pi_\alpha(Q_\infty)$. This completes the proof. □

Since the polynomial q can be selected to be an arbitrary divisor of degree $k = \kappa_\infty(Q_\infty)$ of the polynomial $\widetilde{q}$ in the canonical factorization of Q_∞, the condition (2) in Proposition 6.7 can be satisfied if, for instance, the total number (counting multiplicities) of all generalized zeros in $\mathbb{C}_+$ and all generalized zeros of nonpositive type in $\mathbb{R}$ of Q_∞ which are of type (T1) or (T2) is at least equal to $\kappa_\infty(Q_\infty)$. Therefore, the factorization model in Theorem 4.4 can be minimal, while the canonical factorization model of Q_∞ need not be minimal.

Remark 6.8. Similar facts can be derived for other (localized) factorization models which are build on some (proper) factorization of $Q \in \mathbf{N}_\kappa$, cf. (6.19). The construction of such (localized) factorization models can be based on the (orthogonal) coupling of two minimal models (a method studied in another context in [4]), one of which is a finite-dimensional Pontryagin space model for the rational 2×2-matrix function

$$\widehat{R}(z) = \begin{pmatrix} 0 & \widehat{r}(z) \\ \widehat{r}^\#(z) & 0 \end{pmatrix}, \tag{6.23}$$

whose reproducing kernel space model has been identified in matrix terms with Bezoutians and companion operators in [3, Proposition 3.3]. The other is a minimal Pontryagin space model for the factor $\widehat{Q}_0$. The model for the product $Q = \widehat{r}\widehat{r}^\#\widehat{Q}_0$ in (6.19) is obtained from the orthogonal sum of the models for $\widehat{R}$ and $\widehat{Q}_0$ as a straightforward extension of the model constructed in [3, Theorem 3.3], simply by allowing the factor $\widehat{Q}_0$ to be a generalized Nevanlinna function, too. This procedure can also be described in pure function theoretic terms: perform suitable "block transforms" to the orthogonal sum $\widehat{R} \oplus \widehat{Q}_0$ to get the product $Q = \widehat{r}\widehat{r}^\#\widehat{Q}_0$, cf. [3, Section 3].

Such factorization models are not minimal in general. Non-minimality of such models reflects the fact that some poles and zeros of $\widehat{r}\widehat{r}^\#$ and $\widehat{Q}_0$ may cancel each others when these functions are multiplied. The simplest example here is the function $Q(z) = -z$, which belongs to $\mathbf{N}_1$ and whose canonical factorization is given by

$$Q(z) = -z = z^2 \left(-\frac{1}{z}\right),$$

where $Q_0(z) = -1/z \in \mathbf{N}_0$, cf. [9, Section 3]. Hence a model which is built on the canonical factorization of $Q(z)$ does not produce a minimal model for Q directly. The general characterization for minimality of canonical factorization models was established in [3, Theorem 4.1]. This result is equivalent to the analytic criterion given in Proposition 6.6: Q does not have any generalized zeros or poles in $\mathbb{R} \cup \{\infty\}$ which are of type (T3); cf. also [12, Theorem 4.4].

Finally, it is noted that the construction of a minimal model in the case of the canonical factorization of $Q \in \mathbf{N}_\kappa$ has been recently studied in [12] by using reproducing kernel Pontryagin spaces. In particular, in that paper a detailed analysis concerning the reproducing kernel space model for the matrix function $\widehat{R}$ in (6.23) has been carried out.

References

[1] T.Ya. Azizov and I.S. Iokhvidov, *Foundations of the theory of linear operators in spaces with an indefinite metric*, Nauka, Moscow, 1986 (English translation: Wiley, New York, 1989).

[2] V.A. Derkach, "On generalized resolvents of Hermitian relations", J. Math. Sciences, 97 (1999), 4420–4460.

[3] V.A. Derkach and S. Hassi, "A reproducing kernel space model for $\mathbf{N}_\kappa$-functions", Proc. Amer. Math. Soc., 131 (2003), 3795–3806.

[4] V.A. Derkach, S. Hassi, M.M. Malamud, and H.S.V. de Snoo, "Generalized resolvents of symmetric operators and admissibility", Methods of Functional Analysis and Topology, 6 (2000), 24–55.

[5] V. Derkach, S. Hassi, and H.S.V. de Snoo, "Operator models associated with Kac subclasses of generalized Nevanlinna functions", Methods of Functional Analysis and Topology, 5 (1999), 65–87.

[6] V.A. Derkach, S. Hassi, and H.S.V. de Snoo, "Generalized Nevanlinna functions with polynomial asymptotic behaviour and regular perturbations", Oper. Theory Adv. Appl., 122 (2001), 169–189.

[7] V.A. Derkach, S. Hassi, and H.S.V. de Snoo, "Operator models associated with singular perturbations", Methods of Functional Analysis and Topology, 7 (2001), 1–21.

[8] V.A. Derkach, S. Hassi, and H.S.V. de Snoo, "Rank one perturbations in a Pontryagin space with one negative square", J. Funct. Anal., 188 (2002), 317–349.

[9] V.A. Derkach, S. Hassi, and H.S.V. de Snoo, "A factorization model for the generalized Friedrichs extension in a Pontryagin space", Oper. Theory Adv. Appl., 162 (2005), 117–133.

[10] V. Derkach and M. Malamud, "The extension theory of hermitian operators and the moment problem", J. Math. Sciences, 73 (1995), 141–242.

[11] A. Dijksma, H. Langer, A. Luger, and Yu. Shondin, "A factorization result for generalized Nevanlinna functions of the class $\mathbf{N}_\kappa$", Integral Equations and Operator Theory, 36 (2000), 121–125.

[12] A. Dijksma, H. Langer, A. Luger, and Yu. Shondin, "Minimal realizations of scalar generalized Nevanlinna functions related to their basic factorization", Oper. Theory Adv. Appl., 154 (2004), 69–90.

[13] A. Dijksma, H. Langer, and Yu. Shondin, "Rank one perturbations at infinite coupling in Pontryagin spaces", J. Funct. Anal., 209 (2004), 206–246.

[14] A. Dijksma, H. Langer, Yu.G. Shondin, and C. Zeinstra, "Self-adjoint operators with inner singularities and Pontryagin spaces", Oper. Theory Adv. Appl., 117 (2000), 105–176.

[15] S. Hassi and A. Luger, "Generalized zeros and poles of $\mathcal{N}_\kappa$-functions: on the underlying spectral structure", Methods of Functional Analysis and Topology, 12 no. 2 (2006), 131–150.

[16] S. Hassi, H.S.V. de Snoo, and A.D.I. Willemsma, "Smooth rank one perturbations of selfadjoint operators", Proc. Amer. Math. Soc., 126 (1998), 2663–2675.

[17] P. Jonas, "Operator representations for definitizable functions", Annales Academiae Scientiarum Fennicae, Mathematica, 25 (2000) 41–72.

[18] P. Jonas, H. Langer, and B. Textorius, "Models and unitary equivalence of cyclic selfadjoint operators in Pontryagin spaces", Oper. Theory Adv. Appl., 59 (1992), 252–284.

[19] I.S. Kac and M.G. Kreĭn, "R-functions – analytic functions mapping the upper half-plane into itself", Supplement to the Russian edition of F.V. Atkinson, *Discrete and continuous boundary problems*, Mir, Moscow 1968 (Russian) (English translation: Amer. Math. Soc. Transl. Ser. 2, 103 (1974), 1–18).

[20] M.G. Kreĭn and H. Langer, "Über die Q-function eines π-hermiteschen Operators im Raume Π_κ", Acta. Sci. Math. (Szeged), 34 (1973), 191–230.

[21] M.G. Kreĭn and H. Langer, "Über einige Fortsetzungsprobleme, die eng mit der Theorie hermitescher Operatoren im Raume Π_κ zusammenhängen. 1. Einige Funktionenklassen und ihre Darstellungen", Math. Nachr., 77 (1977), 187–236.

[22] M.G. Kreĭn and H. Langer, "On some extension problems which are closely connected with the theory of hermitian operators in a space Π_κ III. Indefinite analogues of the Hamburger and Stieltjes moment problems, Part (I)", Beiträge Anal., 14 (1979), 25–40.

[23] M.G. Kreĭn and H. Langer, "Some propositions on analytic matrix functions related to the theory of operators in the space Π_κ", Acta Sci. Math. (Szeged), 43 (1981), 181–205.

[24] H. Langer, "A characterization of generalized zeros of negative type of functions of the class $\mathbf{N}_\kappa$", Oper. Theory Adv. Appl., 17 (1986), 201–212.

Vladimir Derkach
Department of Mathematical Analysis
Donetsk National University
Universitetskaya str. 24
83055 Donetsk, Ukraine
e-mail: `derkach.v@gmail.com`

Seppo Hassi
Department of Mathematics and Statistics
University of Vaasa
P.O. Box 700
65101 Vaasa, Finland
e-mail: `sha@uwasa.fi`

Henk de Snoo
Department of Mathematics and Computing Science
University of Groningen
P.O. Box 800
9700 AV Groningen, Nederland
e-mail: `desnoo@math.rug.nl`

Operator Theory:
Advances and Applications, Vol. 175, 89–94

A Necessary Aspect of the Generalized Beals Condition for the Riesz Basis Property of Indefinite Sturm-Liouville Problems

Andreas Fleige

Abstract. For the Sturm-Liouville eigenvalue problem $-f'' = \lambda rf$ on $[-1,1]$ with Dirichlet boundary conditions and with an indefinite weight function r changing it's sign at 0 we discuss the question whether the eigenfunctions form a Riesz basis of the Hilbert space $L^2_{|r|}[-1,1]$. In the nineties the sufficient so called generalized one hand Beals condition was found for this Riesz basis property. Now using a new criterion of Parfyonov we show that already the old approach gives rise to a necessary and sufficient condition for the Riesz basis property under certain additional assumptions.

Mathematics Subject Classification (2000). Primary 34B4; Secondary 34L10.

Keywords. Indefinite Sturm-Liouville problem, Riesz basis, definitizable operator.

1. Introduction

We consider the regular indefinite Sturm-Liouville eigenvalue problem

$$-f'' = \lambda rf \quad \text{on} \quad [-1,1], \qquad f(-1) = f(1) = 0 \tag{1.1}$$

where the real weight function $r \in L^1[-1,1]$ has a single turning point, i.e., sign change, at 0. More precisely we assume

$$r(x) < 0 \quad \text{a.e. on} \quad [-1,0), \qquad\qquad r(x) > 0 \quad \text{a.e. on} \quad (0,1] \tag{1.2}$$

It is well known that this eigenvalue problem has only real and simple eigenvalues ([5], [6], [7]). During the last two decades the question was intensively studied whether the eigenfunctions form a Riesz basis of the Hilbert space $L^2_{|r|}[-1,1]$ with the inner product

$$(f,g) := \int_{-1}^{1} f\bar{g}|r|dx \qquad (f,g \in L^2_{|r|}[-1,1]).$$

(e.g., [2], [5], [12], [6], [7], [13], [8], [1], [10]). This question is motivated by the fact that in case of a positive weight function r the eigenfunctions even form an orthonormal basis. However in the indefinite situation it took some time to find sufficient conditions for this *Riesz basis property of the eigenvalue problem* ([2], [5], [12], [6], [7], [13], [10]). In particular starting from a paper of Beals [2] a technique was developed which led to quite general sufficient conditions in the late eighties and nineties ([5], [6], [7], [13]). For this reason in [7] the author called his result the *generalized one hand Beals condition.* In our days in [4] this technique was still improved and applied to similar problems with λ-dependent boundary conditions.

On the other hand there was a competition to find *counterexamples*, i.e., weight functions r such that the Riesz Basis property fails to hold. After in [13] Volkmer had proved the existence of such counterexamples the first concrete counterexample was presented in [8] and then in [1] counterexamples with continuously differentiable weight functions were given. Moreover, in [3] it was shown that counterexamples also depend on the type of the boundary conditions.

Recently Parfyonov found a connection of the "positive" and the "negative" results in [10] where he gave a sufficient and *necessary* condition for the Riesz Basis property in the special situation of an odd weight function.

We shall see here that already the previous approach using the generalized one hand Beals condition was not too far away from a necessary condition. More precisely we shall show that in the case of an odd weight function and under certain additional assumptions the generalized one hand Beals condition is equivalent to the condition of Parfyonov and hence to the Riesz Basis property of the eigenvalue problem. Moreover, this also leads to a sufficient condition of Parfyonov's type for non odd weight functions. However, the question remains open whether in this general case we also get a *necessary* and sufficient condition for the Riesz Basis property in this way.

In [5] it was detected that the Riesz Basis property is closely connected to the critical points of a certain definitizable operator. Therefore also here we shall make use of Langer's theory of definitizable operators in Krein spaces outlined in [9].

2. Two known conditions for the Riesz basis property

For the real function $r \in L^1[-1,1]$ with (1.2) the space $L^2_r[-1,1]$ of all (equivalence classes of) measurable functions f on $[-1,1]$ with $\int_{-1}^{1} |f|^2 |r| dx < \infty$ is a Krein space with the indefinite inner product

$$[f,g] := \int_{-1}^{1} f\bar{g} r dx \qquad (f,g \in L^2_r[-1,1]).$$

The operator J defined by $Jf := \operatorname{sgn}(r) f$ ($f \in L^2_r[-1,1]$) is a fundamental symmetry and induces the Hilbert space inner product

$$(f,g) = [Jf,g] = \int_{-1}^{1} f\bar{g}|r| dx \qquad (f,g \in L^2_r[-1,1]).$$

It is well known ([5], [6], [7]) that the operator A defined by

$$\begin{aligned} \mathrm{dom}(A) &:= \{f \in L_r^2[-1,1] \mid f, f' \text{ abs. cont.}, \ \frac{1}{r}f'' \in L_r^2[-1,1], \\ &\qquad f(-1) = f(1) = 0\}, \\ Af &:= -\frac{1}{r}f'' \qquad (f \in \mathrm{dom}(A)) \end{aligned}$$

is selfadjoint, nonnegative, boundedly invertible and hence definitizable in the Krein space $(L_r^2[-1,1], [\cdot,\cdot])$. Moreover the spectrum of A consists of a sequence of real and simple eigenvalues accumulating only at $+\infty$ and $-\infty$ and therefore ∞ is the only critical point of A.

According to [7], Sections 3.2 and 3.3, we say that r *satisfies the generalized right-hand Beals condition* if there exist numbers $\mu > 0$, $0 < \xi < \min(1, \frac{1}{\mu})$ and a continuously differentiable real function h on $[0, \xi]$ such that

$$h(t) = \frac{r(t)}{\mu \cdot r(\mu t)} \quad \text{a.e. on } (0, \xi], \qquad h(0) \neq 1. \tag{2.1}$$

Now we recall [7], Theorem 3.7:

Theorem 2.1. *If r satisfies the generalized right-hand Beals condition then ∞ is not a singular critical point of A.*

By [5], Proposition 4.1, the non-singularity of the critical point ∞ of A implies that the Hilbert space $(L_r^2[-1,1], (\cdot,\cdot))$ has a Riesz basis consisting of eigenfunctions of the eigenvalue problem (1.1), i.e., the Riesz basis property. Note that [7] includes a similar result for an analogues generalized *left*-hand Beals condition. Moreover we want to mention that there is a similar result in [13], Corollary 2.7.

Now for $\mu > 0$ we put

$$R_\mu(t) := \int_0^{\mu t} r(x)dx = \int_0^t \mu r(\mu x)dx \quad (t \in [0, \frac{1}{\mu}]).$$

and in view of [10] we say that r *satisfies the Parfyonov condition* if there exists a number $\mu \in (0,1)$ such that for all $\epsilon \in (0,1)$

$$R_\mu(\epsilon) \leq \frac{1}{2} R_1(\epsilon). \tag{2.2}$$

Here we recall a part of [10], Theorem 6:

Theorem 2.2 (Parfyonov). *Assume that the function r is odd. Then r satisfies the Parfyonov condition if and only if the eigenvalue problem* (1.1) *has the Riesz basis property.*

3. A connection of the two conditions

Throughout this section we additionally assume that the weight function r is continuously differentiable on $(0, \xi]$ for a number $0 < \xi \leq 1$. Under some additional assumptions we shall show that the Parfyonov condition implies the generalized right-hand Beals condition. To this end for $\mu \in (0,1)$ we consider the function

$$h_\mu(t) := \frac{r(t)}{\mu r(\mu t)} \quad (t \in (0,\xi])$$

with the derivative

$$h'_\mu(t) = \frac{r(\mu t)r'(t) - \mu r'(\mu t)r(t)}{\mu r(\mu t)^2} \quad (t \in (0,\xi]).$$

Then if r satisfies the Parfyonov condition with $\mu \in (0,1)$ it follows from (2.2) by l'Hôpital's rule that

$$2 \leq \lim_{t \searrow 0} \frac{R_1(t)}{R_\mu(t)} = \lim_{t \searrow 0} \frac{R'_1(t)}{R'_\mu(t)} = \lim_{t \searrow 0} \frac{r(t)}{\mu r(\mu t)} = \lim_{t \searrow 0} h_\mu(t) \tag{3.1}$$

if the last limit exists. Therefore if r is continuously differentiable on $[0,\xi]$ with $r(0) \neq 0$ then so is the function h_μ with $h_\mu(0) \geq 2$ and hence (2.1) is satisfied with $h = h_\mu$. However, since $h_\mu(0) = \frac{1}{\mu} \neq 1$ this is a trivial case of

Theorem 3.1. *Assume that r is continuously differentiable on $(0,\xi]$ for a number $0 < \xi \leq 1$ and r is odd, i.e., $r(-t) = -r(t)$ $(t \in (0,1])$. Moreover assume that for all $\mu \in (0,1)$ the limits*

$$h_{\mu,0} := \lim_{t \searrow 0} \frac{r(t)}{\mu r(\mu t)}, \quad h'_{\mu,0} := \lim_{t \searrow 0} \frac{r(\mu t)r'(t) - \mu r'(\mu t)r(t)}{\mu r(\mu t)^2} \tag{3.2}$$

exist. Then the following statements are equivalent:

(i) *The function r satisfies the generalized right-hand Beals condition.*
(ii) *∞ is not a singular critical point of A.*
(iii) *The eigenvalue problem (1.1) has the Riesz basis property.*
(iv) *The function r satisfies the Parfyonov condition.*
(v) *For at least one $\mu \in (0,1)$ we have $h_{\mu,0} \neq 1$.*

Proof. The proofs of the implications (i) $\Rightarrow$ (ii) $\Rightarrow$ (iii) $\Rightarrow$ (iv) were already mentioned in section 2. The implication (iv) $\Rightarrow$ (v) follows from (3.1). For the implication (v) $\Rightarrow$ (i) we must show that the continuous function

$$h(t) := \begin{cases} h_\mu(t) & (t \in (0,\xi]) \\ h_{\mu,0} & (t = 0) \end{cases}$$

is continuously differentiable. By the mean value theorem for each $t \in (0,\xi]$ there is a number $\xi_t \in (0,t)$ such that

$$\frac{h(t) - h(0)}{t} = h'_\mu(\xi_t) \longrightarrow h'_{\mu,0} \quad (t \longrightarrow 0).$$

Therefore h is differentiable at 0 and the derivative is continuous in 0. $\square$

Now from Theorem 3.1 we easily obtain the following sufficient condition of Parfyonov's type without the restriction to odd weight functions. Note that this was already shown in [11].

Corollary 3.2. *Assume that r is continuously differentiable on $(0,\xi]$ for a number $0<\xi\le 1$ and for all $\mu\in(0,1)$ the limits* (3.2) *exist. Then the eigenvalue problem* (1.1) *has the Riesz basis property if the function r satisfies the Parfyonov condition.*

Proof. Consider the "odd extension" $\tilde{r}$ of r restricted to $(0,1]$

$$\tilde{r}(t) := \begin{cases} r(t) & (t\in(0,1]) \\ -r(-t) & (t\in[-1,0)). \end{cases}$$

Then $\tilde{r}$ satisfies the Parfyonov condition and by Theorem 3.1 also the generalized right-hand Beals condition. Since this is then also true for r the Riesz basis property follows from Theorem 2.1. □

It is well known that the eigenvalue problem (1.1) has the Riesz basis property if $r\in C^2[0,\xi]$ with $r(0)=0$ and $r'(0)\neq 0$ (e.g., [7], Corollary 3.8). This also serves as an example for the existence of the limits (3.2) since in this case we have by l'Hospital's rule

$$h_{\mu,0} = \lim_{t\searrow 0} h_\mu(t) = \lim_{t\searrow 0} \frac{r'(t)}{\mu^2 r'(\mu t)} = \frac{1}{\mu^2}\ (\neq 1),$$

$$h'_{\mu,0} = \lim_{t\searrow 0} h'_\mu(t) = \lim_{t\searrow 0}\left(\frac{r''(t)}{2\mu^2 r'(\mu t)} - \frac{r''(\mu t)}{2r'(\mu t)}\mu h_\mu(t)\right) = \frac{r''(0)}{2\mu^2 r'(0)} - \frac{r''(0)}{2\mu r'(0)}.$$

References

[1] N.L. Abasheeva, S.G. Pyatkov, *Counterexamples in indefinite Sturm-Liouville problems.* Siberian Adv. Math. **7**, No. 4 (1997), 1–8.

[2] R. Beals, *Indefinite Sturm-Liouville problems and half range completeness.* J. Differential Equations **56** (1985), 391–407.

[3] P. Binding, B. Curgus, *A counterexample in Sturm-Liouville completeness theory.* Proc. Roy. Soc. Edinburgh Sect. A **134** (2004), 244–248.

[4] P. Binding, B. Curgus, *Riesz Bases of Root Vectors of Indefinite Sturm-Liouville Problems with Eigenparameter Dependent Boundary Conditions, I.* To appear.

[5] B. Curgus, H. Langer, *A Krein space approach to symmetric ordinary differential operators with an indefinite weight function.* J. Differential Equations **79** (1989), 31–61.

[6] A. Fleige, *The "turning point condition" of Beals for indefinite Sturm-Liouville problems.* Math. Nachr. **172** (1995), 109–112.

[7] A. Fleige, *Spectral Theory of Indefinite Krein-Feller Differential Operators.* Mathematical Research **98**. Akademie Verlag, Berlin, 1996.

[8] A. Fleige, *A counterexample to completeness properties for indefinite Sturm-Liouville problems.* Math. Nachr. **190** (1998), 123–128.

[9] H. Langer, *Spectral functions of definitizable operators in Krein spaces* In: D. Butkovic, H. Kraljevic and S. Kurepa (eds.): Functional analysis. Conf. held at Dubrovnik, Yugoslavia, November 2–14, 1981. Lecture Notes in Mathematics, Vol. **948**, Springer-Verlag, Berlin, Heidelberg, New York, 1982, 1–46.

[10] A.I. Parfyonov, *On an Embedding Criterion for Interpolation Spaces and Application to Indefinite Spectral Problems.* Siberian Mathematical Journal, Vol. **44**, No. 4 (2003), 638–644.

[11] A.I. Parfyonov, *On Curgus' condition in indefinite Sturm-Liouville problems.* Matem. Trudy., Vol. **7**, No. 1 (2004), 153–188. (In Russian, translated in Sib. Adv. Math.)

[12] S.G. Pyatkov, *Elliptic eigenvalue problems with an indefinite weight function.* Siberian Adv. Math. **4**, No. 2 (1994), 87–121.

[13] H. Volkmer, *Sturm-Liouville problems with indefinite weights and Everitt's inequality.* Proc. Roy. Soc. Edinburgh Sect. A **126**, No. 5, (1996), 1097–1112.

Andreas Fleige
Baroper Schulstraße 27a
D-44225 Dortmund, Germany
e-mail: andreas.fleige@lycos.de

Operator Theory:
Advances and Applications, Vol. 175, 95–109

On Reducible Nonmonic Matrix Polynomials with General and Nonnegative Coefficients

K.-H. Förster and B. Nagy

Abstract. We consider nonmonic quadratic polynomials acting on a general or on a finite-dimensional linear space as a continuation of our work in [7,8]. Conditions are given for the existence of right roots, if the coefficient operators have lower block triangular representations. In the finite-dimensional case we consider (in a certain sense, entrywise) nonnegative coefficient matrices in the general (reducible) case, and extend several earlier results from the case of irreducible coefficients. In particular, we generalize results of Gail, Hantler and Taylor [9]. We show that our general methods are sufficiently strong to prove a remarkable result by Butler, Johnson and Wolkowicz [3], proved there by ingenious ad hoc methods.

Mathematics Subject Classification (2000). Primary 15A22; Secondary 15A48, 47A56, 15A18.

Keywords. Reducible matrix polynomials, block triangular operator coefficients, (entrywise) nonnegative matrix coefficients, nonnegative matrix roots.

1. Introduction

We consider quadratic (in general nonmonic) polynomials $Q(\cdot)$ of the form

$$\lambda^2 M + \lambda L + K.$$

In Section 2 we study such polynomials when the coefficients M, L and K are lower block triangular linear maps (with the same block size). Such polynomials can have (right) roots, but none of these roots need to be lower block triangular; see Example 2.2. In Theorem 2.6 we give a necessary and sufficient condition for the existence of a lower block triangular (of the same block size as the coefficients) right root of this matrix polynomial.

This work was completed with partial support of the Hungarian National Science Grant OTKA No T-047276 and partial support of the DAAD, the Technical University of Berlin and the Budapest University of Technology and Economics.

The main topic of this paper is the matrix polynomial

$$Q(\lambda) = \lambda I_{n\times n} - (\lambda^2 A + \lambda B + C)$$

with entrywise nonnegative $n \times n$-matrices A, B and C. This class of matrix polynomials play an important role in the theory of quasi-birth-and-death Markov chains (see [2, 5.6]) and has been studied very intensively; for a rather extensive review on this topic see [2] and [14].

In [7] the spectral properties of such matrix polynomials were studied. In the case when $S = A + B + C$ is irreducible a complete set of necessary and sufficient conditions for the existence of a nonnegative right root W of $Q(\cdot)$ were given and the spectral properties of the matrices W and V such that

$$Q(\lambda) = (V - \lambda A)(\lambda I_{n\times n} - W) \text{ and } W \geq 0_{n\times n}$$

were studied; see [7, Section 4].

Results in this direction for the case when S is reducible can be found in [2, 4.7], [9, Theorem 5] and [3, Theorem 2.3]. In Section 3 we study systematically the existence of nonnegative right roots of $Q(\cdot)$ when the coefficients are nonnegative and reducible. In contrast to the example in Section 2 mentioned above here $Q(\cdot)$ has nonnegative lower triangular right roots if and only if it has nonnegative right roots (Theorem 3.1). With the help of the Frobenius normal form of S and the function $0 \leq \rho \mapsto$ spectral radius of $S(\rho) = \rho^2 A + \rho B + C$ we give several sufficient conditions for the existence of nonnegative roots of $Q(\cdot)$, and we generalize the results mentioned above. In particular, we give a completely different proof of [3, Theorem 2.3].

Our terminology is mostly traditional. For standard facts in the theory of linear bounded operators in Banach spaces we refer to [4], concerning (matrix or operator) polynomials to [10] or [15], concerning nonnegative matrices to [11], [1] or [16].

2. Roots of polynomials with lower triangular block operator coefficients

Let E be a (finite- or infinite-dimensional complex) vector space. We consider the polynomial

$$Q(\lambda) = \lambda^2 M + \lambda L + K, \ \lambda \in \mathbb{C}, \tag{2.1}$$

where M, L and K are linear maps in E. We assume that $E = E_1 \times E_2$ and that the coefficients have lower block triangular representations; for example

$$M = \begin{pmatrix} M_{11} & 0 \\ M_{21} & M_{22} \end{pmatrix}, \tag{2.2}$$

where M_{rs} is a linear map form E_s into E_r for $1 \leq s \leq r \leq 2$ and 0 is the zero map from E_2 into E_1.

Then $Q(\lambda)$ has the lower block triangular representation

$$Q(\lambda) = \begin{pmatrix} Q_{11}(\lambda) & 0 \\ Q_{21}(\lambda) & Q_{22}(\lambda) \end{pmatrix}. \tag{2.3}$$

Note that if the dimension of the vector space E is finite, and the $Q's$ are matrices, then the form (2.3) shows that $Q(\cdot)$ is reducible in the sense of [11].Note also that the irreducible case (for nonnegative coefficients) was treated in [7, Section 4].

Assume that $Q(\cdot)$ has a right root W, i.e., W is a linear map in E such that $Q(W) = MW^2 + LW + K = 0$, and let V be a linear map in E such that

$$Q(\lambda) = (\lambda M - V)(\lambda I_E - W), \tag{2.4}$$

where I_E denotes the identity operator in E.

If V and W have the block representation

$$V = \begin{pmatrix} V_{11} & V_{12} \\ V_{21} & V_{22} \end{pmatrix} \quad \text{and} \quad W = \begin{pmatrix} W_{11} & W_{12} \\ W_{21} & W_{22} \end{pmatrix}, \tag{2.5}$$

direct computations show that (2.4) implies the following system of matrix equations

$$\begin{aligned} M_{11}W_{12} + V_{12} &= 0 \\ V_{11}W_{12} + V_{12}W_{22} &= 0, \\ V_{21}W_{11} + V_{22}W_{21} &= K_{21}, \\ M_{21}W_{11} + V_{21} + M_{22}W_{21} &= -L_{21}. \end{aligned} \tag{2.6}$$

From the first two equations in (2.6) it follows immediately: If W is lower block triangular, then V is lower block triangular (in this case the second equation is automatically fulfilled), the diagonal polynomials $Q_{rr}(\cdot)$ have the right roots W_{rr}, and $Q_{rr}(\lambda) = (\lambda M_{rr} - V_{rr})(\lambda I_{E_r} - W_{rr})$ for $r = 1, 2$.
Let the linear map X of E_1 into E_2 be a solution of the equation

$$V_{22}X - M_{22}XW_{11} = K_{21} + (L_{21} + M_{21}W_{11})W_{11}, \tag{2.7}$$

and define

$$W_{21} := X \qquad \text{and} \qquad V_{21} := -L_{21} - M_{21}W_{11} - M_{22}X. \tag{2.8}$$

Then the last two equations of (2.6) are fulfilled.

On the other hand, from the last two equations of (2.6) it follows: if we define $X := W_{21}$, then (2.7) and (2.8) are satisfied.

Therefore we have proved that 1. implies 2. in the following proposition.

Proposition 2.1. *With the assumptions, definitions and notations above the following assertions are equivalent:*

1. *The polynomial $Q(\cdot)$ has a lower block triangular right root (of the same block size as the coefficients).*
2. *For $r = 1, 2$ the polynomial $Q_{rr}(\cdot)$ has a right root W_{rr}, and equation (2.7) has a solution X, which is a linear map of E_1 into E_2, where V_{22} satisfies $Q_{22}(\lambda) = (\lambda M_{22} - V_{22})(\lambda I_{E_2} - W_{22})$.*

Proof. 2. implies 1.: Let $Q_{rr}(\lambda) = (\lambda M_{rr} - V_{rr})(\lambda I_{E_r} - W_{rr})$ for $r = 1, 2$, let X be a solution of (2.7), and W_{21} and V_{21} be defined as in (2.8). Then these maps are the lower left blocks, and W_{rr} and V_{rr} are the diagonal blocks of a lower block triangular right root W of $Q(\cdot)$ and of a lower block diagonal V, respectively, such that $Q(\lambda) = (\lambda M - V)(\lambda I_E - W)$. □

The next example exhibits a reducible matrix polynomial which has right roots, but none of its right roots is lower triangular.

Example 2.2. Consider the matrix polynomial

$$Q(\lambda) = \lambda^2 M + K = \lambda^2 \begin{pmatrix} 0 & 0 \\ 1 & 0 \end{pmatrix} + \begin{pmatrix} 0 & 0 \\ -2 & -2 \end{pmatrix}$$
$$= \left(\lambda \begin{pmatrix} 0 & 0 \\ 1 & 0 \end{pmatrix} - \begin{pmatrix} 0 & 0 \\ -1 & -1 \end{pmatrix}\right)\left(\lambda I_{2\times 2} - \begin{pmatrix} 1 & 1 \\ 1 & 1 \end{pmatrix}\right).$$

Let $Q(\lambda) = (\lambda M - V)(\lambda I - W)$ with 2×2-matrices V and $W = (w_{rs})$. Then

$$V = -MW = -\begin{pmatrix} 0 & 0 \\ w_{11} & w_{12} \end{pmatrix} \quad \text{and}$$

$$\begin{pmatrix} 0 & 0 \\ -2 & -2 \end{pmatrix} = K = VW = -\begin{pmatrix} 0 & 0 \\ w_{11}^2 + w_{12}w_{21} & w_{12}(w_{11} + w_{22}) \end{pmatrix},$$

and this implies $w_{12} \neq 0$.

Remark. Assume that the coefficients M, L, K of $Q(\cdot)$ are lower triangular, that the pair (V, W) factorizes $Q(\cdot)$ (as in (2.4)), and let (V_0, W_0) denote the pair obtained by substituting the blocks V_{12}, W_{12} in (2.5) by zero blocks. We have seen that if the pair (V_0, W_0) also factorizes $Q(\cdot)$, then the diagonal blocks (V_{rr}, W_{rr}) factorize the corresponding diagonal block $Q_{rr}(\cdot)$. We show that if the diagonal blocks (V_{rr}, W_{rr}) factorize the corresponding diagonal block $Q_{rr}(\cdot)$ and the block M_{21} is an injective operator, then $(V_0, W_0) = (V, W)$, hence it factorizes $Q(\cdot)$.

Indeed, if the diagonal blocks factorize as well as the pair (V, W), then we have

$$V_{jj}W_{jj} = K_{jj}, \quad M_{jj}W_{jj} + V_{jj} = -L_{jj} \quad (j = 1, 2),$$
$$V_{j1}W_{1j} + V_{j2}W_{2j} = K_{jj}, \quad M_{j1}W_{1j} + M_{j2}W_{2j} + V_{jj} = -L_{jj} \quad (j = 1, 2).$$

Hence we obtain $M_{21}W_{12} = 0$, and the injectivity of M_{21} implies $W_{12} = 0$. It means that $W = W_0$, and we have seen that then $V_{12} = 0$ follows. Hence $(V_0, W_0) = (V, W)$ factorizes the polynomial $Q(\cdot)$.

Corollary 2.3. *Let the coefficients of $Q(\cdot)$ be block diagonal (i.e., $M_{12} = L_{12} = K_{12} = 0$ and $M_{21} = L_{21} = K_{21} = 0$). Then $Q(\cdot)$ has a (block diagonal) right root if and only if all diagonal entries of $Q(\cdot)$ have right roots.*

Assume that V_{22} is invertible. Then equation (2.7) is equivalent to

$$X - V_{22}^{-1}M_{22}XW_{11} = V_{22}^{-1}(K_{21} + (L_{21} + M_{21}W_{11})W_{11}). \qquad (2.9)$$

Let $L(E_1, E_2)$ denote the vector space of all linear maps from E_1 into E_2. We define the linear map

$$\diamond({V_{22}}^{-1}M_{22}, W_{11}) : L(E_1, E_2) \to L(E_1, E_2), \textit{ with } X \mapsto {V_{22}}^{-1}M_{22}XW_{11}. \quad (2.10)$$

Corollary 2.4. *Assume that* $Q_{rr}(\lambda) = (\lambda M_{rr} - V_{rr})(\lambda I_{E_r} - W_{rr})$ *for* $r = 1, 2$. *If* V_{22} *and* $I_{L(E_1,E_2)} - \diamond({V_{22}}^{-1}M_{22}, W_{11})$ *are invertible, then* $Q(\cdot)$ *has a lower triangular right root.*

Now let E_1 and E_2 be Banach spaces, E their Banach product space and the coefficients of $Q(\cdot)$ and their entries in the representation (2.2) be linear and bounded operators between the corresponding Banach spaces. By $\sigma(T)$ we denote the spectrum of a bounded linear operator T in a Banach space. Then we have:

Corollary 2.5. *Assume that* $Q_{rr}(\lambda) = (\lambda M_{rr} - V_{rr})(\lambda I_{E_r} - W_{rr})$ *for* $r = 1, 2$. *If*

$$1 \notin \sigma(V_{22}^{-1}M_{22}) \cdot \sigma(W_{11}) = \{\lambda \cdot \mu : \lambda \in \sigma(V_{22}^{-1}M_{22}) \textit{ and } \mu \in \sigma(W_{11})\}, \quad (2.11)$$

then $Q(\cdot)$ *has a lower triangular right root.*

The proof follows from the last corollary and the equality

$$\sigma(\diamond(T, S)) = \sigma(T) \cdot \sigma(S),$$

see [20, Theorem].

The next theorem follows from Proposition 2.1 by constructing inductively the blocks in the subdiagonals of W and V.

Theorem 2.6. *Let the polynomial* $Q(\cdot)$ *be as in (2.1),* $E = E_1 \times E_2 \times \cdots \times E_m$, *and assume that the coefficients* M, L *and* K *have lower block triangular representations, for example*

$$M = \begin{pmatrix} M_{11} & 0 & 0 & \cdots & 0 \\ M_{21} & M_{22} & 0 & \vdots & 0 \\ \vdots & \vdots & \ddots & \ddots & \vdots \\ M_{m-1,1} & \cdots & \cdots & M_{m-1,m-1} & 0 \\ M_{m1} & \cdots & \cdots & \cdots & M_{mm} \end{pmatrix},$$

where M_{pq} *are linear maps form* E_q *into* E_p *for* $1 \le q \le p \le m$. *Then the following assertions are equivalent:*

1. *The polynomial* $Q(\cdot)$ *has a right root which is lower block triangular (of the same block size as the coefficients).*
2. *For* $r = 1, 2, \ldots, m$ *the polynomial* $Q_{rr}(\cdot)$ *has a right root* W_{rr} *and for* $1 \le q < p \le m$ *the equation*

$$V_{pp}X - M_{pp}XW_{qq} = K_{pq} + (L_{pq} + \sum_{q \le r < p} M_{pr}W_{rq})W_{qq} - \sum_{q < r < p} V_{pr}W_{rq} \quad (2.12)$$

has a solution X *(of the same size as* M_{pq}*) which is a linear map from* E_q *into* E_p. *Here* V_{pp} *is a linear map in* E_p *such that* $Q_{pp}(\lambda) = (\lambda M_{pp} - V_{pp})(\lambda I_{E_p} - W_{pp})$.

For the proof we note the following: Let $Q_{rr}(\lambda) = (\lambda M_{rr} - V_{rr})(\lambda I_{E_r} - W_{rr})$ for $r = 1, 2, \dots, m$, and let X_{pq} be a solution of (2.12) for $1 \le q < p \le m$. Define for $1 \le q < p \le m$

$$W_{pq} := X_{pq} \qquad \text{and} \qquad V_{pq} := -L_{pq} - \sum_{q \le r \le p} M_{pr} W_{rq}. \tag{2.13}$$

Then the first of these maps are the lower left blocks, and W_{rr} are the diagonal blocks of a lower block triangular right root W of $Q(\cdot)$. The second of these maps are the lower left blocks, and V_{rr} are the diagonal blocks of a lower block triangular V such that $Q(\lambda) = (\lambda M - V)(\lambda I_E - W)$.

We can generalize Corollaries 2.3–2.4 in an obvious way to the more general case of the last theorem.

3. Nonnegative roots of nonmonic matrix polynomials with nonnegative coefficients

In this section we consider the matrix polynomial

$$Q(\lambda) = \lambda I_{n\times n} - (\lambda^2 A + \lambda B + C) \tag{3.1}$$

with entrywise nonnegative $n \times n$-matrices A,B and C. Here and in the remainder $I_{n\times n}$ and $0_{n\times n}$ denote the $n \times n$ identity matrix and the $n \times n$ zero matrix, respectively; sometimes we write for short I and 0 for an appropriate (square) identity matrix and a (not necessarily square) zero matrix, respectively.

As mentioned in the introduction, we are interested in the existence of nonnegative right roots of $Q(\cdot)$ in the case when $S = A + B + C$ is *reducible*; for the simpler case when S is irreducible see [7, Sections 3 and 4].

The nonnegative matrix S is cogredient to its (lower triangular) Frobenius normal form (see [18, Section 2]), i.e., there exists a permutation matrix P such that

$$PSP^t = \begin{pmatrix} S_{11} & 0 & 0 & \cdots & 0 \\ S_{21} & S_{22} & 0 & \vdots & 0 \\ \vdots & \vdots & \ddots & \ddots & \vdots \\ S_{m-1,1} & \cdots & \cdots & S_{m-1,m-1} & 0 \\ S_{m1} & \cdots & \cdots & \cdots & S_{mm} \end{pmatrix} \tag{3.2}$$

where the diagonal blocks S_{rr} are irreducible square matrices. For our purpose we can and will assume, that S is in its Frobenius form.

Since $0_{n\times n} \le A \le S$, and a corresponding inequality for the other coefficients B and C holds, all coefficients and $Q(\cdot)$ are lower block triangular matrices of the same block size as S.

If W is a right root of $Q(\cdot)$, then there exists a (unique) matrix V such that

$$Q(\lambda) = (V - \lambda A)(\lambda I_{n\times n} - W), \qquad \lambda \in \mathbb{C}, \tag{3.3}$$

and (3.3) is equivalent to

$$V = I_{n\times n} - B - AW \quad \text{and} \quad C = VW \tag{3.4}$$

In the next two theorems it is not essential that (3.2) is the Frobenius normal form of S; only the lower block triangular form of S is needed, and the fact that the nonnegative coefficients A, B and C have a corresponding lower block triangular form.

If $Q(\cdot)$ has a nonnegative right root, we know from [7, Proposition 3.1] that $Q(\cdot)$ has a smallest (= minimal) nonnegative right root.

Theorem 3.1. *Let $Q(\lambda) = \lambda I_{n\times n} - (\lambda^2 A + \lambda B + C)$ with nonnegative $n \times n$-matrices A, B and C and let $S = A + B + C$ be in lower block triangular form (3.2). Then the following assertions are equivalent:*

1. *$Q(\cdot)$ has a nonnegative right root.*
2. *$Q(\cdot)$ has a nonnegative right root which is lower block triangular (of the same block sizes as S); in particular, the smallest nonnegative right root of $Q(\cdot)$ is lower block triangular.*
3. *For $r = 1, 2, \ldots, m$ the polynomial $Q_{rr}(\cdot)$ has a nonnegative right root W_{rr} and for $1 \le q < p \le m$ the equation*

$$V_{pp}X - A_{pp}XW_{qq} = C_{pq} + \left(B_{pq} + \sum_{q\le r<p} A_{pr}W_{rq} \right) W_{qq} - \sum_{q<r<p} V_{pr}W_{rq} \tag{3.5}$$

has a nonnegative matrix X (of the same size as S_{pq}) as a solution. Here V_{pp} is a Z-matrix such that $Q_{pp}(\lambda) = (V_{pp} - \lambda A_{pp})(\lambda I - W_{pp})$.

Proof. 1. implies 2.: From [7, Proposition 3.1] we know that the fixed point iteration

$$W_{l+1} = AW_l^2 + BW_l + C \quad \text{with} \quad 0_{n\times n} \le W_0 \le C,$$

$l = 0, 1, \ldots,$ converges to the smallest nonnegative right root of $Q(\cdot)$. In our case W_0, and then all matrices of the sequence (W_l) are lower block triangular (of the same block sizes as S). Therefore its limit is also lower block triangular (of the same block sizes as S).

That 2. implies 1. is clear.

For the equivalence of 2. and 3. apply Theorem 2.5 (*note the sign change for V between the first factors in* (2.4) *and* (3.3)!) and results of [7, Section 3]. □

Note that in the case of Theorem 3.1 the matrices V_{pq} for $1 \le q < p \le m$ defined in (2.13) are given by

$$V_{pq} = -B_{pq} - \sum_{q\le r<p} A_{pr}W_{rq} - A_{pp}X. \tag{3.6}$$

Therefore they are nonpositive, and the block matrix V is a Z-matrix (see (3.4)).

In the following $r(T)$ denotes the spectral radius of the square matrix T, and $S(\lambda)$ denotes the polynomial $\lambda^2 A + \lambda B + C$. Thus $Q(\lambda) = \lambda I - S(\lambda)$.

Theorem 3.2. *Let $Q(\lambda) = \lambda I_{n\times n} - (\lambda^2 A + \lambda B + C)$ with nonnegative $n \times n$-matrices A, B and C and let $S = A + B + C$ be in the lower block triangular form* (3.2). *Assume that for $k = 1, 2, \dots, m$ there exists a positive ρ_k such that*

$$r(S_{kk}(\rho_k)) = r(\rho_k{}^2 A_{kk} + \rho_k B_{kk} + C_{kk}) < \rho_k, \tag{3.7}$$

and

$$\rho_1 \le \rho_2 \le \cdots \le \rho_m. \tag{3.8}$$

Then $Q(\cdot)$ has a nonnegative right root; more precisely, there exist lower block triangular matrices W and V (with the same block sizes as S) such that $Q(\lambda) = (\lambda M - V)(\lambda I_{n\times n} - W)$, where W is a nonnegative matrix and V is a nonsingular M-matrix.

Proof. By [7, Theorem 3.4] there exists for $k = 1, 2, \dots, m$ a nonnegative matrix W_{kk} with $r(W_{kk}) < \rho_k$ and a nonsingular M-matrix V_{kk} with $r(V_{kk}{}^{-1} A_{kk}) < 1/\rho_k$ such that $Q_{kk}(\lambda) = \lambda I - (\lambda^2 A_{kk} + \lambda B_{kk} + C_{kk}) = (\lambda M_{kk} - V_{kk})(\lambda I_{kk} - W_{kk})$. This and (3.8) imply that $r(\diamond(V_{pp}{}^{-1} A_{pp}, W_{qq})) < 1$ for $1 \le q < p \le m$, see the remark to the proof of Corollary 2.4. The right-hand side of equation (3.5) is a nonnegative matrix. Using the Neumann series expansion of the inverse of $I - \diamond(V_{pp}{}^{-1} A_{pp}, W_{qq})$ we obtain that for $1 \le q < p \le m$ the unique solution of (3.5) is a nonnegative matrix. Theorem 3.1 completes the proof of the first assertion. Note that the diagonal blocks of V are nonsingular M-matrices, therefore they have nonnegative inverses. Further the left lower blocks of V are nonpositive (see (3.6)). Therefore $V = I - B - AW$ is a nonsingular Z-matrix with a nonnegative inverse, thus it is a M-matrix (see [1, (N_{38}), p. 137]). □

By Theorem 5.1 of [8] it is possible to generalize both theorems above to polynomials with coefficients which are more general than matrices, for example operators in ordered Banach spaces.

Note that the assumptions of Theorem 3.2 about the spectral radii do not imply that $r(\rho^2 A + \rho B + C) < \rho$ for some positive ρ (see [7, Example 3.3]). Therefore one cannot directly apply [7, Theorem 3.4]. The following example shows that the above theorem does not hold if we assume only (3.7), but not (3.8).

Example 3.3. Let a be a positive number and define

$$Q(\lambda) = \lambda I_{2\times 2} - \left(\lambda^2 \begin{pmatrix} 0 & 0 \\ 0 & a \end{pmatrix} + \lambda \begin{pmatrix} 0 & 0 \\ a & 0 \end{pmatrix} + \begin{pmatrix} 1 & 0 \\ 1 & 0 \end{pmatrix}\right).$$

Then equation (3.5) becomes $(1-a)x = 1 + a$. Theorem 3.1 shows that $Q(\cdot)$ has no nonnegative (triangular) right root for $a > 1$. For $a > 1$ and positive ρ_1 and ρ_2

$$1 = r(S_{11}(\rho_1)) < \rho_1 \quad \textit{and} \quad a\rho_2{}^2 = r(S_{22}(\rho_2)) < \rho_2$$

imply $\rho_1 > 1 > 1/a > \rho_2$.

As a preparation for the next theorem we introduce the following notation. Let $Q(\lambda) = \lambda I_{n\times n} - (\lambda^2 A + \lambda B + C)$ with nonnegative $n \times n$-matrices A, B and

C and let $S = A + B + C$ be in (lower triangular) Frobenius normal form (3.2). Assume that

$$r(S(\hat{\rho})) = r(\,\hat{\rho}^2 A + \hat{\rho} B + C\,) \le \hat{\rho} \quad \text{for some positive } \hat{\rho}. \tag{3.9}$$

Since for $k = 1, 2, \ldots, m$ the matrix $S_{kk}(1)$ is irreducible, the function $]0, \infty[\to [0, \infty[,\ \rho \mapsto r(S_{kk}(\rho))$ is real analytic; see [7, Section 3]. By $r'(S_{kk}(\cdot))$ we denote its derivative. Now we define the following sets:

$$\begin{aligned}
\mathbb{M} &:= \{k \in \{1, 2, \ldots, m\} : r(S_{kk}(\hat{\rho})) = \hat{\rho}\},\\
\mathbb{M}_- &:= \{k \in \mathbb{M} : r'(S_{kk}(\hat{\rho})) < 1\},\\
\mathbb{M}_1 &:= \{k \in \mathbb{M} : r'(S_{kk}(\hat{\rho})) = 1\} \qquad \text{and}\\
\mathbb{M}_+ &:= \{k \in \mathbb{M} : r'(S_{kk}(\hat{\rho})) > 1\}.
\end{aligned}$$

For $k = 1, 2, \ldots, m$ we denote by d_k the *index of phase imprimitivity* (see [7]) of the graph $G(C_{kk}]B_{kk}[A_{kk})$. We recall (see [7, Proposition 4.5]): $d_k \neq 0$ implies that for a positive ρ with $r(S_{kk}(\rho)) = \rho$ and for a $\theta \in [0, 2\pi[$ the complex number $\rho e^{i\theta}$ is an eigenvalue of $Q_{kk}(\cdot)$ if and only if $\theta \in \{0, \frac{2\pi}{d_k}, 2 \cdot \frac{2\pi}{d_k}, \ldots, (d_k - 1)\frac{2\pi}{d_k}\}$.

The following theorem generalizes [9, Theorem 5] for quadratic matrix polynomials considered in this paper.

Theorem 3.4. *Let $Q(\lambda) = \lambda I_{n\times n} - (\lambda^2 A + \lambda B + C)$ with nonnegative $n \times n$-matrices A, B and C and let $S = A+B+C$ be in its Frobenius normal form* (3.2). *Assume that there exists a positive $\hat{\rho}$ such that*

$$r(S(\hat{\rho})) = r(\,\hat{\rho}^2 A + \hat{\rho} B + C\,) \le \hat{\rho}. \tag{3.10}$$

With the notation above the following assertions hold:

1. *The set of the eigenvalues of $Q(\cdot)$ is the whole plane $\mathbb{C}$ if and only if there exists a $k \in \mathbb{M}$ such that $r(S_{kk}(\rho)) = \rho$ for all nonnegative ρ.*
2. *Assume that the set of the eigenvalues of $Q(\cdot)$ is not the whole plane $\mathbb{C}$. Then*
   (a) *$Q(\cdot)$ has exactly $n - \sum_{k\in\mathbb{M}_-\cup\mathbb{M}_1} d_k$ eigenvalues (counting algebraic multiplicities) in the open disc $\mathbb{D}_{\hat{\rho}} = \{\lambda \in \mathbb{C} : |\lambda| < \hat{\rho}\}$.*
   (b) *$Q(\cdot)$ has exactly $\sum_{k\in\mathbb{M}_-\cup\mathbb{M}_+} d_k + 2\sum_{k\in\mathbb{M}_1} d_k$ eigenvalues (counting algebraic multiplicities) on the circle $\mathbb{T}_{\hat{\rho}} = \{\lambda \in \mathbb{C} : |\lambda| = \hat{\rho}\}$.*

Proof. The polynomial $Q(\cdot)$ is block triangular, therefore the set of its eigenvalues is the union of the sets of the eigenvalues of its diagonal blocks $Q_{kk}(\cdot), k = 1, 2, \ldots, m$. Each diagonal block in the Frobenius normal form is irreducible. Therefore we can apply to each of them [7, Theorem 4.10] and will obtain the assertions. □

In [9, Theorem 5] the main assumption is that S is substochastic. If $r(S) = 1$, this implies (see [9, Equations (55) and (56)]) that there exists a $p \in \{1, 2, \ldots, m\}$

such that $\mathbb{M} = \{1, 2, \dots, p\}$ and the Frobenius normal form of $S(\rho)$ is

$$\begin{pmatrix} S_{11}(\rho) & 0 & \cdots & 0 & 0 & \cdots & & 0 \\ 0 & \ddots & \ddots & 0 & 0 & \cdots & & 0 \\ \vdots & \ddots & \ddots & \vdots & \vdots & \cdots & & \vdots \\ 0 & & & 0 & 0 & & & 0 \\ 0 & \cdots & 0 & S_{pp}(\rho) & 0 & \cdots & & 0 \\ S_{p+1,1}(\rho) & \cdots & \cdots & S_{p+1,p}(\rho) & S_{p+1,p+1}(\rho) & 0 & \cdots & 0 \\ \vdots & & & \vdots & \vdots & \ddots & \ddots & \vdots \\ \vdots & & & \vdots & \vdots & & \ddots & 0 \\ S_{m1}(\rho) & \cdots & \cdots & S_{mp}(\rho) & S_{m,p+1}(\rho) & \cdots & \cdots & S_{mm}(\rho) \end{pmatrix}. \tag{3.11}$$

Corollary 3.5. *Let $Q(\lambda) = \lambda I_{n\times n} - (\lambda^2 A + \lambda B + C)$ with nonnegative $n \times n$-matrices A, B and C. Assume that there exists a positive $\hat{\rho}$ such that $r(S(\hat{\rho})) \leq \hat{\rho}$. Then $Q(\cdot)$ has a nonnegative right root if one of the following assumptions is satisfied:*

1. *$\mathbb{M} := \{k \in \{1, 2, \dots, m\} : r(S_{kk}(\hat{\rho})) = \hat{\rho}\}$ is empty.*
2. *There exists a $p \in \{1, 2, \dots, m\}$ such that*

$$\mathbb{M} := \{k \in \{1, 2, \dots, m\} : r(S_{kk}(\hat{\rho})) = \hat{\rho}\} = \{1, 2, \dots, p\},$$

and (3.11) is the Frobenius normal form of $S(\rho)$.

Proof. Assumption 1. is equivalent to $r(S(\hat{\rho})) < \hat{\rho}$. The assertion follows from [7, Theorem 3.4] or Theorem 3.2 with $m = 1$.
Let Assumption 2. hold. Then $r(S(\hat{\rho})) = \hat{\rho}$. For $k = 1, 2, \dots, p$ each $Q_{kk}(\cdot)$ has a nonnegative right root with spectral radius equal to $\hat{\rho}$ (see [7, Proposition 4.6]). Since the upper left corner of $Q(\cdot)$ (of the first $p \times p$ blocks) is diagonal, it has a nonnegative right root with spectral radius equal to $\hat{\rho}$ (see Corollary 2.3). The right lower corner of $S(\cdot)$ has a spectral radius strictly less than $\hat{\rho}$. Now an analogous argument as in the proof of Theorem 3.2 proves the assertion. □

As a completion to [9, Theorem 5] we obtain in the case of a quadratic matrix polynomial, the following:

Corollary 3.6. *Let $Q(\lambda) = \lambda I_{n\times n} - (\lambda^2 A + \lambda B + C)$ with nonnegative $n \times n$-matrices A, B and C. If $S = A + B + C$ is substochastic, then $Q(\cdot)$ has a nonnegative right root.*

The proof follows from the preceding corollary as the case $\hat{\rho} = 1$.

In [3] the authors consider the polynomial $-Q(\cdot)$ where

$$Q(\lambda) = \lambda I_{n\times n} - (\lambda^2 I_{n\times n} + C). \tag{3.12}$$

and C is a nonnegative $n \times n$-matrix. In [3, Theorem 2.3] they give necessary and sufficient conditions for the existence of nonnegative roots of $Q(\cdot)$. Their proof uses analytical methods, for example the convergence of the binomial series of $\frac{1}{4}I - C$.

We will give a different proof of their results by methods in the spirit of this paper (see Theorem 3.10 below). We need some preparations.

Let $Q(\cdot)$ be as in (3.12). Then $r(S(\rho)) = \rho^2 + r(C)$ for positive ρ. This implies that both equalities $r(S(\hat{\rho})) = \hat{\rho}$ and $r'(S(\hat{\rho})) = 1$ hold if and only if both equalities $r(C) = 1/4$ and $\hat{\rho} = 1/2$ hold.

From [7, Theorem 3.4] it follows

Proposition 3.7. *Let $Q(\cdot)$ be as in* (3.12) *with a nonnegative square matrix C with $r(C) < 1/4$. Then $Q(\cdot)$ has a nonnegative right root.*
In addition, there exist a nonnegative matrix W and a nonsingular M-matrix V such that

$$Q(\lambda) = \lambda I_{n\times n} - (\lambda^2 I_{n\times n} + C) = (V - \lambda I)(\lambda I - W), \tag{3.13}$$

$r(W) < 1/2$ *and* $r(V^{-1}) < 2$.

From [7, Theorems 3.4 and 4.10] it follows

Proposition 3.8. *Let $Q(\cdot)$ be as in* (3.12) *with a nonnegative irreducible square matrix C. Then $Q(\cdot)$ has a nonnegative right root if and only if $r(C) \leq 1/4$.*
In addition, there exist a nonnegative matrix W and a nonsingular M-matrix V such that (3.13) *holds, and*

1. $r(W) = 1/2$ *and* $r(V^{-1}) = 2$, *if* $r(C) = 1/4$,
2. $r(W) < 1/2$ *and* $r(V^{-1}) < 2$, *if* $r(C) < 1/4$.

Note that for a nonnegative irreducible matrix its spectral radius is a simple root of its characteristic polynomial (see [16, Theorem 1.4.3]). This implies that it is in the terminology of [3] an eigenvalue value of degree 1, i.e., it is a semisimple eigenvalue. Therefore the first part of the last proposition is exactly [3, Theorem 2.3] for the case of an *irreducible* matrix C.

Lemma 3.9.

1. *Let $Q(\cdot)$ be as in* (3.12) *with a nonnegative irreducible square matrix C and $Q(\lambda) = \lambda I_{n\times n} - (\lambda^2 I_{n\times n} + C) = (V - \lambda I)(\lambda I - W)$, with nonnegative W and a nonsingular M-matrix V. Then W is irreducible and $V^{-1} = (I - W)^{-1}$ is strictly positive, i.e., all entries of V^{-1} are positive.*
2. *Let W be nonnegative and irreducible, V have a strictly positive inverse, $r(W) = r(V^{-1})^{-1}$ and $B \geq 0_{n\times n}$. Then the equation $VX - XW = B$ has a nonnegative solution if and only if $B = 0_{n\times n}$; here $W, V, B, 0_{n\times n}$ and X are matrices of the size $n \times n$.*

Proof. 1. The irreducibility of W is easily proved by contradiction using $C = VW$ and $V = I - W$. The second assertion follows from [16, Theorem 2.3].

2. The equation $VX - XW = B$ is equivalent to $X - \diamond(V^{-1}, W)X = X - V^{-1}XW = V^{-1}B$. By the description of the spectrum of $\diamond(V^{-1}, W)$ following Corollary 2.5, we obtain $r(\diamond(V^{-1}, W)) = r(V^{-1})r(W) = 1$. Now we will use some results from the local spectral theory of nonnegative operators in finite-dimensional spaces, see [5, 6]. In the following we write for short $\diamond$ instead of $\diamond(V^{-1}, W)$.

Set $\diamondsuit = \sum_{0\le j\le n} \diamond^j$. Then the linear maps $\diamondsuit$ and $\diamond$ (acting in the linear space of all complex $n\times n$-matrices) commute. Since W is irreducible and V^{-1} is strictly positive, it follows that $\diamondsuit V^{-1}B$ is strictly positive when B is nonnegative and nonzero. This implies that the local spectral radius $r_\diamond(\diamondsuit V^{-1}B)$ of $\diamondsuit V^{-1}B$ with respect to $\diamond$ is equal to the spectral radius of $\diamond$, see [6, Proposition 1]. Now $1 = r_\diamond(\diamondsuit V^{-1}B) \le r_\diamond(V^{-1}B) \le r(\diamond) = 1$. We are in finite-dimensional spaces, therefore, by [5, Theorem 12] or [19, Theorem 3.1], the equation $X - \diamond X = X - V^{-1}XW = V^{-1}B$ with nonnegative B has a nonnegative solution if and only if $B = 0_{n\times n}$. □

The following theorem was proved in [3, Theorem 2.3] by completely different methods.

Theorem 3.10. *Let*

$$Q(\lambda) = \lambda I_{n\times n} - (\lambda^2 I_{n\times n} + C). \tag{3.14}$$

where C is a nonnegative $n\times n$-matrix. Then

1. *$r(C) < 1/4$ implies that there is a nonnegative right root of $Q(\cdot)$.*
2. *$r(C) > 1/4$ implies that there is no nonnegative right root of $Q(\cdot)$.*
3. *If $r(C) = 1/4$, then $Q(\cdot)$ has a nonnegative right root if and only if $1/4$ is a semisimple eigenvalue of C.*

Proof. For $S(\lambda) = \lambda^2 I_{n\times n} + C$ we have $r(S(\rho)) = \rho^2 + r(C)$ for nonnegative ρ. Then $r(C) < 1/4$ implies that $r(S(1/2)) < 1/2$. From [7, Theorem 3.4] we obtain that $Q(\cdot)$ has a nonnegative right root, which proves 1.

Let $r(C) > 1/4$. In the Frobenius normal form of C there exists an irreducible diagonal block C_{kk} with $r(C_{kk}) > 1/4$. Then $r(S_{kk}(\rho)) = \rho^2 + r(C_{kk}) \ge \rho + r(C_{kk}) - 1/4$ for all nonnegative ρ. By [7, Theorem 4.10(i)] $Q_{kk}(\cdot)$ has no nonnegative right root, here $Q_{kk}(\lambda) = \lambda - (\lambda^2 + C_{kk})$. By Theorem 3.1, $Q(\cdot)$ has no nonnegative right root, this proves 2.

Now assume that $r(C) = 1/4$ and $1/4$ is an eigenvalue of C of degree 1. By [17, Theorem 3.1] or [13, Proposition II.1.], C is (cogredient to)

$$\tilde{C} = \begin{pmatrix} C_{11} & 0 & 0 \\ C_{21} & C_{22} & 0 \\ C_{31} & C_{32} & C_{33} \end{pmatrix},$$

where

1. C_{11} are C_{33} are possibly empty, but $r(C_{kk}) < 1/4$ for $k = 1, 3$ if this is not the case,
2. C_{22} is block diagonal,

$$C_{22} = \mathrm{Diag}(C_1, C_2, \ldots, C_p)$$

where p is the dimension of the eigenspace $Ker(1/4I - C)$, C_l is irreducible and $r(C_l) = 1/4$ for $l = 1, 2, \ldots, p$.

From Proposition 3.8.1. we obtain

$$Q_{22}(\lambda) = \lambda I - (\lambda^2 I + C_{22}) = (V_{22} - \lambda I)(\lambda I - W_{22})$$

with nonnegative $W_{22} = \mathrm{Diag}(W_1, W_2, \ldots, W_p)$, an invertible M-matrix $V_2 = \mathrm{Diag}(V_1, V_2, \ldots, V_p)$ and $r(W_{22}) = r(W_l) = 1/2 = r({V_{22}}^{-1})^{-1} = r({V_l}^{-1})^{-1}$ for $l = 1, 2, \ldots, p$.
From Proposition 3.8.2. we obtain for $k = 1, 3$

$$Q_{kk}(\lambda) = \lambda I - (\lambda^2 I + C_{kk}) = (V_{kk} - \lambda I)(\lambda I - W_{kk})$$

with nonnegative W_{kk}, invertible M-matrices V_{kk}, $r(W_{kk}) < 1/2$ and $r({V_{kk}}^{-1}) < 2$. This implies $r(\diamond({V_{pp}}^{-1}, W_{qq})) < 1$ for $1 \leq p < q \leq 3$. As in the proof of Theorem 3.2, it follows that for $1 \leq p < q \leq 3$ the equations (3.5) have nonnegative solutions (note that in our case $A_{pp} = I$, and $B_{rs} = A_{rs} = 0$ for $r \neq s$). Therefore $Q(\cdot)$ has a nonnegative right root, by Theorem 3.1.
Now assume that $r(C) = 1/4$, and $Q(\cdot)$ has a nonnegative right root W. Then there exists a matrix V such that (3.13), and therefore $V = I_{n\times n} - W$ and $C = VW = WV$ hold. By Theorem 3.1, we can and will assume that W and V have the same block structure as the Frobenius normal form of C.
Further, assume that $1/4$ is an eigenvalue of C of degree greater than 1. By [17, Theorem 3.1] or [13, (2.28)], there exists a principal submatrix $\tilde{C}$ of C such that

$$\tilde{C} = \begin{pmatrix} C_{11} & 0 & 0 \\ C_{21} & C_{22} & 0 \\ C_{31} & C_{33} & C_{33} \end{pmatrix}, \tag{3.15}$$

where

1. C_{11} and C_{33} are irreducible,
2. $r(C_{kk}) = 1/4$ for $k = 1, 3$,
3. C_{22} may be empty (then C_{21} and C_{32} are also empty), but $r(C_{22}) < 1/4$ if this is not the case,
4. $$C_{31} + C_{32}C_{21} > 0, \tag{3.16}$$
   i.e., the matrix of the left-hand side is nonnegative and nonzero.

Then $\tilde{Q}(\lambda) = \lambda I - (\lambda^2 I + \tilde{C}) = (\tilde{V} - \lambda I)(\lambda I - \tilde{W})$, where $\tilde{W}$ and $\tilde{V}$ are the corresponding principal submatrices of W and V, respectively. Further $\tilde{W}$ and $\tilde{V}$ have the same block structure as $\tilde{C}$ in (3.15). By Theorem 3.1, W_{31} is a nonnegative solution of

$$V_{33}X - XW_{11} = C_{31} - V_{32}W_{21}.$$

Now $V_{rs} = -W_{rs}$ are nonpositive for $1 \leq s < r \leq 3$, since $\tilde{W} \geq 0$ and $\tilde{V} = I - \tilde{W}$. From Lemma 3.9.1 we obtain that W_{11} and W_{33} are nonnegative and irreducible, and V_{33} has a strictly positive inverse. By Proposition 3.8.1, we have $r(W_{11}) = 1/2$ and $r({V_{33}}^{-1}) = 2$. Then, by Lemma 3.9.2, the last equation has a nonnegative solution if and only if its right-hand side is zero, which is equivalent to $C_{31} = 0$ and $V_{32}W_{21} = 0$; note that C and W are nonnegative, and V is an M-matrix.

From $\tilde{C} = \tilde{V}\tilde{W} = \tilde{W}\tilde{V}$, $\tilde{W} \geq 0$ and $\tilde{V} = I - \tilde{W}$ we obtain $0 \leq C_{21} = W_{21}V_{11} + W_{22}V_{21} \leq W_{21}V_{11}$. Now the inverse of V_{11} exists and is nonnegative, therefore $0 \leq C_{21}{V_{11}}^{-1} \leq W_{21}$; in a similar way we obtain from $0 \leq C_{32} = V_{32}W_{22} + V_{33}W_{32}$ that $0 \leq {V_{33}}^{-1}C_{32} \leq W_{32} = -V_{32}$. But then $0 \leq {V_{33}}^{-1}C_{32}C_{21}{V_{11}}^{-1} \leq -V_{32}W_{21} = 0$, which is equivalent to $C_{32}C_{21} = 0$. Together with $C_{31} = 0$ from above, we get a contradiction to (3.16). Therefore $1/4$ is not an eigenvalue of C of degree greater than 1. □

Acknowledgment

The authors wish to thank the referee for his/her careful reading and for several suggestions.

References

[1] Berman, A., Plemmons, R.: Nonnegative Matrices in the Mathematical Sciences. SIAM: Philadelphia, 1994.

[2] Bini, D.A., Latouche, G., Meini, B.: Numerical Methods for Structured Markov Chains, Oxford University Press, Oxford-New York, 2005.

[3] Butler, G.J., Johnson, C.R., Wolkowicz, H.: Nonnegative Solutions of a Quadratic Matrix Equation Arising from Comparison Theorems in Ordinary Differential Equations. SIAM J. Alg. Disc. Meth. **6**, 47–53 (1985).

[4] Conway, J.B.: A Course in Functional Analysis, Springer: New York-Berlin-Heidelberg-Tokyo, 1985.

[5] Förster, K.-H., Nagy, B.: On the Local Spectral Theory of Positive Operators. Operator Theory, Advances and Applications **28**, 71–81, Birkhäuser: Basel-Boston-Berlin, 1988.

[6] Förster, K.-H., Nagy, B.: On the Local Spectral Radius of a Nonnegative Element with Respect to an Irreducible Operator, Acta Sci. Math. **55**, 155–166 (1991).

[7] Förster, K.-H., Nagy, B.: On Nonmonic Quadratic Matrix Polynomials with Nonnegative Coefficients. Operator Theory, Advances and Applications **162**, 145–163, Birkhäuser: Basel-Boston-Berlin, 2005.

[8] Förster, K.-H., Nagy, B.: Spectral Properties of Operator Polynomials with Nonnegative Coefficients. Operator Theory, Advances and Applications **163**, 147–162, Birkhäuser: Basel-Boston-Berlin, 2005.

[9] Gail, H.R., Hantler, S.L., Taylor, B.A.: Spectral Analysis of $M/G/1$ and $G/M/1$ Type Markov Chains, Adv. Appl. Prob. **28**, 114–165 (1996).

[10] Gohberg, I., Lancaster, P., Rodman, L.: Matrix Polynomials, Academic Press: New York, 1982.

[11] Horn, R.A., Johnson, C.R.: Matrix Analysis, Cambridge University Press: Cambridge, 1985.

[12] Horn, R.A., Johnson, C.R.: Topics in Matrix Analysis, Cambridge University Press: Cambridge, 1991.

[13] Jang, R., Victory, Jr., H.D.: On the Ideal Structure of Positive, Eventually Compact Operators on Banach Lattices, Pacific J. Math. **157**, 57–85 (1993).

[14] Latouche, G., Ramaswami, S.: Introduction to Matrix Analytic Methods in Stochastic Modeling. ASA-SIAM Series on Stochastics and Applied Probability. SIAM: Philadelphia, 1999.

[15] Marcus, A.S.: Introduction to the Spectral Theory of Polynomial Operator Pencils. Translation of Mathematical Monographs, Vol. **71**, Amer. Math. Soc., Providence, 1988.

[16] Minc, H.: Nonnegative Matrices, Wiley: New York, 1988.

[17] Rothblum, U.G.: Algebraic Eigenspaces of Nonnegative Matrices, Linear Algebra Appl. **12**, 281–291 (1975).

[18] Schneider, H.: The Influence of the Marked Reduced Graph of a Nonnegative Matrix on the Jordan Form and on Related Properties: A Survey. Linear Algebra Appl. **84**, 169–189 (1986).

[19] Tam, Bit-Shun, Schneider, H. Linear Equations over Cones and Collatz-Wielandt Numbers. Linear Algebra Appl. **363**, 295–332 (2003).

[20] Trampus, A.: A Spectral Mapping Theorem for Functions of Two Commuting Linear Operators, Proc. Amer. Math. Soc. **14**, 893–895 (1963).

K.-H. Förster
Technische Universität Berlin
Institut für Mathematik, MA 6-4
D-10623 Berlin, Germany
e-mail: `foerster@math.tu-berlin.de`

B. Nagy
University of Technology and Economics
Department of Analysis
Institute of Mathematics
H-1521 Budapest, Hungary
e-mail: `bnagy@math.bme.hu`

Operator Theory:
Advances and Applications, Vol. 175, 111–120

On Exceptional Extensions Close to the Generalized Friedrichs Extension of Symmetric Operators

Seppo Hassi, Henk de Snoo and Henrik Winkler

Abstract. If the Q-function Q corresponding to a closed symmetric operator S with defect numbers $(1,1)$ and one of its selfadjoint extensions belongs to the Kac class $\mathbf{N}_1$ then it is known that all except one of the Q-functions of S belong to $\mathbf{N}_1$, too. In this note the situation that the given Q-function does not belong to the class $\mathbf{N}_1$ is considered. If $Q \in \mathbf{N}_p$, i.e., if the restriction of the spectral measure of Q on the positive or the negative axis corresponds to an $\mathbf{N}_1$-function, then Q itself is the Q-function of the exceptional extension, and, hence, it is associated with the generalized Friedrichs extension of S. If Q or, equivalently, the spectral measure of Q is symmetric, or if the difference of Q and a symmetric Nevanlinna function belongs to the class $\mathbf{N}_1$ or $\mathbf{N}_p$, then Q is still exceptional in a wider sense. Similar results hold for the generalized Kreĭn-von Neumann extension of the symmetric operator.

Mathematics Subject Classification (2000). Primary 47A06, 47B25; Secondary 47A57 .

Keywords. Q-function, generalized Friedrichs extension, generalized Kreĭn-von Neumann extension, Kac class.

1. Introduction

Let S be a not necessarily densely defined closed symmetric relation in a Hilbert space $(\mathfrak{H},\ (\cdot,\cdot))$ with defect numbers $(1,1)$ and let A be some canonical (i.e., $\operatorname{dom} A \subset \mathfrak{H}$) selfadjoint extension of S with resolvent set $\rho(A)$. Let $\chi(\mu)$, $\mu \in \mathbb{C}\backslash\mathbb{R}$, be a nontrivial element of $\ker(S^* - \mu)$, where S^* denotes the adjoint of S. Then the element

$$\chi(z) := (I + (z-\mu)(A-z)^{-1})\chi(\mu), \quad z \in \rho(A),$$

The research was partially supported by the Research Institute for Technology at the University of Vaasa. H. Winkler was supported by the "Fond zur Förderung der wissenschaftlichen Forschung" (FWF, Austria), grant number P15540-N05.

belongs to $\ker(S^* - z)$ and for each $z \in \rho(A)$ the symmetric relation S can be recovered from A and $\chi(z)$ by

$$S = \{ \{f, g\} \in A : (g - zf, \chi(\bar{z})) = 0 \}.$$

Let $\mathbf{N}$ be the set of Nevanlinna functions, i.e., the set of all functions which are analytic on $\mathbb{C}\setminus\mathbb{R}$, satisfy $\overline{Q(z)} = Q(\bar{z})$, and map the open upper half-plane $\mathbb{C}^+$ into $\mathbb{C}^+ \cup \mathbb{R}$. Recall that the Q-function associated with S and A is a solution (unique up to a real constant) of the equation

$$\frac{Q(z) - \overline{Q(\mu)}}{z - \bar{\mu}} = (\chi(z), \chi(\mu)), \quad z, \mu \in \rho(A),$$

which implies that $Q \in \mathbf{N}$. Each Nevanlinna function $Q \in \mathbf{N}$ which is not equal to a real constant is the Q-function of a closed symmetric relation S with defect numbers $(1, 1)$ and a canonical selfadjoint extension A of S, see, e.g., [4]. Moreover, for a given $Q \in \mathbf{N}$ the relations S and A are uniquely determined up to isometric isomorphisms if S is completely nonselfadjoint, i.e., if there exists no nontrivial orthogonal decomposition of S such that one of the summands is selfadjoint. Note that a completely nonselfadjoint closed symmetric relation is automatically an operator. The canonical selfadjoint extensions of S can be uniquely parametrized by $\alpha \in (-\pi/2, \pi/2]$:

$$(A(\alpha) - z)^{-1} = (A - z)^{-1} - \chi(z) \frac{1}{Q(z) + \tan \alpha} (\cdot, \chi(\bar{z})), \tag{1.1}$$

where Q is the Q-function corresponding to S and $A = A(\pi/2)$.

If S is nonnegative, among all selfadjoint extensions of S there are two extremal nonnegative selfadjoint extensions, the so-called Friedrichs extension and the Kreĭn-von Neumann extension. In [4] a generalization of the Friedrichs extension was introduced for a class of non-semibounded symmetric operators with defect numbers (1,1), which was further studied in, e.g., [1], cf. [2]. If one of the Q-functions of S belongs to the so-called Kac class $\mathbf{N}_1$ (see below), then all but one of the Q-functions belong to $\mathbf{N}_1$. The Q-function which does not belong to $\mathbf{N}_1$ corresponds to the generalized Friedrichs extension. A corresponding characterization of the Kreĭn-von Neumann extension is investigated in [3]. For an extension of such results to a class of Nevanlinna functions wider than $\mathbf{N}_1$ which is related to Sturm-Liouville equations, see [5]. A generalization of the above results in a different direction can be found in [9] and will be considered in the present paper.

In the above papers it is usually assumed that the Q-function corresponding to a given extension belongs to the class $\mathbf{N}_1$ or $\widehat{\mathbf{N}}_1$, respectively. In this note some results are presented for the cases that the given Q-function does not belong to the class $\mathbf{N}_1$ or $\widehat{\mathbf{N}}_1$ (the precise definitions of various classes are given in Section 2). If Q belongs to the class $\mathbf{N}_p$ or $\widehat{\mathbf{N}}_p$ then it corresponds to the generalized Friedrichs or to the generalized Kreĭn-von Neumann extension, respectively; see Theorems 4.2 and 5.2 below. A new situation occurs if Q is symmetric or if the difference between Q and a symmetric Nevanlinna function is a function from one of the above classes. Then there are at most two extensions which do not belong to the classes $\mathbf{N}_{1+}$ or $\widehat{\mathbf{N}}_{1+}$, see Theorems 4.3 and 5.3 below, cf. [9]. Nevertheless, it is also possible that there is no exceptional extension.

2. Preliminaries

Let A be a canonical selfadjoint extension of S with a corresponding Q-function denoted by Q and let $A(\alpha)$ be the canonical selfadjoint extension of S which is determined by the identity (1.1). For $\alpha \in (-\pi/2, \pi/2)$ the Q-function Q_α corresponding to $A(\alpha)$ and S satisfies

$$Q_\alpha(z) = \frac{Q(z)\tan\alpha - 1}{Q(z) + \tan\alpha} = \tan\alpha - \frac{1 + \tan^2\alpha}{Q(z) + \tan\alpha}, \quad z \in \mathbb{C} \setminus \mathbb{R}, \tag{2.1}$$

(see [6], Proposition 4.4 and what follows). It follows from (2.1) that

$$\operatorname{Im} Q_\alpha(z) = \frac{1 + \tan^2\alpha}{|Q(z) + \tan\alpha|^2} \operatorname{Im} Q(z). \tag{2.2}$$

Each function $Q \in \mathbf{N}$ has an integral representation of the form

$$Q(z) = bz + a + \int_{\mathbb{R}} \left(\frac{1}{\lambda - z} - \frac{\lambda}{1 + \lambda^2} \right) d\sigma(\lambda), \tag{2.3}$$

with $a \in \mathbb{R}$, $b \geq 0$, and a measure σ with the property that

$$\int_{\mathbb{R}} \frac{d\sigma(\lambda)}{1 + \lambda^2} < +\infty.$$

The following identities are immediate from (2.3)

$$\operatorname{Im} Q(iy) = by + y \int_{\mathbb{R}} \frac{d\sigma(\lambda)}{\lambda^2 + y^2}, \quad y > 0, \tag{2.4}$$

and

$$\operatorname{Re} Q(iy) = a + (1 - y^2) \int_{\mathbb{R}} \frac{\lambda d\sigma(\lambda)}{(\lambda^2 + y^2)(1 + \lambda^2)}, \quad y > 0. \tag{2.5}$$

Let $\mathbf{N}_\gamma$, $\gamma \in [0, 2)$, be the class of all functions $Q \in \mathbf{N}$ for which $b = 0$ in the representation (2.3) and for which the inequality

$$\int_{\mathbb{R}} \frac{d\sigma(\lambda)}{1 + |\lambda|^\gamma} < +\infty \tag{2.6}$$

is satisfied (see [7]). If $Q \in \mathbf{N}_1$, the so-called Kac class, it has an integral representation of the form

$$Q(z) = a_1 + \int_{\mathbb{R}} \frac{d\sigma(\lambda)}{\lambda - z} \quad \text{with} \quad a_1 = a - \int_{\mathbb{R}} \frac{\lambda}{1 + \lambda^2} d\sigma(\lambda). \tag{2.7}$$

Let $\widehat{\mathbf{N}}_\gamma$, $\gamma \in [0, 2)$, be the class of all functions $Q \in \mathbf{N}$ with the property that

$$\int_{-1}^{1} \frac{d\sigma(\lambda)}{|\lambda|^{2-\gamma}} < +\infty. \tag{2.8}$$

If $Q \in \widehat{\mathbf{N}}_1$, it has an integral representation of the form (see [3])

$$Q(z) = bz + \hat{a}_1 + \int_{\mathbb{R}} \frac{z}{\lambda(\lambda - z)}\, d\sigma(\lambda) \quad \text{with} \quad \hat{a}_1 = a + \int_{\mathbb{R}} \frac{1}{\lambda(1+\lambda^2)}\, d\sigma(\lambda). \quad (2.9)$$

Clearly, if $0 \le \gamma_1 < \gamma_2 < 2$ then $\mathbf{N}_{\gamma_1} \subset \mathbf{N}_{\gamma_2}$ and $\widehat{\mathbf{N}}_{\gamma_1} \subset \widehat{\mathbf{N}}_{\gamma_2}$. Moreover, with $\gamma = 0$ one has $Q \in \mathbf{N}_0$ if and only if $\lim_{y\to\infty} y \operatorname{Im} Q(iy) < \infty$ and $Q \in \widehat{\mathbf{N}}_0$ if and only if $\lim_{y\to 0+} y^{-1} \operatorname{Im} Q(iy) < \infty$. The remaining subclasses with $\gamma \in (0,2)$ can be characterized as follows (see [7], [8]):

Proposition 2.1. *Let $\gamma \in (0,2)$. A Nevanlinna function Q belongs to the class $\mathbf{N}_\gamma$ if and only if*

$$\int_1^\infty \frac{\operatorname{Im} Q(iy)}{y^\gamma}\, dy < +\infty, \quad (2.10)$$

and it belongs to the class $\widehat{\mathbf{N}}_\gamma$ if and only if

$$\int_0^1 \frac{\operatorname{Im} Q(iy)}{y^{2-\gamma}}\, dy < +\infty. \quad (2.11)$$

In [9] a generalization of results from [4] and [3] led to the following two propositions:

Proposition 2.2. *Let $Q \in \mathbf{N}_\gamma$ for some $\gamma \in (0,1]$, and let Q_α be given by the identity (2.1). Let $\alpha_0 \in (-\pi/2, \pi/2)$ be the solution of $\tan \alpha_0 + a_1 = 0$, where a_1 is the corresponding constant in the representation (2.7) of Q. Then $Q_\alpha \in \mathbf{N}_\gamma$ for each $\alpha \in (-\pi/2, \pi/2) \setminus \{\alpha_0\}$, whereas Q_{α_0}, which corresponds to the generalized Friedrichs extension, does not belong to the class $\mathbf{N}_{2-\gamma}$: $Q_{\alpha_0} \notin \mathbf{N}_{2-\gamma}$. If $Q \notin \mathbf{N}_{\tilde{\gamma}}$ for some $\tilde{\gamma} \in (0,\gamma)$ then $Q_\alpha \notin \mathbf{N}_{\tilde{\gamma}}$ for each $\alpha \in (-\pi/2, \pi/2) \setminus \{\alpha_0\}$.*

Proposition 2.3. *Let $Q \in \widehat{\mathbf{N}}_\gamma$ for some $\gamma \in (0,1]$, and let Q_α be given by the identity (2.1). Let $\alpha_0 \in (-\pi/2, \pi/2)$ be the solution of $\tan \alpha_0 + \hat{a}_1 = 0$, where $\hat{a}_1$ is the corresponding constant in the representation (2.9) of Q. Then $Q_\alpha \in \widehat{\mathbf{N}}_\gamma$ for each $\alpha \in (-\pi/2, \pi/2) \setminus \{\alpha_0\}$, whereas Q_{α_0}, which corresponds to the generalized Kreĭn-von Neumann extension, does not belong to the class $\mathbf{N}_{2-\gamma}$: $Q_{\alpha_0} \notin \widehat{\mathbf{N}}_{2-\gamma}$. If $Q \notin \widehat{\mathbf{N}}_{\tilde{\gamma}}$ for some $\tilde{\gamma} \in (0,\gamma)$, then $Q_\alpha \notin \widehat{\mathbf{N}}_{\tilde{\gamma}}$ for each $\alpha \in (-\pi/2, \pi/2) \setminus \{\alpha_0\}$.*

3. Some new subclasses of Nevanlinna functions

For the sequel of the paper it is necessary to introduce some more subclasses of the class of all Nevanlinna functions $\mathbf{N}$. First the classes $\mathbf{N}_{1+}$ and $\widehat{\mathbf{N}}_{1+}$ will be defined:

$$\mathbf{N}_{1+} := \bigcap_{1<\gamma<2} \mathbf{N}_\gamma \quad \text{and} \quad \widehat{\mathbf{N}}_{1+} := \bigcap_{1<\gamma<2} \widehat{\mathbf{N}}_\gamma.$$

Then $Q \in \mathbf{N}_{1+}$ if and only if

$$\int_1^\infty \frac{\operatorname{Im} Q(iy)}{y^{1+\delta}}\, dy < +\infty \quad (3.1)$$

for each $\delta > 0$, and $Q \in \widehat{\mathbf{N}}_{1+}$ if and only if

$$\int_0^1 \frac{\text{Im } Q(iy)}{y^{1-\delta}} dy < +\infty \tag{3.2}$$

for each $\delta > 0$. In particular,

$$\mathbf{N}_1 \subset \mathbf{N}_{1+} \quad \text{and} \quad \widehat{\mathbf{N}}_1 \subset \widehat{\mathbf{N}}_{1+},$$

and these inclusions are proper.

Now the subclasses $\mathbf{N}_p$ and $\widehat{\mathbf{N}}_p$ will be introduced. Let $Q \in \mathbf{N}$ have the integral representation (2.3). Then there is a decomposition of Q in terms of the corresponding spectral function: define the functions Q_+ and Q_- by

$$Q_+(z) := bz + a + \int_{[0,\infty)} \left(\frac{1}{\lambda - z} - \frac{\lambda}{1+\lambda^2} \right) d\sigma(\lambda)$$

and

$$Q_-(z) := Q(z) - Q_+(z).$$

Observe that $Q \in \mathbf{N} \setminus \mathbf{N}_1$ if and only if either $Q_+ \in \mathbf{N} \setminus \mathbf{N}_1$ or $Q_- \in \mathbf{N} \setminus \mathbf{N}_1$. The class of all Nevanlinna functions Q with the property that either $Q_+ \in \mathbf{N}_1$ and $Q_- \in \mathbf{N} \setminus \mathbf{N}_1$ or $Q_- \in \mathbf{N}_1$ and $Q_+ \in \mathbf{N} \setminus \mathbf{N}_1$ is denoted by $\mathbf{N}_p$. Note that $\mathbf{N}_p \subset \mathbf{N} \setminus \mathbf{N}_1$. Similarly, observe that $Q \in \mathbf{N} \setminus \widehat{\mathbf{N}}_1$ if and only if either $Q_+ \in \mathbf{N} \setminus \widehat{\mathbf{N}}_1$ or $Q_- \in \mathbf{N} \setminus \widehat{\mathbf{N}}_1$. The class of all Nevanlinna functions Q with the property that either $Q_+ \in \widehat{\mathbf{N}}_1$ and $Q_- \in \mathbf{N} \setminus \widehat{\mathbf{N}}_1$ or $Q_- \in \widehat{\mathbf{N}}_1$ and $Q_+ \in \mathbf{N} \setminus \widehat{\mathbf{N}}_1$ is denoted by $\widehat{\mathbf{N}}_p$.

Finally, $\mathbf{N}_s$ stands for the class of all symmetric Nevanlinna functions Q, that is, functions $Q \in \mathbf{N}$ for which the identity $Q(z) = -Q(-z)$ holds for all $z \in \mathbb{C} \setminus \mathbb{R}$. In particular, if $Q \in \mathbf{N_s}$, then $\text{Re } Q(iy) = 0$, $y > 0$.

4. The case that $Q \in \mathbf{N} \setminus \mathbf{N}_1$

In this section some functions in $\mathbf{N} \setminus \mathbf{N}_1$ will be identified as exceptional functions for a family of Nevanlinna functions of the form (2.1). A first observation concerns the limiting behavior of functions in $\mathbf{N} \setminus \mathbf{N}_1$.

Lemma 4.1. *Let $Q \in \mathbf{N}$. If $Q_+ \in \mathbf{N} \setminus \mathbf{N}_1$ then $\lim_{y \to +\infty} \text{Re } Q_+(iy) = -\infty$, and if $Q_- \in \mathbf{N} \setminus \mathbf{N}_1$ then $\lim_{y \to +\infty} \text{Re } Q_-(iy) = +\infty$.*

Proof. Assume that $Q_+ \in \mathbf{N} \setminus \mathbf{N}_1$ and define

$$d\sigma_1(\lambda) := \frac{\lambda}{1+\lambda^2} d\sigma(\lambda), \quad \lambda > 0.$$

Then the identity $\int_0^\infty d\sigma_1(\lambda) = +\infty$ holds. Choose $K > 0$ and a corresponding $l > 0$ such that $\int_0^l d\sigma_1(\lambda) \geq K$. Then for $y \geq l$ one has

$$\int_0^\infty \frac{y^2}{\lambda^2 + y^2} d\sigma_1(\lambda) \geq \int_0^l \frac{y^2}{l^2 + y^2} d\sigma_1(\lambda) \geq \frac{K}{2}.$$

It follows from the identity (2.5) that for $y \geq \max\{\sqrt{2}, l\}$ the inequality

$$\text{Re } Q_+(iy) \leq a - \frac{1}{2}\int_0^\infty \frac{y^2}{\lambda^2+y^2}\, d\sigma_1(\lambda) \leq a - \frac{K}{4}$$

holds. Hence $\lim_{y\to\infty} \text{Re } Q_+(iy) = -\infty$, since K can be chosen arbitrary large. If $Q_- \in \mathbf{N} \setminus \mathbf{N}_1$, the statement of the lemma can be shown in a similar way. □

Theorem 4.2. *Let $Q \in \mathbf{N}_p$ and let Q_α be given by (2.1). Then $Q_\alpha \in \mathbf{N}_1$ for each $\alpha \in (-\pi/2, \pi/2)$.*

Proof. Assume that $Q_- \in \mathbf{N}_1$ and $Q \in \mathbf{N} \setminus \mathbf{N}_1$. Let

$$Q_{+\alpha}(z) := \tan\alpha - \frac{1+\tan^2\alpha}{Q_+(z)+\tan\alpha} \tag{4.1}$$

As in [4], for $x < -1$ one has

$$Q_+(x) - Q_+(-1) = \int_{[0,\infty)} \left(\frac{1}{\lambda - x} - \frac{1}{\lambda+1}\right) d\sigma(\lambda) < 0.$$

It follows that the Nevanlinna function $Q_1(z) := -(Q_+(z) - Q_+(-1))^{-1}$ is positive on $(-\infty, -1)$. Since $\lim_{y\to+\infty} y^{-1}Q_1(iy) = \lim_{y\to+\infty}(yQ_+(iy))^{-1} = 0$ one can assume that Q_1 has a representation of the form

$$Q_1(z) = a + \int_{-1}^\infty \left(\frac{1}{\lambda - z} - \frac{\lambda}{1+\lambda^2}\right) d\sigma_1(\lambda).$$

As in [7], for each $\lambda \geq -1$ the function $h_\lambda(x) := (\lambda - x)^{-1}$ is increasing on $(-\infty, -1)$. By monotone convergence one concludes

$$\lim_{x\to-\infty} Q_1(x) = a - \int_{-1}^\infty \frac{\lambda}{1+\lambda^2}\, d\sigma_1(\lambda) \geq 0,$$

hence $Q_1 \in \mathbf{N}_1$. Let $\tan\alpha_1 := -Q_+(-1)$. It follows that $Q_{\alpha_1+} \in \mathbf{N}_1$, which implies by Proposition 2.2 that $Q_{+\alpha} \in \mathbf{N}_1$ for each $\alpha \in (-\pi/2, \pi/2)$ (see also [4]). Let $c := \lim_{y\to\infty} \text{Re } Q_-(iy)$ and let $\tan\beta := \tan\alpha + c + 1$ for $\alpha \in (-\pi/2, \pi/2)$. Since $\lim_{y\to+\infty} \text{Re } Q_+(iy) = -\infty$ by Lemma 4.1, it follows from (2.2) that for $\alpha \in (-\pi/2, \pi/2)$ and sufficiently large y

$$\begin{aligned}
\frac{\text{Im } Q_\alpha(iy)}{1+\tan^2\alpha} &= \frac{\text{Im } Q_+(iy) + \text{Im } Q_-(iy)}{(\text{Re } Q_+(iy) + \text{Re } Q_-(iy) + \tan\alpha)^2 + (\text{Im } Q_-(iy) + \text{Im } Q_+(iy))^2} \\
&\leq \frac{\text{Im } Q_+(iy)}{(\text{Re } Q_+(iy) + c + 1 + \tan\alpha)^2 + (\text{Im } Q_+(iy))^2} + \text{Im } Q_-(iy) \\
&< \frac{\text{Im } Q_{+\beta}(iy)}{1+\tan^2\beta} + \text{Im } Q_-(iy).
\end{aligned} \tag{4.2}$$

Hence $Q_\alpha \in \mathbf{N}_1$, since $Q_{+\beta} \in \mathbf{N}_1$ and $Q_- \in \mathbf{N}_1$. If $Q_+ \in \mathbf{N}_1$, the statement of the theorem can be shown in a similar way. □

Theorem 4.3. *Assume that* $Q \in \mathbf{N} \setminus \mathbf{N}_1$ *is of the form* $Q = Q_s + Q_p$ *such that* $Q_p \in \mathbf{N}_p \cup \mathbf{N}_1$ *and that for some* $x_q \in \mathbb{R}$ *the shifted function* $\hat{Q}_s(z) := Q_s(z - x_q)$ *belongs to* $\mathbf{N}_s$. *Let* Q_α *be given by the identity* (2.1). *Then:*

(i) *if* $Q_p \in \mathbf{N}_1$, *there is at most one value* $\alpha_0 \in (-\pi/2, \pi/2)$ *such that* $Q_\alpha \in \mathbf{N}_{1+}$ *for each* $\alpha \in (-\pi/2, \pi/2) \setminus \{\alpha_0\}$;

(ii) *if* $Q_p \in \mathbf{N}_p$, *then* $Q_\alpha \in \mathbf{N}_{1+}$ *for each* $\alpha \in (-\pi/2, \pi/2)$.

Proof. Let $\hat{Q}(z) := Q(z - x_q)$ and $\hat{Q}_p(z) := Q_p(z - x_q)$. Then also $\hat{Q}_p \in \mathbf{N}_p \cup \mathbf{N}_1$. Let $\hat{Q}_\alpha$ be given by (2.1) with $\hat{Q}$ instead of Q. Since $\hat{Q}_s \in \mathbf{N}_s$, one has $\mathrm{Re}\ \hat{Q}_s(iy) = 0$ and the identity (2.2) gives

$$\begin{aligned}\mathrm{Im}\ \hat{Q}_\alpha(iy) &= \frac{(1+\tan^2\alpha)\mathrm{Im}\ \hat{Q}(iy)}{(\mathrm{Im}\ \hat{Q}(iy))^2 + (\tan\alpha + \mathrm{Re}\ \hat{Q}_p(iy))^2} \\ &\le \frac{1+\tan^2\alpha}{2|\tan\alpha + \mathrm{Re}\ \hat{Q}_p(iy)|}.\end{aligned} \tag{4.3}$$

(i) Assume $Q_p \in \mathbf{N}_1$ or equivalently that $\hat{Q}_p \in \mathbf{N}_1$. Let

$$a_1 := \lim_{y \to +\infty} \mathrm{Re}\ \hat{Q}_p(iy),$$

and let $\alpha_0 \in (-\pi/2, \pi/2)$ be the solution of $\tan\alpha_0 + a_1 = 0$. For $\alpha \neq \alpha_0$ it follows from (4.3) that $\mathrm{Im}\ \hat{Q}_\alpha$ is uniformly bounded on the interval $(i, i\infty)$, hence $\hat{Q}_\alpha \in \mathbf{N}_{1+}$.

(ii) Assume that $Q_p \in \mathbf{N}_p$ or equivalently $\hat{Q}_p \in \mathbf{N}_p$. Then Lemma 4.1 shows that

$$\lim_{y \to +\infty} |\mathrm{Re}\ \hat{Q}_p(iy)| = +\infty,$$

and it follows from (4.3) that $\mathrm{Im}\ \hat{Q}_\alpha$ is uniformly bounded on the interval $(i, i\infty)$. Hence $\hat{Q}_\alpha \in \mathbf{N}_{1+}$ for each $\alpha \in (-\pi/2, \pi/2)$.

Finally, observe that $Q_\alpha(z - x_q) = \hat{Q}_\alpha$ and therefore $Q_\alpha \in \mathbf{N}_\gamma$ if and only if $\hat{Q}_\alpha \in \mathbf{N}_\gamma$. This completes the proof. □

In particular, if $Q \in \mathbf{N} \setminus \mathbf{N_1}$ is symmetric, i.e., if $Q \in \mathbf{N_s}$ then assertion (i) of Theorem 4.3 holds. However, it should be observed that there are functions in $\mathbf{N} \setminus \mathbf{N}_1$ which do not serve as exceptional functions for a family of Nevanlinna functions in $\mathbf{N}_{1+}$ of the form (2.1). Consider, for example, $Q = i$. Then $Q \in \mathbf{N_s}$ and it is easy to see that $Q \in \mathbf{N}_{1+} \setminus \mathbf{N}_1$. Since $Q_\alpha = i$ for each $\alpha \in (-\pi/2, \pi/2)$, there is no exceptional extension.

5. The case that $Q \in \mathbf{N} \setminus \widehat{\mathbf{N}}_1$

In this section some functions in $\mathbf{N} \setminus \widehat{\mathbf{N}}_1$ will be identified as exceptional functions for a family of Nevanlinna functions of the form (2.1). Again, a first observation concerns the limiting behavior of functions in $\mathbf{N} \setminus \widehat{\mathbf{N}}_1$.

Lemma 5.1. *Let* $Q \in \mathbf{N}$. *If* $Q_+ \in \mathbf{N} \setminus \widehat{\mathbf{N}}_1$ *then* $\lim_{y \to +0} \operatorname{Re} Q_+(iy) = +\infty$, *and if* $Q_- \in \mathbf{N} \setminus \widehat{\mathbf{N}}_1$ *then* $\lim_{y \to +0} \operatorname{Re} Q_-(iy) = -\infty$.

Proof. Assume that $Q_+ \in \mathbf{N} \setminus \widehat{\mathbf{N}}_1$. Choose $K > 0$ and a corresponding real l with $0 < l < 1/\sqrt{2}$ such that $\int_l^1 d\sigma(\lambda)/\lambda \geq K$. If $0 \leq y \leq l$ then using

$$\frac{\lambda}{\lambda^2 + y^2} \geq \frac{1}{2\lambda}, \quad \lambda \geq l,$$

one obtains the estimate

$$(1 - y^2) \int_0^\infty \frac{\lambda}{(\lambda^2 + y^2)(1 + \lambda^2)} \, d\sigma(\lambda) \geq \int_l^1 \frac{d\sigma(\lambda)}{8\lambda} \geq \frac{K}{8}.$$

This together with (2.5) gives

$$\operatorname{Re} Q_+(iy) \geq a + \frac{K}{8}, \quad 0 \leq y \leq l.$$

Hence $\lim_{y \to +0} \operatorname{Re} Q_+(iy) = +\infty$, since K can be chosen arbitrary large. In case $Q_- \in \mathbf{N} \setminus \widehat{\mathbf{N}}_1$, the statement of the lemma can be shown in a similar way. □

Of course, the result in Lemma 5.1 can be derived also from Lemma 4.1 by means of the transform $\tilde{Q}(z) = -Q(1/z)$, which connects the classes $\mathbf{N}_1$ and $\widehat{\mathbf{N}}_1$, see [3].

Theorem 5.2. *Let* $Q \in \widehat{\mathbf{N}}_p$ *and let* Q_α *be given by* (2.1). *Then* $Q_\alpha \in \widehat{\mathbf{N}}_1$ *for each* $\alpha \in (-\pi/2, \pi/2)$.

Proof. Assume that $Q_- \in \widehat{\mathbf{N}}_1$ and $Q \in \mathbf{N} \setminus \widehat{\mathbf{N}}_1$. Let $Q_{+\alpha}$ be given by (4.1). For $-1 < x < 0$ one has

$$Q_+(x) - Q_+(-1) = \int_{[0,\infty)} \left(\frac{1}{\lambda - x} - \frac{1}{\lambda + 1} \right) d\sigma(\lambda) > 0.$$

Then the Nevanlinna function $Q_1(z) := -(Q_+(z) - Q_+(-1))^{-1}$ is nonpositive on $(-1, 0)$. Since the function Q_1 is nonnegative on $(-\infty, -1)$, the support of the corresponding spectral measure σ_1 on the interval $(-\infty, 0)$ must be concentrated at the point -1. Therefore Q_1 has a representation of the form

$$Q_1(z) = bz + a - \frac{\sigma_1(\{-1\})}{1 + z} + \int_0^\infty \left(\frac{1}{\lambda - z} - \frac{\lambda}{1 + \lambda^2} \right) d\sigma_1(\lambda).$$

For each $\lambda \geq 0$ the function $h_\lambda(x) := (\lambda - x)^{-1}$ is increasing on $(-1, 0)$. By monotone convergence one concludes

$$\lim_{x \to 0-} Q_1(x) = a - \sigma_1(\{-1\}) + \int_0^\infty \frac{d\sigma_1(\lambda)}{\lambda(1 + \lambda^2)} \leq 0.$$

Hence $Q_1 \in \widehat{\mathbf{N}}_1$. Let $\tan \alpha_1 := -Q_+(-1)$. It follows that $Q_{\alpha_1 +} \in \widehat{\mathbf{N}}_1$, which implies by Proposition 2.3 that $Q_{\alpha +} \in \widehat{\mathbf{N}}_1$ for each $\alpha \in (-\pi/2, \pi/2)$. Let $c :=$

$\lim_{y\to+0}\mathrm{Re}\,Q_-(iy)$ and let $\tan\beta := \tan\alpha + c + 1$ for $\alpha \in (-\pi/2, \pi/2)$. Since $\lim_{y\to+0}\mathrm{Re}\,Q_+(iy) = +\infty$ by Lemma 5.1, it is seen as in (4.2) that

$$\frac{\mathrm{Im}\,Q_\alpha(iy)}{1+\tan^2\alpha} < \frac{\mathrm{Im}\,Q_{+\beta}(iy)}{1+\tan^2\beta} + \mathrm{Im}\,Q_-(iy)$$

for $\alpha \in (-\pi/2, \pi/2)$ and sufficiently small y. It follows that $Q_\alpha \in \widehat{\mathbf{N}}_1$, since $Q_{+\beta} \in \widehat{\mathbf{N}}_1$ and $Q_- \in \widehat{\mathbf{N}}_1$. Again if $Q_+ \in \widehat{\mathbf{N}}_1$, the statement of the theorem can be shown in a similar way. □

Theorem 5.3. *Assume that $Q \in \mathbf{N} \setminus \widehat{\mathbf{N}}_1$ is of the form $Q = Q_s + Q_p$ such that $Q_s \in \mathbf{N}_s$ and $Q_p \in \widehat{\mathbf{N}}_p \cup \widehat{\mathbf{N}}_1$. Let Q_α be given by the identity* (2.1). *Then:*

(i) *if $Q_p \in \widehat{\mathbf{N}}_1$, then there is at most one value $\alpha_0 \in (-\pi/2, \pi/2)$ such that $Q_\alpha \in \widehat{\mathbf{N}}_{1+}$ for each $\alpha \in (-\pi/2, \pi/2) \setminus \{\alpha_0\}$;*

(ii) *if $Q_p \in \widehat{\mathbf{N}}_p$, then $Q_\alpha \in \widehat{\mathbf{N}}_{1+}$ for each $\alpha \in (-\pi/2, \pi/2)$.*

Proof. Since $\mathrm{Re}\,Q_s(iy) = 0$, the estimate (4.3) still holds:

$$\mathrm{Im}\,\hat{Q}_\alpha(iy) \le \frac{1+\tan^2\alpha}{2|\tan\alpha + \mathrm{Re}\,\hat{Q}_p(iy)|}. \tag{5.1}$$

Hence, if $Q_p \in \widehat{\mathbf{N}}_1$, $\hat{a}_1 := \lim_{y\to+0}\mathrm{Re}\,Q_p(iy)$, and $\alpha_0 \in (-\pi/2, \pi/2)$ is the solution of $\tan\alpha_0 + \hat{a}_1 = 0$, then with $\alpha \neq \alpha_0$ (5.1) implies that $\mathrm{Im}\,Q_\alpha$ is uniformly bounded on the interval $(0, i)$, so that $Q_\alpha \in \widehat{\mathbf{N}}_{1+}$. On the other hand, if $Q_p \in \widehat{\mathbf{N}}_p$ then Lemma 5.1 implies that $\lim_{y\to+0}|\mathrm{Re}\,Q_p(iy)| = +\infty$. Now $\mathrm{Im}\,\hat{Q}_\alpha$ is uniformly bounded on the interval $(0, i)$ and $Q_\alpha \in \widehat{\mathbf{N}}_{1+}$ for each $\alpha \in (-\pi/2, \pi/2)$ by (5.1). □

Here again one observes that assertion (i) of Theorem 5.3 holds for symmetric functions in $\mathbf{N} \setminus \widehat{\mathbf{N}}_1$. Moreover, there are functions in $\mathbf{N} \setminus \widehat{\mathbf{N}}_1$ which do not serve as exceptional functions for a family of Nevanlinna functions in $\widehat{\mathbf{N}}_{1+}$ of the form (2.1).

References

[1] S. Hassi, M. Kaltenbäck, and H.S.V. de Snoo, "Triplets of Hilbert spaces and Friedrichs extensions associated with the subclass $\mathbf{N}_1$ of Nevanlinna functions", J. Operator Theory, 37 (1997), 155–181.

[2] S. Hassi, M. Kaltenbäck, and H.S.V. de Snoo, "A characterization of semibounded selfadjoint operators", Proc. Amer. Math. Soc., 125 (1997), 2681–2692.

[3] S. Hassi, M. Kaltenbäck, and H.S.V. de Snoo, "Generalized Kreĭn-von Neumann extensions and associated operator models", Acta Sci. Math. (Szeged), 64 (1998), 627–655.

[4] S. Hassi, H. Langer, and H.S.V. de Snoo, "Selfadjoint extensions for a class of symmetric operators with defect numbers (1,1)", 15th OT Conference Proceedings, (1995), 115–145.

[5] S. Hassi, M. Möller, and H.S.V. de Snoo, "A class of Nevanlinna functions associated with Sturm-Liouville operators", Proc. Amer. Math. Soc., 134 (2006), 2885–2893.

[6] S. Hassi, H.S.V. de Snoo, and H. Winkler, "Boundary-value problems for two-dimensional canonical systems", Integral Equations Operator Theory, 36 (2000), 445–479.

[7] I.S. Kac and M.G. Kreĭn, "R-functions-analytic functions mapping the upper half-plane into itself", Amer. Math. Soc. Transl. (2), 103 (1974), 1–18.

[8] H. Winkler, "Spectral estimations for canonical systems", Math. Nachr., 220 (2000), 115–141.

[9] H. Winkler, "On generalized Friedrichs and Kreĭn-von Neumann extensions and canonical systems", Math. Nachr., 236 (2002), 175–191.

Seppo Hassi
Department of Mathematics and Statistics
University of Vaasa
P.O. Box 700
65101 Vaasa, Finland
e-mail: sha@uwasa.fi

Henk de Snoo
Department of Mathematics and Computing Science
University of Groningen
P.O. Box 800
9700 AV Groningen, Nederland
e-mail: desnoo@math.rug.nl

Henrik Winkler
Institut für Mathematik, MA 6-4
Technische Universität Berlin
Strasse des 17. Juni 136
D-10623 Berlin, Germany
e-mail: winkler@math.tu-berlin.de

Operator Theory:
Advances and Applications, Vol. 175, 121–158

On the Spectrum of the Self-adjoint Extensions of a Nonnegative Linear Relation of Defect One in a Krein Space

P. Jonas and H. Langer

Abstract. A nonnegative symmetric linear relation A_0 with defect one in a Krein space $\mathcal{H}$ has self-adjoint extensions which are not nonnegative. If the resolvent set of such an extension A is not empty, A has a so-called exceptional eigenvalue α. For $\alpha \neq 0, \infty$ this means that α is an eigenvalue in the open upper half-plane, or a positive eigenvalue with a nonpositive eigenvector, or a negative eigenvalue with a nonnegative eigenvector. In this paper we study these exceptional eigenvalues and their dependence on a parameter if the self-adjoint extensions of A_0 are parametrized according to M. G. Krein's resolvent formula. An essential tool is a family of generalized Nevanlinna functions of the class $\mathcal{N}_1$ and their zeros or generalized zeros of nonpositive type.

Mathematics Subject Classification (2000). Primary 47B50; Secondary 47A20, 47A55.

Keywords. Linear relations in Krein spaces, nonnegative operators, operators with one negative square, selfadjoint extensions, generalized Nevanlinna functions.

1. Introduction

It is well known that a nonnegative symmetric operator A_0 in a Hilbert space $\mathcal{H}$ with defect one has self-adjoint extensions in $\mathcal{H}$. They can be described, e.g., by Krein's formula by means of a real parameter $\gamma \in \overline{\mathbb{R}}$ $(= \mathbb{R} \cup \{\infty\})$. If a suitable such parametrization is fixed and the self-adjoint extensions of A_0 are denoted by $A_{(\gamma)}$, $\gamma \in \overline{\mathbb{R}}$, then there exists a nonempty open interval $(\gamma_-, \gamma_+) \subset \mathbb{R}$, such that for $\gamma \in (\gamma_-, \gamma_+)$ the symmetric form $\big(A_{(\gamma)}\cdot, \cdot\big)$ on $\operatorname{dom} A_{(\gamma)}$ has one negative square or, equivalently, the operator $A_{(\gamma)}$ has one negative eigenvalue, and for $\gamma \in \overline{\mathbb{R}} \backslash (\gamma_-, \gamma_+)$

The first author was supported by the Hochschul- und Wissenschaftsprogramm des Bundes und der Länder of Germany.

the operator $A_{(\gamma)}$ is nonnegative. If for $\gamma \in (\gamma_-, \gamma_+)$ the negative eigenvalue of $A_{(\gamma)}$ is denoted by $\alpha(\gamma)$, then $\alpha(\gamma)$ is a strictly monotonous function of γ on (γ_-, γ_+) and tends to zero or $-\infty$ if γ tends to the boundary points γ_- or γ_+.

We shall illustrate this for the simple example of the nonnegative symmetric operator A_0 in $L^2(0,1)$ generated by $-\dfrac{d^2}{dx^2}$ and the boundary conditions $y(0) = y'(0) = 0$, $y(1) = 0$. Its self-adjoint extensions $A_{(\gamma)}$ are given by the same differential expression, and the boundary conditions $y(1) = 0$, $y(0) - \gamma y'(0) = 0$ with $\gamma \in \overline{\mathbb{R}}$; for $\gamma = \infty$ this boundary condition reads as $y'(0) = 0$. Then the extension $A_{(\gamma)}$ has a negative eigenvalue $\alpha(\gamma)$ if and only if $\gamma \in (-1, 0)$, and $\alpha(\gamma)$ is a strictly decreasing function on $(-1, 0)$.

Now let A be a self-adjoint relation in some Krein space $\big(\mathcal{H}, [\cdot,\cdot]\big)$ with the property that the Hermitian sesquilinear form $[\cdot,\cdot]_A$ on A, defined by

$$\left[\begin{pmatrix} f \\ f' \end{pmatrix}, \begin{pmatrix} g \\ g' \end{pmatrix}\right]_A := [f', g], \quad \begin{pmatrix} f \\ f' \end{pmatrix}, \begin{pmatrix} g \\ g' \end{pmatrix} \in A, \tag{1.1}$$

has one negative square on A. Under some regularity assumptions this is equivalent to the fact that the self-adjoint relation A has exactly one so-called exceptional eigenvalue α; this is an eigenvalue of A in the open upper half-plane $\mathbb{C}^+$ or a negative eigenvalue with a nonnegative eigenelement or a positive eigenvalue with a nonpositive eigenelement, or $\alpha \in \{0, \infty\}$ and in this case to α there corresponds a neutral eigenelement with an 'approximate associated vector' with some sign property (see (3.3)).

The main object of our studies is a nonnegative symmetric relation A_0 of defect one in some Krein space $\mathcal{H}$ and its self-adjoint extensions in $\mathcal{H}$. We make use of the parametrization $A_{(\gamma)}$ of the self-adjoint extensions of A_0 by a parameter $\gamma \in \overline{\mathbb{R}}$ considered in [JL3]. In this paper there was found an interval (γ_-, γ_+) with the property that γ belongs to this interval if and only if the Hermitian sesquilinear form $[\,\cdot\,,\,\cdot\,]_{A_{(\gamma)}}$ has one negative square. Under an additional assumption all extensions have non-empty resolvent sets and then, according to what was said above, for $\gamma \in (\gamma_-, \gamma_+)$ the self-adjoint relation $A_{(\gamma)}$ has one exceptional eigenvalue $\alpha(\gamma)$. The main results of this paper concern the dependence of this exceptional eigenvalue $\alpha(\gamma)$ on γ: we show that the function $\gamma \mapsto \alpha(\gamma)$ is continuous and that there exist two more numbers γ_0, γ_∞ such that

$$\gamma_- \le \gamma_0 \le \gamma_\infty \le \gamma_+, \tag{1.2}$$

and $\alpha(\gamma) = 0$ for $\gamma \in (\gamma_-, \gamma_0)$, $\alpha(\gamma) = \infty$ for $\gamma \in (\gamma_\infty, \gamma_+)$, and that on the interval $(\gamma_0, \gamma_\infty)$ the function $\gamma \mapsto |\alpha(\gamma)|$ is nondecreasing and there is no subinterval of $(\gamma_0, \gamma_\infty)$ on which the function $\gamma \mapsto \alpha(\gamma)$ is constant. We also describe the root subspace of $A_{(\gamma)}$ at a real exceptional eigenvalue $\neq 0, \infty$.

Our study of the exceptional eigenvalues of the operators $A_{(\gamma)}$ is based on the study of the zeros of nonpositive type and their dependence on γ of the generalized Nevanlinna functions

$$F_\gamma(z) = \gamma z + G_1(z) + z^2 G_2(z), \quad \gamma \in \mathbb{R}, \tag{1.3}$$

where G_1 and G_2 are Nevanlinna functions. These results may be of independent interest.

For particular cases of families of generalized Nevanlinna functions the same questions were studied in [JL2] and [DaL2]. In the first mentioned paper the results were applied to the description of the eigenvalues of nonpositive type of a family of self-adjoint operators in Pontryagin spaces with negative index one. In [DaL2] a periodic Sturm-Liouville problem with an indefinite weight was studied. There a family of self-adjoint operators $A_{(\gamma)}$ in some Krein space as above arises from a self-adjoint boundary condition, which depends on some parameter, and the dependence of the exceptional eigenvalue on this parameter was described in the same way as in the more general abstract framework of the present paper.

A brief synopsis is as follows. In the next section we consider the family of functions F_γ from (1.3). In Section 3 the exceptional eigenvalue of a self-adjoint relation A with the property that the Hermitian sesquilinear form $[\cdot,\cdot]_A$ has one negative square is characterized in different ways. Section 4 contains the main result about the existence of the four numbers γ_-, γ_0, γ_∞, γ_+, the monotonicity of the modulus of the exceptional eigenvalue $\alpha(\gamma)$ on the interval (γ_0,γ_∞), and the further properties of the curve $\gamma\mapsto\alpha(\gamma)$ as mentioned above. Finally, we describe the root subspace of $A_{(\gamma)}$ at a real exceptional eigenvalue, and we give some criteria for the coincidence of the numbers γ_- and γ_0 as well as of γ_∞ and γ_+.

2. A family of holomorphic functions

For $\kappa\in\mathbb{N}_0$ we denote by $\mathcal{N}_\kappa$ the set of all complex functions Q which are meromorphic in $\mathbb{C}\setminus\mathbb{R}$, such that $Q(\overline{z})=\overline{Q(z)}$, $z\in\operatorname{hol}Q$ (here $\operatorname{hol}Q$ denotes the domain of holomorphy of Q), and for which the kernel

$$\frac{Q(z)-\overline{Q(\zeta)}}{z-\overline{\zeta}},\qquad z,\zeta\in\operatorname{hol}Q,\ z\neq\overline{\zeta},$$

has κ negative squares. Now let Q be a function of the class $\mathcal{N}_1$. Recall that $z_0\in\overline{\mathbb{R}}:=\mathbb{R}\cup\{\infty\}$ is called a *generalized zero* (*generalized pole*, respectively) *of nonpositive type of* Q if for each neighborhood $\mathcal{U}$ of z_0 in $\overline{\mathbb{C}}$ there exists a $\delta_{\mathcal{U}}>0$ such that for $0<\nu<\delta_{\mathcal{U}}$ ($\delta_{\mathcal{U}}<\nu<\infty$, respectively) the equation $Q(z)=-i\nu$ has a solution $z\in\mathcal{U}\cap\mathbb{C}^+$. If the function Q is holomorphic in a deleted neighborhood of its generalized zero (generalized pole, respectively) z_0 of nonpositive type, then z_0 is a zero (pole, respectively) of Q. Every function $Q\in\mathcal{N}_1$ has either a generalized zero of nonpositive type in $\overline{\mathbb{R}}$ or a simple zero in the open upper half-plane $\mathbb{C}^+$, see [KL]. This point, which is uniquely determined, is denoted by ζ_Q. Similarly, $Q\in\mathcal{N}_1$ has either a generalized pole of nonpositive type in $\overline{\mathbb{R}}$ or a simple pole in the open upper half-plane $\mathbb{C}^+$. This point, which is also uniquely determined, is denoted by $\widehat{\zeta}_Q$; clearly, $\widehat{\zeta}_Q=\zeta_{-Q^{-1}}$. We mention that if $\widehat{\zeta}_Q\in\overline{\mathbb{R}}$ then it coincides with the uniquely determined point $\widehat{\zeta}\in\overline{\mathbb{R}}$ for which there exists a sequence $(\zeta_n)\subset\mathbb{C}^+$ converging to $\widehat{\zeta}$ in $\overline{\mathbb{C}}$ such that $\liminf_{n\to\infty}\operatorname{Im}Q(\zeta_n)<0$. This is a direct consequence of the definition of $\widehat{\zeta}_Q$ mentioned above and well-known decomposition properties of functions of the class $\mathcal{N}_1$.

By $\mathcal{N}_1^\infty$ we denote the class of all functions $Q \in \mathcal{N}_1$ such that $\widehat{\zeta}_Q = \infty$. This class will play a special role in this section. It was studied (with one replaced by any natural number) also in the papers [DLShZ], [DLSh], [DHdS].

The first lemma is an easy consequence of results from [DaL1] and [L2], therefore the proof is omitted.

Lemma 2.1. *Let the function Q be holomorphic at zero. Then $Q \in \mathcal{N}_1^\infty$ if and only if it has a representation of the form*

$$Q(z) = z^{2\rho} \int_{\mathbb{R}} \frac{1+tz}{t-z}\, d\sigma(t) + \sum_{\nu=0}^{\ell} a_\nu z^\nu, \tag{2.1}$$

where $\rho \in \{0,1\}$, σ is a bounded positive measure on $\mathbb{R}$ with $0 \notin \operatorname{supp}\sigma$ and $\int_{\mathbb{R}} t^2\, d\sigma(t) = \infty$ if $\rho = 1$, $\ell \in \{0,1,2,3\}$, and $a_\nu \in \mathbb{R}$, $\nu = 0,\dots,\ell$, $a_\ell \neq 0$ if $\ell > 0$, such that one of the following conditions is fulfilled:

(a) $\rho = 0$, $\ell = 1$, $a_1 < 0$; (b) $\rho = 0$, $\ell = 2$;

(c) $\rho \in \{0,1\}$, $\ell = 3$, $a_3 > 0$; (d) $\rho = 1$, $\ell \in \{0,1,2\}$.

The representation of a function $Q \in \mathcal{N}_1^\infty$ in the form (2.1) is unique.

In the following, let G_1, $G_2 \in \mathcal{N}_0$, and for $j = 1,2$, we write G_j in the form

$$G_j(z) = \alpha_j + \beta_j z + \int_{\mathbb{R}} \frac{1+tz}{t-z}\, d\sigma_j(t), \tag{2.2}$$

where $\alpha_j \in \mathbb{R}$, $\beta_j \geq 0$, and σ_j is a bounded positive measure on $\mathbb{R}$ (cf., e.g., [KK]).

Theorem 2.2. *The union $\mathcal{N}_0 \cup \mathcal{N}_1^\infty$ is the set of all functions F of the form*

$$F(z) = \gamma z + G_1(z) + z^2 G_2(z) \text{ with } \gamma \in \mathbb{R},\ G_1, G_2 \in \mathcal{N}_0. \tag{2.3}$$

This function F belongs to the class $\mathcal{N}_0$ if and only if

$$\beta_2 = 0,\ \int_{\mathbb{R}} (1+t^2) d\sigma_2(t) < \infty,\ \alpha_2 = \int_{\mathbb{R}} t\, d\sigma_2(t),\ \gamma \geq \int_{\mathbb{R}} (1+t^2)\, d\sigma_2(t) - \beta_1.$$

Proof. Denote the set of functions F of the form (2.3) by $\widetilde{\mathcal{N}}$. Evidently, $\mathcal{N}_0 \subset \widetilde{\mathcal{N}}$. Every function $F \in \mathcal{N}_1^\infty$ can be written as a sum $F = F_0 + F_1$ where $F_0 \in \mathcal{N}_0$, $F_1 \in \mathcal{N}_1^\infty$, and F_1 is holomorphic at zero ([L2, §2]). By Lemma 2.1, the function F_1 belongs to $\widetilde{\mathcal{N}}$, hence $\mathcal{N}_0 \cup \mathcal{N}_1^\infty \subset \widetilde{\mathcal{N}}$.

Now let $F \in \widetilde{\mathcal{N}}$, i.e., $F(z) = \gamma z + G_1(z) + z^2 G_2(z)$ where $\gamma \in \mathbb{R}$ and G_1, $G_2 \in \mathcal{N}_0$ with integral representations (2.2). Choose $\Delta_0 := (-1,1)$, $\Delta_\infty := \mathbb{R} \setminus \Delta_0$ and define for $j = 1,2$

$$G_{j,0}(z) := \alpha_j + \int_{\Delta_0} \frac{1+tz}{t-z}\, d\sigma_j(t), \quad G_{j,\infty}(z) := \beta_j z + \int_{\Delta_\infty} \frac{1+tz}{t-z}\, d\sigma_j(t).$$

Then

$$\begin{aligned} z^2G_{2,0}(z) &= -\int_{\Delta_0} t\,d\sigma_2(t) - z\int_{\Delta_0}(1+t^2)\,d\sigma_2(t) \\ &\quad + z^2\left(\alpha_2 - \int_{\Delta_0} t\,d\sigma_2(t)\right) + \int_{\Delta_0}\frac{1+tz}{t-z}\,t^2\,d\sigma_2(t) \end{aligned}$$

and

$$\begin{aligned} F(z) &= \gamma z + G_{1,0}(z) + z^2G_{2,0}(z) + G_{1,\infty}(z) + z^2G_{2,\infty}(z) \\ &= -\int_{\Delta_0} t\,d\sigma_2(t) + z\left(\gamma - \int_{\Delta_0}(1+t^2)\,d\sigma_2(t)\right) + z^2\left(\alpha_2 - \int_{\Delta_0} t\,d\sigma_2(t)\right) \\ &\quad + G_{1,0}(z) + \int_{\Delta_0}\frac{1+tz}{t-z}\,t^2\,d\sigma_2(t) + G_{1,\infty}(z) + z^2\,G_{2,\infty}(z). \end{aligned}$$

Hence F belongs to $\mathcal{N}_0$ or $\mathcal{N}_1^\infty$ if and only if the following function F_∞ belongs to $\mathcal{N}_0$ or $\mathcal{N}_1^\infty$, respectively:

$$\begin{aligned} F_\infty(z) :&= z\left(\gamma - \int_{\Delta_0}(1+t^2)\,d\sigma_2(t)\right) + z^2\left(\alpha_2 - \int_{\Delta_0} t\,d\sigma_2(t)\right) \\ &\quad + G_{1,\infty}(z) + z^2G_{2,\infty}(z) \\ &= z\left(\gamma + \beta_1 - \int_{\Delta_0}(1+t^2)\,d\sigma_2(t)\right) + z^2\left(\alpha_2 - \int_{\Delta_0} t\,d\sigma_2(t)\right) \\ &\quad + z^3\,\beta_2 + \int_{\Delta_\infty}\frac{1+tz}{t-z}\,d\sigma_1(t) + z^2\int_{\Delta_\infty}\frac{1+tz}{t-z}\,d\sigma_2(t). \end{aligned} \tag{2.4}$$

If $\int_{\mathbb{R}}(1+t^2)d\sigma_2(t) = \infty$ then we replace the second last integral in (2.4):

$$\begin{aligned} \int_{\Delta_\infty}\frac{1+tz}{t-z}d\sigma_1(t) &= \int_{\Delta_\infty} t^{-1}\,d\sigma_1(t) + z\int_{\Delta_\infty} t^{-2}(1+t^2)d\sigma_1(t) \\ &\quad + z^2\int_{\Delta_\infty} t^{-1}\,d\sigma_1(t) + z^2\int_{\Delta_\infty}\frac{1+tz}{t-z}t^{-2}d\sigma_1(t). \end{aligned}$$

It follows that F_∞ is of the form (2.1) with $\rho = 1$ and $a_3 \geq 0$, therefore $F_\infty \in \mathcal{N}_1^\infty$.
If $\int_{\mathbb{R}}(1+t^2)\,d\sigma_2(t) < \infty$, then

$$\begin{aligned} z^2\int_{\Delta_\infty}\frac{1+tz}{t-z}\,d\sigma_2(t) &= -\int_{\Delta_\infty} t\,d\sigma_2(t) - z\int_{\Delta_\infty}(1+t^2)\,d\sigma_2(t) \\ &\quad - z^2\int_{\Delta_\infty} t\,d\sigma_2(t) + \int_{\Delta_\infty}\frac{t^2(1+tz)}{t-z}\,d\sigma_2(t). \end{aligned}$$

Inserting this expression for the last integral into (2.4) we see that F_∞ has the form (2.1) with $\rho = 0$. Then, by Lemma 2.1, $F_\infty \in \mathcal{N}_1^\infty$ if $\beta_2 \neq 0$.

It remains to consider the case $\beta_2 = 0$. Then $F_\infty \in \mathcal{N}_1^\infty$ if $\alpha_2 \neq \int_{\mathbb{R}} t\, d\sigma_2(t)$. If $\beta_2 = 0$ and $\alpha_2 = \int_{\mathbb{R}} t\, d\sigma_2(t)$ then $F_\infty \in \mathcal{N}_1^\infty$ if $\gamma + \beta_1 - \int_{\mathbb{R}} (1+t^2) d\sigma_2(t) < 0$. If $\beta_2 = 0$, $\alpha_2 = \int_{\mathbb{R}} t\, d\sigma_2(t)$ and $\gamma + \beta_1 - \int_{\mathbb{R}} (1+t^2)\, d\sigma_2(t) \geq 0$, then $F_\infty \in \mathcal{N}_0$. □

For the rest of this section we fix two functions G_1, $G_2 \in \mathcal{N}_0$ and consider the family of functions F_γ, $\gamma \in \mathbb{R}$, as on the right-hand side of (2.3):

$$F_\gamma(z) = \gamma z + G_1(z) + z^2 G_2(z).$$

Define

$$\gamma_\infty := \begin{cases} \int_{\mathbb{R}} (1+t^2)\, d\sigma_2(t) - \beta_1 & \text{if } \int_{\mathbb{R}} (1+t^2)\, d\sigma_2(t) < \infty,\ \beta_2 = 0, \\ & \qquad \text{and } \alpha_2 = \int_{\mathbb{R}} t\, d\sigma_2(t), \\ \infty & \text{otherwise.} \end{cases}$$

Then it follows that

$$F_\gamma \in \begin{cases} \mathcal{N}_1 & \text{if } \gamma \in (-\infty, \gamma_\infty), \\ \mathcal{N}_0 & \text{if } \gamma \in [\gamma_\infty, \infty). \end{cases}$$

If $\gamma \in (-\infty, \gamma_\infty)$ we denote the zero of nonpositive type of F_γ in $\overline{\mathbb{R}} \cup \mathbb{C}^+$ by $\zeta(\gamma) : \zeta(\gamma) := \zeta_{F_\gamma}$. Observe that $\gamma_\infty < \infty$ holds if and only if

$$\int_{\mathbb{R}} (1+t^2)\, d\sigma_2(t) < \infty \quad \text{and} \quad G_2(z) = \int_{\mathbb{R}} \frac{1+t^2}{t-z}\, d\sigma_2(t),$$

which in the notation of [KK, 4] means that G_2 belongs to the class (R_0).

In the following, if σ is a measure with $\sigma(\{\omega\}) > 0$ for some $\omega \in \mathbb{R}$ then we set

$$\int_{\mathbb{R}} (t-\omega)^{-2}\, d\sigma(t) = \infty.$$

Lemma 2.3. *If $\gamma \in (-\infty, \gamma_\infty)$, then $\zeta(\gamma) \neq \infty$ and the following statements hold:*

(i) $\zeta(\gamma) = 0$ *if and only if*

$$\int_{\mathbb{R}} t^{-2}(1+t^2)\, d\sigma_1(t) < \infty, \quad \gamma \leq \sigma_2(\{0\}) - \beta_1 - \int_{\mathbb{R}} t^{-2}(1+t^2)\, d\sigma_1(t),$$

$$\alpha_1 = -\int_{\mathbb{R}} t^{-1}\, d\sigma_1(t). \tag{2.5}$$

(ii) $\zeta(\gamma) = \omega \in \mathbb{R} \setminus \{0\}$ *if and only if*

$$\int_{\mathbb{R}} |t-\omega|^{-2}\, d\sigma_1(t) < \infty, \qquad \int_{\mathbb{R}} |t-\omega|^{-2}\, d\sigma_2(t) < \infty \tag{2.6}$$

and

$$\begin{aligned}\gamma\omega + \alpha_1 + \beta_1\omega &+ \int_{\mathbb{R}} \frac{1+t\omega}{t-\omega}\, d\sigma_1(t) \\ &+ \omega^2\left(\alpha_2 + \beta_2\omega + \int_{\mathbb{R}} \frac{1+t\omega}{t-\omega}\, d\sigma_2(t)\right) = 0,\end{aligned} \tag{2.7}$$

$$\begin{aligned}\gamma + \beta_1 + 2\omega\alpha_2 &+ 3\omega^2\beta_2 + \textstyle\int_{\mathbb{R}} |t-\omega|^{-2}\big(1+t^2\big)\, d\sigma_1(t) \\ &+ \int_{\mathbb{R}} \frac{2t\omega - 2t\omega^3 - \omega^2 + 3t^2\omega^2}{(t-\omega)^2}\, d\sigma_2(t) \le 0.\end{aligned} \tag{2.8}$$

The inequality (2.8) means that $F'_\gamma(\omega) \le 0$, where the derivative is to be understood in the Carathéodory sense; the relation (2.7) means that $F_\gamma(\omega) = 0$ if $F_\gamma(\omega)$ is the nontangential boundary value of F_γ at ω.

Proof. By [L2, Corollary 2.1] the point ∞ cannot be a zero of nonpositive type of F_γ. It follows from [L2, Remark 3.2 and proof of Theorem 3.1] that $\zeta(\gamma) = \omega \in \mathbb{R}$ is a generalized zero of nonpositive type of F_γ if and only if

$$\lim_{\eta \downarrow 0} \eta^{-1} \operatorname{Im} F_\gamma(\omega + i\eta) \text{ exists and is nonpositive} \tag{2.9}$$

and

$$\lim_{\eta \downarrow 0} \eta^{-1} \operatorname{Re} F_\gamma(\omega + i\eta) = 0. \tag{2.10}$$

For arbitrary $z = \omega + i\eta$ we find

$$\begin{aligned}\eta^{-1} \operatorname{Im} F_\gamma(z) &= \gamma + \beta_1 + 2\omega\alpha_2 + (3\omega^2 - \eta^2)\beta_2 + \int_{\mathbb{R}} |t-z|^{-2}\big(1+t^2\big)\, d\sigma_1(t) \\ &\quad + \int_{\mathbb{R}} \big(2\omega t - \omega^2 + 3\omega^2 t^2 - 2t\omega^3\big)|t-z|^{-2}\, d\sigma_2(t) \\ &\quad - \eta^2 \int_{\mathbb{R}} \big(1 + t^2 - 2\omega t\big)|t-z|^{-2}\, d\sigma_2(t).\end{aligned} \tag{2.11}$$

If $\omega = 0$ and (2.9) holds then the last term on the right-hand side of (2.11) converges to $-\sigma_2(\{0\})$ if $\eta \to 0$. Moreover, by Fatou's lemma,

$$\sigma_1(\{0\}) = 0, \quad \int_{\mathbb{R}} t^{-2}\, d\sigma_1(t) < \infty, \quad \gamma + \beta_1 + \int_{\mathbb{R}} t^{-2}\big(1+t^2\big)\, d\sigma_1(t) - \sigma_2(\{0\}) \le 0.$$

Now it is easy to see that the relation (2.10) can be written as (2.5), that the inequalities in (i) imply (2.9) and that (2.5) implies (2.10).

If $\zeta(\gamma) = \omega \in \mathbb{R} \setminus \{0\}$, then the last term in (2.11) converges for $\eta \to 0$ to $\big(\omega^2 - 1\big)\sigma_2(\{\omega\})$, and we conclude from (2.9) and Fatou's lemma that

$$\int_{\mathbb{R}} |t-\omega|^{-2}\, d\sigma_1(t) < \infty, \quad \int_{\mathbb{R}} |t-\omega|^{-2}\, d\sigma_2(t) < \infty,$$

in particular, $\sigma_1(\{\omega\}) = \sigma_2(\{\omega\}) = 0$. Now it is easy to see that (2.9) implies (2.8) and that (2.10) implies (2.7), and also that, conversely, (2.6), (2.7) and (2.8) imply (2.9) and (2.10). □

We define γ_0 as follows:

$$\gamma_0 := \begin{cases} \sigma_2(\{0\}) - \displaystyle\int_{\mathbb{R}} \frac{1+t^2}{t^2}\, d\sigma_1(t) - \beta_1 & \text{if } \displaystyle\int_{\mathbb{R}} \frac{d\sigma_1(t)}{t^2} < \infty,\ \alpha_1 + \int_{\mathbb{R}} \frac{d\sigma_1(t)}{t} = 0, \\ -\infty & \text{otherwise.} \end{cases}$$

The numbers γ_0, γ_∞ satisfy the following inequalities:

$$-\infty \le \gamma_0 \le \gamma_\infty \le \infty, \quad \gamma_0 < \infty, \quad -\infty < \gamma_\infty,$$

and Lemma 2.3 implies that for $\gamma \in (-\infty, \gamma_\infty)$ the relation $\zeta(\gamma) = 0$ holds if and only if $\gamma \le \gamma_0$. Evidently, if

$$F_\gamma(z) = \gamma z - cz \text{ with some } c \in \mathbb{R}$$

then F_γ belongs to N_1 if $\gamma < c$ and to N_0 if $\gamma \ge c$, hence, for this family F_γ, $\gamma \in \mathbb{R}$, we have $\gamma_0 = \gamma_\infty = c$. In the next theorem, where we study the dependence of the zero $\zeta(\gamma)$ of nonpositive type of F_γ on γ in the interval $(-\infty, \gamma_\infty)$, it will be shown that this is the only case where $\gamma_0 = \gamma_\infty$.

Theorem 2.4. *The equality $\gamma_0 = \gamma_\infty$ holds if and only if $F_\gamma(z) = \gamma z - cz$ with some $c \in \mathbb{R}$; in this case $\gamma_0 = \gamma_\infty = c$.*

If $\gamma_0 \ne \gamma_\infty$ then the function ζ on $(-\infty, \gamma_\infty)$ has the following properties:

(1) *ζ is a continuous function, $\zeta(\gamma) = 0$ if and only if $\gamma \in (-\infty, \gamma_0]$, and $\zeta(\gamma_1) \ne \zeta(\gamma_2)$ if $\gamma_1 \ne \gamma_2$, $\gamma_1, \gamma_2 \in (\gamma_0, \gamma_\infty)$.*
(2) $\lim_{\gamma \downarrow \gamma_0} \zeta(\gamma) = 0, \quad \lim_{\gamma \uparrow \gamma_\infty} \zeta(\gamma) = \infty.$
(3) *ζ is real analytic at points $\gamma \in (-\infty, \gamma_\infty)$ with $\zeta(\gamma) \notin (\operatorname{supp} \sigma_1 \cup \operatorname{supp} \sigma_2)$ with possible exception of branching points of order ≤ 3 for which $\zeta(\gamma) \in \mathbb{R}$.*
(4) $\sigma_j(\{\zeta(\gamma) : \gamma \in (\gamma_0, \gamma_\infty)\} \cap \mathbb{R}) = 0,\ j = 1, 2.$
(5) *The function $|\zeta|$: $|\zeta|(\gamma) := |\zeta(\gamma)|$ is nondecreasing; it is increasing on $(\gamma_0, \gamma_\infty)$ if and only if F_γ is not of the form*

$$F_\gamma(z) = \gamma z + \alpha_1 + \beta_1 z + \alpha_2 z^2$$

with $\alpha_1 \alpha_2 > 0$.

Proof. If $\gamma_0 = \gamma_\infty$, then the numbers γ_0 and γ_∞ are finite, $\beta_2 = 0$, $\int_{\mathbb{R}} (1+t^2)\, d\sigma_2(t)$ and $\int_{\mathbb{R}} t^{-2}(1+t^2)\, d\sigma_1(t)$ are finite and $\alpha_1 = -\int_{\mathbb{R}} t^{-1}\, d\sigma_1(t)$, $\alpha_2 = \int_{\mathbb{R}} t\, d\sigma_2(t)$. Moreover,

$$\int_{\mathbb{R}} (1+t^2)\, d\sigma_2(t) = \sigma_2(\{0\}) - \int_{\mathbb{R}} t^{-2}(1+t^2)\, d\sigma_1(t) \le \sigma_2(\{0\}) \le \int_{\mathbb{R}} (1+t^2)\, d\sigma_2(t),$$

hence $\sigma_1 = 0$, $\sigma_2 = c_0 \delta_0$, where $c_0 \in \mathbb{R}$ and δ_0 is the unit mass at 0. It follows that $F_\gamma(z) = \gamma z + \beta_1 z - c_0 z$. Conversely, evidently, $F_\gamma(z) = \gamma z - cz$, $c \in \mathbb{R}$, implies $\gamma_0 = \gamma_\infty$.

For the rest of the proof we assume that $\gamma_0 \ne \gamma_\infty$. The continuous dependence of $\zeta(\gamma)$ on γ is an easy consequence of [J1, Corollary 3.2 and Proposition 3.5]. The second assertion of (1) follows from Lemma 2.3: Suppose that for $\gamma_1, \gamma_2 \in (\gamma_0, \gamma_\infty)$ we have $\zeta(\gamma_1) = \zeta(\gamma_2) =: \zeta_0$. Then $\zeta_0 \neq 0$ and the relations

$$\gamma_j \zeta_0 + G_1(\zeta_0) + \zeta_0^2 G_2(\zeta_0) = 0 \quad j = 1, 2,$$

(understood as in (2.7)) imply $\gamma_1 = \gamma_2$.

In order to prove the first relation of (2) we can suppose that $\gamma_0 = -\infty$ and consider only $\gamma < 0$. Then $\zeta(\gamma)$ is the generalized zero of nonpositive type of the function $\frac{1}{|\gamma|} F_\gamma$, and the first relation of (2) is again a consequence of [J1, Corollary 3.2 and Proposition 3.5].

In order to prove the second relation of (2) we introduce the function

$$\widetilde{F}_\gamma(z) := z^2 F_{-\gamma}\big(-z^{-1}\big) = \gamma z + \widetilde{G}_1(z) + z^2 \widetilde{G}_2(z)$$

with $\widetilde{G}_1(z) := G_2\big(-z^{-1}\big)$, $\widetilde{G}_2(z) := G_1\big(-z^{-1}\big)$, and we write

$$\widetilde{G}_j(z) = \widetilde{\alpha}_j + \widetilde{\beta}_j z + \int_{\mathbb{R}} (1 + tz)(t - z)^{-1} \, d\widetilde{\sigma}_j(t), \; j = 1, 2,$$

where $\widetilde{\alpha}_j$, $\widetilde{\beta}_j$ and $\widetilde{\sigma}_j$ have the same properties as α_j, β_j and σ_j. Define $\widetilde{\gamma}_0$, $\widetilde{\gamma}_\infty$ for $\widetilde{F}_\gamma$ in the same way as γ_0 and γ_∞ were defined for F_γ. Then [ADo, Section 2] yields the relations

$$\widetilde{\alpha}_1 = \alpha_2, \; \widetilde{\beta}_1 = \sigma_2(\{0\}), \; \widetilde{\sigma}_1(\{0\}) = \beta_2, \; d\widetilde{\sigma}_1(t) = d\sigma_2(-t^{-1}) \text{ on } \mathbb{R} \setminus \{0\},$$

and also the same relations with indices 1 and 2 interchanged. Further, an easy computation shows that $\widetilde{\gamma}_0 = -\gamma_\infty$, $\widetilde{\gamma}_\infty = -\gamma_0$, and if $\gamma \in (\gamma_0, \gamma_\infty)$ the generalized zero $\widetilde{\zeta}(\gamma)$ of nonpositive type of $\widetilde{F}_\gamma$ is $\widetilde{\zeta}(\gamma) = -\zeta(-\gamma)^{-1}$. Therefore, the relation $\lim_{\gamma \downarrow \gamma_0} \widetilde{\zeta}(\widetilde{\gamma}) = 0$ implies $\lim_{\gamma \uparrow \gamma_\infty} \zeta(\gamma) = +\infty$.

Statement (3) follows from (2.7), (4) is a consequence of the relations (2.6) and [JL2, Lemma 2.1]. It remains to prove (5). We first show that the function $|\zeta|$ is nondecreasing in $(\gamma_0, \gamma_\infty)$. To this end we observe that there exist a sequence of real numbers b_n and a sequence of compactly supported positive measures τ_n on $\mathbb{R}$, such that the sequence of functions H_n:

$$H_n(z) := b_n z + \int_{\mathbb{R}} (t - z)^{-1} \, d\tau_n(t), \quad n = 1, 2, \ldots,$$

converges for $n \to \infty$ locally uniformly on $\mathbb{C}^+$ to $G_1(z) + z^2 G_2(z)$. Indeed, it is easy to see that the functions

$$\alpha_1 + \beta_1 z + \int_{-n}^{n} \frac{1 + tz}{t - z} \, d\sigma_1(t) + z^2 \bigg(\alpha_2 + \beta_2 z + \int_{-n}^{n} \frac{1 + tz}{t - z} \, d\sigma_2(t) \bigg), \; n = 1, 2, \ldots,$$

converge for $m \to \infty$ locally uniformly on $\mathbb{C}^+$ to $G_1(z) + z^2 G_2(z)$ for $n \to \infty$. Since for arbitrary $\alpha \in \mathbb{R}$ and $\beta \geq 0$ the functions

$$|\alpha| \, m \left((\operatorname{sign} \alpha) m - z \right)^{-1} + \frac{\beta}{2} m^2 \left((m - z)^{-1} + (-m - z)^{-1} \right), \quad m = 1, 2, \ldots,$$

converge locally uniformly to the function $\alpha+\beta z$, there exist sequences of compactly supported positive measures $\sigma'_{1,n}(t)$, $\sigma'_{2,n}(t)$, such that the functions

$$\begin{aligned}&\int_{\mathbb{R}}(t-z)^{-1}\,d\sigma'_{1,n}(t)+z^2\int_{\mathbb{R}}(t-z)^{-1}\,d\sigma'_{2,n}(t)\\&=\int_{\mathbb{R}}(t-z)^{-1}\,d\sigma'_{1,n}(t)-z\int_{\mathbb{R}}d\sigma'_{2,n}(t)-\int_{\mathbb{R}}t\,d\sigma'_{2,n}(t)+\int_{\mathbb{R}}(t-z)^{-1}t^2\,d\sigma'_{2,n}(t)\end{aligned}$$

converge locally uniformly to $G_1(z)+z^2G_2(z)$. If we approximate the constants as above by Nevanlinna functions we get a sequence of functions H_n with the desired properties.

Consider $\gamma'\in(\gamma_0,\gamma_\infty)$. Since $F_{\gamma'}\in\mathcal{N}_1$, from the definition of the class $\mathcal{N}_1$ it follows that there exist a natural number n_0 and an $\varepsilon>0$ with $\Delta:=[\gamma'-\varepsilon,\gamma'+\varepsilon]\subset(\gamma_0,\gamma_\infty)$, such that for $\gamma\in\Delta$ and $n>n_0$ the functions $F_{\gamma,n}$:

$$F_{\gamma,n}(z):=\gamma z+H_n(z)$$

belong to $\mathcal{N}_1$. Denote the zero of nonpositive type of $F_{\gamma,n}$ by $\zeta_n(\gamma)$. Then, according to [JL2, Theorem 3.1], for $n>n_0$ the functions $|\zeta_n|$ are nondecreasing on Δ. By [J1, Corollary 3.2 and Proposition 3.5], $\lim_{n\to\infty}\zeta_n(\gamma)=\zeta(\gamma)$, $\gamma\in\Delta$, therefore also the function $|\zeta|$ is nondecreasing on Δ.

For the proof of the second statement of (5) it is sufficient to consider the function $|\zeta|$ on an interval (γ_1,γ_2) such that $\zeta(\gamma)\in\mathbb{C}^+$ if $\gamma\in[\gamma_1,\gamma_2]$. Suppose first that $\sigma_1\neq0$ or that $(\operatorname{supp}\sigma_2)\setminus\{0\}\neq\emptyset$. Then, by the above construction of H_n, we may assume that for each bounded interval Δ there exists an n_1 such that for $n\geq n_1$ the measure τ_n, restricted to Δ, is not zero and does not depend on n. Applying [JL2, (3.6)] and the reasoning which follows that relation, we see that there exists an $a>0$ such that

$$\inf_{\gamma\in[\gamma_1,\gamma_2]}\frac{d|\zeta_n(\gamma)|}{d\gamma}\geq a\quad\text{if } n\geq n_1,$$

which implies that the function $|\zeta|$ is increasing.

Let now $\sigma_1=0$ and $\operatorname{supp}\sigma_2\subset\{0\}$. Then

$$F_\gamma(z)=\gamma z+\alpha_1+\big(\beta_1-\sigma_2(\{0\})\big)z+\alpha_2z^2+\beta_2z^3.$$

Set $\beta':=\beta_1-\sigma_2(\{0\})$, suppose $\beta_2>0$ and let (γ_1,γ_2) be as above. Then a simple computation, starting from $F_\gamma(\zeta(\gamma))=0$, yields

$$\begin{aligned}\frac{d\,\log|\zeta(\gamma)|}{d\gamma}&=\operatorname{Im}\frac{d\,\zeta(\gamma)}{d\gamma}\,\zeta(\gamma)^{-1}\\&=\beta_2\,\big|\beta_2\operatorname{Im}\,\zeta(\gamma)-\mathrm{i}\big(\alpha_2+3\beta_2\operatorname{Re}\,\zeta(\gamma)\big)\big|^{-2}>0\end{aligned}$$

for all $\gamma\in(\gamma_1,\gamma_2)$. Therefore the function $|\zeta|$ is strictly increasing on (γ_0,γ_∞). Finally, the case

$$F_\gamma(z)=\gamma(z)+\alpha_1+\beta'z+\alpha_2z^2$$

can be considered by straightforward calculations. □

For $\gamma_1 \in (\gamma_0, \gamma_\infty)$ and

$$\zeta_1 = \zeta(\gamma_1) \in \mathbb{R} \setminus (\operatorname{supp} \sigma_1 \cup \operatorname{supp} \sigma_2 \cup \{0\}),$$

in a neighborhood of ζ_1 the graph of the function ζ can be described similarly to [JL2, Proposition 3.2]. To this end we denote by $a_\pm$ the directions of the (one-sided) tangents of the curve $\gamma \mapsto \zeta(\gamma)$ at γ_1:

$$a_\pm := \pm \lim_{\gamma \to \gamma_1 \pm 0} \frac{\zeta(\gamma) - \zeta(\gamma_1)}{|\zeta(\gamma) - \zeta(\gamma_1)|}. \tag{2.12}$$

Then one of the following four cases prevails for some $\varepsilon > 0$:

(rr) $\zeta(\gamma) \in \mathbb{R}$ if $|\gamma - \gamma_1| < \varepsilon$.

(cr) $\zeta(\gamma) \in \mathbb{C}^+$ if $\gamma \in (\gamma_1 - \varepsilon, \gamma_1)$, $\zeta(\gamma) \in \mathbb{R}$ if $\gamma \in (\gamma_1, \gamma_1 + \varepsilon)$, and $a_- = -i$.

(rc) $\zeta(\gamma) \in \mathbb{R}$ if $\gamma \in (\gamma_1 - \varepsilon, \gamma_1)$, $\zeta(\gamma) \in \mathbb{C}^+$ if $\gamma \in (\gamma_1, \gamma_1 + \varepsilon)$, and $a_+ = i$.

(cc) $\zeta(\gamma) \in \mathbb{C}^+$ if $0 < |\gamma - \gamma_1| < \varepsilon$ and
$$a_- = (\operatorname{sign} \gamma_0) \exp\left(-(\operatorname{sign} \gamma_0) i\pi/3\right), \; a_+ = (\operatorname{sign} \gamma_0) \exp\left((\operatorname{sign} \gamma_0) i\pi/3\right).$$

If we denote the inverse function of ζ by γ, that is, $\gamma(\zeta(\gamma)) = \gamma$ for $\gamma \in (\gamma_0, \gamma_\infty)$, then

$$\gamma(\zeta) = -\zeta^{-1}\left(G_1(\zeta) + \zeta^2 G_2(\zeta)\right).$$

The case (rr) holds if and only if $\gamma'(\zeta_1) \neq 0$, in the other cases we have $\gamma'(\zeta_1) = 0$ and, additionally, $\gamma''(\zeta_1) > 0$, < 0, $= 0$, respectively, in the case (cr), (rc) and (cc), respectively.

The graph of the function ζ 'visits' an open interval I with $0 \notin I$ and such that G_1 and G_2 are holomorphic on I at most once. This is a consequence of the following theorem.

Theorem 2.5. *If I is a component of the open set $\mathbb{R} \setminus (\operatorname{supp} \sigma_1 \cup \operatorname{supp} \sigma_2 \cup \{0\})$, then the set $I \cap \zeta\big((\gamma_0, \gamma_\infty)\big)$ is empty or connected.*

Proof. If

$$F_\gamma(z) = \gamma z + \alpha_1 + \beta' z + \alpha_2 z^2, \; \beta' := \beta_1 - \sigma_2(\{0\}), \tag{2.13}$$

the theorem can be proved by a straightforward computation. Suppose that F_γ is not of the form (2.13). Then the third derivative of the function

$$F(\zeta) := -\zeta\gamma(\zeta) = G_1(\zeta) + \zeta^2 G_2(\zeta)$$

is positive in real points of holomorphy of F. We consider the case $I \subset (0, \infty)$, the case $I \subset (-\infty, 0)$ can be treated similarly. Assume that $I \cap \zeta\big((\gamma_0, \gamma_\infty)\big)$ is not connected. Then there exist points ζ_1, $\zeta_2 \in I \cap \zeta\big((\gamma_0, \gamma_\infty)\big)$, $\zeta_1 < \zeta_2$, such that

$$\gamma'(\zeta_1) = \gamma'(\zeta_2) = 0, \quad \gamma''(\zeta_1) \leq 0, \quad \gamma''(\zeta_2) \geq 0.$$

therefore $F''(\zeta_1) \geq 0$, $F''(\zeta_2) \leq 0$. But this is impossible since $F'''(\zeta)$ is positive on (ζ_1, ζ_2). □

3. The exceptional eigenvalue of a self-adjoint relation with one negative square

Let $(\mathcal{H}, [\cdot,\cdot])$ be a Krein space, and let $\|\cdot\|$ be a Hilbert norm on $\mathcal{H}$ such that $[\cdot,\cdot]$ is $\|\cdot\|$-continuous. A *closed linear relation* A in $\mathcal{H}$ is a closed linear subspace of $\mathcal{H}^2$. In the sequel a closed linear operator A in $\mathcal{H}$ is identified with the closed linear relation given by its graph $\left\{\begin{pmatrix} f \\ Af \end{pmatrix} : f \in \mathcal{D}(A)\right\}$. For the definitions of the domain $\mathcal{D}(A)$, the kernel $\ker A$, the range $\operatorname{ran} A$, the adjoint A^+, the resolvent set $\rho(A)$, the spectrum $\sigma(A)$ and the point spectrum $\sigma_p(A)$ we refer to [DdS1], [DdS2], [JL3]. The *extended spectrum* $\widetilde{\sigma}(A)$ is defined as follows:

$$\widetilde{\sigma}(A) := \begin{cases} \sigma(A) & \text{if} \quad 0 \notin \sigma(A^{-1}), \\ \sigma(A) \cup \{\infty\} & \text{if} \quad 0 \in \sigma(A^{-1}). \end{cases}$$

If A is not an operator, that is $A(0) = \left\{g : \begin{pmatrix} 0 \\ g \end{pmatrix} \in A\right\} \neq \{0\}$, then ∞ is called an *eigenvalue of* A with the nonzero elements of $A(0)$ as corresponding *eigenvectors.* The set

$$\widetilde{\sigma}_p(A) := \begin{cases} \sigma_p(A) & \text{if} \quad A(0) = \{0\}, \\ \sigma_p(A) \cup \{\infty\} & \text{if} \quad A(0) \neq \{0\} \end{cases}$$

is called the *extended point spectrum of* A.

The linear relation A in the Krein space $\mathcal{H}$ is called *symmetric* if the sesquilinear form $[\cdot,\cdot]_A$ (see (1.1)) is Hermitian, that is

$$[f', g] = [f, g'], \quad \begin{pmatrix} f \\ f' \end{pmatrix}, \begin{pmatrix} g \\ g' \end{pmatrix} \in A,$$

and A is called *self-adjoint* if for elements $g, g' \in \mathcal{H}$ the relation

$$[f', g] = [f, g'] \quad \text{for all} \quad \begin{pmatrix} f \\ f' \end{pmatrix} \in A$$

yields $\begin{pmatrix} g \\ g' \end{pmatrix} \in A$, see, e.g., [DdS1], [DdS2]. The symmetric linear relation A in $\mathcal{H}$ is called *nonnegative* if

$$\left[\begin{pmatrix} f \\ f' \end{pmatrix}, \begin{pmatrix} f \\ f' \end{pmatrix}\right]_A \equiv [f', f] \geq 0, \quad \begin{pmatrix} f \\ f' \end{pmatrix} \in A,$$

and it is said to have *one negative square* if the Hermitian sesquilinear form

$$\left[\begin{pmatrix} f \\ f' \end{pmatrix}, \begin{pmatrix} g \\ g' \end{pmatrix}\right]_A \equiv [f', g], \quad \begin{pmatrix} f \\ f' \end{pmatrix}, \begin{pmatrix} g \\ g' \end{pmatrix} \in A, \tag{3.1}$$

has this property.

Let A be a self-adjoint relation with $\rho(A) \neq \emptyset$, let $\mu \in \rho(A)$ and denote by $\{\cdot,\cdot\}$ the following Hermitian sesquilinear form on $\mathcal{H}$:

$$\{f, g\} := \left[\left(I + \mu(A-\mu)^{-1}\right) f, (A-\mu)^{-1} g\right], \quad f, g \in \mathcal{H}. \tag{3.2}$$

For $\kappa \in \mathbb{N}_0$ and a Krein space $\mathcal{H}$, by $\mathcal{N}_\kappa(\mathcal{H})$ we denote the set of all functions Q with values in $\mathcal{L}(\mathcal{H})$ (the set of bounded linear operators in $\mathcal{H}$), which are meromorphic in $\mathbb{C} \setminus \mathbb{R}$ and such that $Q(\overline{z}) = Q(z)^+$ for all $z \in \operatorname{hol}(Q)$ and that the kernel

$$\frac{Q(z) - Q(\zeta)^+}{z - \overline{\zeta}}, \quad z, \zeta \in \operatorname{hol}(Q),$$

has κ negative squares; here $^+$ denotes the Krein space adjoint. Then the following statements are equivalent:

(a) A is nonnegative (has one negative square, respectively).

(b) For some (and hence for all) $\mu \in \rho(A)$ the Hermitian sesquilinear form $\{\cdot, \cdot\}$ on $\mathcal{H}$ is nonnegative (has one negative square, respectively).

(c) The Krein space operator function $z \mapsto z(A-z)^{-1}$ belongs to $\mathcal{N}_0(\mathcal{H})$ ($\mathcal{N}_1(\mathcal{H})$, respectively).

In particular, the fact that for a nonnegative self-adjoint relation A with $\rho(A) \neq \emptyset$ the operator function in (c) is a Nevanlinna function implies that in this case $\mathbb{C} \setminus \mathbb{R} \subset \rho(A)$.

The main result of this section is the following theorem.

Theorem 3.1. *Let A be a self-adjoint relation with one negative square in the Krein space $\mathcal{H}$ such that $\rho(A) \neq \emptyset$. Then exactly one of the following two statements holds true:*

(i) *The non-real spectrum of A consists of one pair of complex conjugate points $\alpha \in \mathbb{C}^+$ and $\overline{\alpha}$ which are simple normal eigenvalues and such that the two-dimensional linear span of the corresponding eigenvectors is a non-degenerate indefinite subspace. In the orthogonal complement of this span the relation A induces a nonnegative self-adjoint linear relation with non-empty resolvent set.*

(ii) *The spectrum of A is real and there exists a unique point $\alpha \in \overline{\mathbb{R}}$ with the following properties:*
If $\alpha \neq 0, \infty$ then it is an eigenvalue of A with an eigenelement e_0 such that

$$(\operatorname{sgn} \alpha)\ [e_0, e_0] \leq 0;$$

if $\alpha = 0$ then α is an eigenvalue of A with a neutral eigenelement e_0 and there exists a sequence $\left(\begin{pmatrix} e_n \\ e_n' \end{pmatrix}\right) \subset A$ such that $\|e_n' - e_0\| \to 0$ and

$$[e_n' - e_m', e_n - e_m] \to 0, \quad \lim_{n\to\infty} [e_n', e_n] \leq 0 \quad \text{if } m, n \to \infty; \tag{3.3}$$

if $\alpha = \infty$ there exists a sequence $\left(\begin{pmatrix} e_n \\ e_n' \end{pmatrix}\right) \subset A$ such that $\|e_n - e_0'\| \to 0$ if $n \to \infty$ for some nonzero neutral element $e_0' \in A(0)$ and (3.3) holds.

In the proof of this theorem we need the notion of a definitizable self-adjoint relation in a Krein space and its spectral function. The self-adjoint relation A in the Krein space $\mathcal{H}$ is called *definitizable* if $\rho(A) \neq \emptyset$ and if there exists a real polynomial p such that the self-adjoint relation $p(A)$ is nonnegative (see [DdS2]). Obviously, a nonnegative self-adjoint relation A with $\rho(A) \neq \emptyset$ is definitizable. The self-adjoint relation A with $\mu \in \rho(A)$, $\mu \neq \overline{\mu}$, is definitizable if and only if the unitary operator U in the Krein space $\mathcal{H}$ defined by

$$U := -I + (\mu - \overline{\mu})(A - \overline{\mu})^{-1}, \tag{3.4}$$

i.e., the Cayley transform of A, is definitizable in the sense of [L1]. The spectrum of a definitizable unitary operator U lies on the unit circle $\mathbb{T}$ with possible exception of a finite set of eigenvalues, which is symmetric with respect to $\mathbb{T}$. On $\mathbb{T}$ the operator U has a spectral function with possibly a finite number of critical points, see [L1]; the set of these critical points is denoted by $c(U)$.

The Cayley transform U of the definitizable self-adjoint relation A with $\mu \in \rho(A)$ can be written as $U = \psi(A)$ with

$$\psi(z) := -(z - \mu)(z - \overline{\mu})^{-1}. \tag{3.5}$$

According to [DdS2, Proposition 3.2] we have $\psi(\widetilde{\sigma}(A)) = \sigma(U)$ and hence $\widetilde{\sigma}(A)$ lies on $\mathbb{R}$ with possible exception of a finite set of complex eigenvalues, which is symmetric with respect to $\mathbb{R}$. If E_U denotes the spectral function of U then the spectral function E_A of A is defined by the relation

$$E_A(\Delta) = E_U\left(\psi(\Delta)\right),$$

where $\Delta \in \mathcal{B}(A)$. Here $\mathcal{B}(A)$ denotes the Boolean algebra generated by the connected subsets of $\overline{\mathbb{R}}$ ($\overline{\mathbb{R}}$ is considered to be homeomorphic to the unit circle) for which the endpoints are not in $c(A) := \psi^{-1}(c(U))$; this set $c(A)$ is called the set of *critical points of* A. The spectral function E_A defines a homomorphism from $\mathcal{B}(A)$ into the set of self-adjoint projections in $\mathcal{H}$ such that for all $\Delta \in \mathcal{B}(A)$ the following holds:

1. $E_A\left(\overline{\mathbb{R}}\right) = 1 - E_0$, where E_0 is the Riesz–Dunford projection corresponding to the set of nonreal eigenvalues of A.
2. A is the direct sum of the three subspaces $A \cap (E_0\mathcal{H})^2$, $A \cap \left(E_A(\Delta)\mathcal{H}\right)^2$, and $A \cap \left(E_A(\overline{\mathbb{R}} \setminus \Delta)\mathcal{H}\right)^2$ of $\mathcal{H}^2$.
3. $\widetilde{\sigma}\left(A \cap \left(E_A(\Delta)\mathcal{H}\right)^2\right) \subset \overline{\Delta}$.

The point $\lambda \in \sigma(A)$ is said to be a *spectral point of positive* (*negative*, respectively) *type* of the definitizable self-adjoint relation A if for some open set $\Delta \in \mathcal{B}(A)$ such that $\lambda \in \Delta$ the set $E_A(\Delta)\mathcal{H} \setminus \{0\}$ consists of positive (negative, respectively) elements or, equivalently, the space $(E_A(\Delta)\mathcal{H}, [\,\cdot\,,\,\cdot\,])$ is a Hilbert space (anti-Hilbert space, respectively). The spectral function E_A has also the following properties:

4. If $\lambda \in \sigma(A)$ and for a definitizing polynomial p of A we have $p(\lambda) > 0$ (< 0, respectively) then λ is a spectral point of positive (negative, respectively) type of A.
5. $\lambda \in c(A)$ if and only if for all $\Delta \in \mathcal{B}(A)$ with $\lambda \in \Delta$ the range $E_A(\Delta)\mathcal{H}$ contains positive as well as negative elements.

In the following lemma we set

$$\varphi(U) := (2\operatorname{Im}\mu)^{-2}\left(\overline{\mu}(1+U)+\mu(1+U^{-1})\right) = (A-\overline{\mu})^{-1}A(A-\mu)^{-1}. \quad (3.6)$$

Then the Hermitian sesquilinear form in (3.2) becomes

$$\{f,g\} = [\varphi(U)f,g], \quad f,g\in\mathcal{H}.$$

Lemma 3.2. *Let A be a self-adjoint relation with one negative square in the Krein space $\mathcal{H}$, suppose that $\rho(A)\neq\emptyset$, and choose $\mu\in\rho(A)\cap\mathbb{C}^+$. Then A and also its Cayley transform $U=\psi(A)$ (see (3.4)) are definitizable. There exists a unique point $\beta\in\mathbb{C}$, $0<|\beta|\leq 1$, such that with the operator $\varphi(U)$ from (3.6) the relation*

$$\left[\varphi(U)(U-\beta)(U^{-1}-\overline{\beta})x,x\right]\geq 0 \text{ for all } x\in\mathcal{H} \quad (3.7)$$

holds.

The point β is an eigenvalue of U with a $[\varphi(U)\,\cdot\,,\,\cdot\,]$-nonpositive eigenelement e, and $\alpha=\psi^{-1}(\beta)$ is an eigenvalue of the self-adjoint relation A with the same eigenelement e.

If $|\beta|<1$, that is $\alpha\in\mathbb{C}^+$, then $\sigma(U)\setminus\mathbb{T}=\left\{\beta,\overline{\beta}^{-1}\right\}$ and $\sigma(A)\setminus\mathbb{R}=\{\alpha,\overline{\alpha}\}$, β and α are simple eigenvalues of U and A, respectively, the corresponding eigenvectors of U and A coincide and the two-dimensional subspace

$$\ker(U-\beta)+\ker\left(U-\overline{\beta}^{-1}\right)=\ker(A-\alpha)+\ker(A-\overline{\alpha})$$

is non-degenerated and indefinite.

If $|\beta|=1$, that is if $\alpha\in\overline{\mathbb{R}}$ then $\sigma(U)\subset\mathbb{T}$ and $\sigma(A)\subset\mathbb{R}$. For $\Delta\in\mathcal{B}(A)$ it holds $\alpha\in\Delta$ ($\alpha\notin\Delta$, respectively) if and only if the Hermitian sesquilinear form $[\cdot,\cdot]_A$ has one negative square (is positive semi-definite, respectively) on $A\cap\left(E_A(\Delta)\mathcal{H}\right)^2$.

Proof. We consider the factor space $\widehat{\mathcal{H}}:=\mathcal{H}/\ker\varphi(U)$. Since $\ker\varphi(U)$ is the isotropic subspace of the Hermitian sesquilinear form (3.2), the space $\left(\widehat{\mathcal{H}},\{\,\cdot\,,\,\cdot\,\}\right)$ is a non-degenerated inner product space with negative index one. We complete it to a Pontryagin space $\left(\widetilde{\mathcal{H}},\{\,\cdot\,,\,\cdot\,\}\right)$ with one negative square. The operator U induces a bounded operator $\widehat{U}$ in $\widehat{\mathcal{H}}$, which maps $\widehat{\mathcal{H}}$ onto itself and leaves the inner product $\{\,\cdot\,,\,\cdot\,\}$ invariant. The closure $\widetilde{U}$ of $\widehat{U}$ in $\widetilde{\mathcal{H}}$ is a unitary operator. According to Pontryagin's Theorem there exists a point $\beta\in\sigma_p(\widetilde{U})$, $|\beta|\leq 1$, with a nonpositive eigenelement $\widetilde{e}$. The relation $\left\{\operatorname{ran}(\widetilde{U}^{-1}-\overline{\beta}),\widetilde{e}\right\}=\{0\}$ implies

$$\left\{(\widetilde{U}-\beta)(\widetilde{U}^{-1}-\overline{\beta})\widetilde{x},\widetilde{x}\right\}\geq 0 \text{ for all } \widetilde{x}\in\widetilde{\mathcal{H}},$$

consequently, the relation (3.7) holds and U is a definitizable operator. The point β is an eigenvalue of U since otherwise the range of $U-\beta$ would be dense in $\mathcal{H}$ and hence $[\varphi(U)x,x]\geq 0$ for all $x\in\mathcal{H}$, which is impossible since this form has one negative square. On $\ker(U-\beta)$ we have $[\varphi(U)\,\cdot\,,\,\cdot\,]=\varphi(\beta)[\,\cdot\,,\,\cdot\,]$. If this form would be positive definite there, then $\ker(U-\beta)+(\ker(U-\beta))^{[\perp]}$ would be dense in $\mathcal{H}$. Since, by (3.7), $[\varphi(U)\cdot,\cdot]$ is positive semi-definite on $(\ker(U-\beta))^{[\perp]}$ there exists an $x \in \ker(U-\beta)$ with $[\varphi(U)x,x] < 0$, a contradiction. Therefore there exists a $[\varphi(U)\cdot,\cdot]$-nonpositive nonzero element $e \in \ker(U-\beta)$.

The point β, $|\beta| \le 1$, with the property (3.7) is uniquely determined. Indeed, if there would be a point $\beta' \ne \beta$ with the same properties as β, then also $\ker(U-\beta')$ would contain a nonpositive element and this element is orthogonal to e, which is impossible since the inner product $[\varphi(U)\cdot,\cdot]$ has only one negative square. This argument also proves that β with $|\beta| < 1$ is a simple eigenvalue of U. According to the spectral mapping theorem $\alpha = \psi^{-1}(\beta)$ is an eigenvalue of A with the same root subspace as that of U at β.

For $\beta \in \mathbb{T}$ or $\alpha \in \mathbb{R}$ it follows as above that $\sigma(U) \subset \mathbb{T}$ and $\sigma(A) \subset \mathbb{R}$. If $\Delta \in \mathcal{B}(A)$ such that $\alpha \notin \overline{\Delta}$, then $\beta = \psi(\alpha) \notin \overline{\psi(\Delta)}$, $E_U(\psi(\Delta))\mathcal{H} = (U-\beta)E_U(\psi(\Delta))\mathcal{H}$ and we get from (3.7)

$$\left[(A-\overline{\mu})^{-1}A(A-\mu)^{-1}E_A(\Delta)x, E_A(\Delta)x\right] = \left[\varphi(U)E_U(\psi(\Delta))x, E_U(\psi(\Delta))x\right] \ge 0.$$

From this relation the last claim of the lemma follows easily. □

Proof of Theorem 3.1. First we shall show that the point α of Lemma 3.2 has the properties stated in the theorem. Evidently, if $\alpha \in \mathbb{C}^+$ then the case (i) prevails. If $\alpha \in \mathbb{R}\setminus\{0\}$, it follows from Lemma 3.2 that the corresponding point $\beta = \psi(\alpha)$ is an eigenvalue of U and α is an eigenvalue of A with the same $[\varphi(U)\cdot,\cdot]$-nonpositive eigenelement e, and hence

$$0 \ge [\psi(U)e,e] = \left[(A-\overline{\mu})^{-1}A(A-\mu)e,e\right] = |\alpha-\mu|^2\,\alpha\,[e,e].$$

It follows that $[e,e]$ is nonpositive if $\alpha > 0$ and nonnegative if $\alpha < 0$. It remains to consider the cases $\alpha = 0$ and $\alpha = \infty$. We restrict ourselves to the first case, for $\alpha = \infty$ a similar reasoning applies.

If $\alpha = 0$ we choose a bounded open interval $\Delta \in \mathcal{B}(A)$ with $0 \in \Delta$ and consider in the Krein space $\mathcal{H}_\Delta := E(\Delta)\mathcal{H}$ the restriction $A_\Delta := A|_{\mathcal{H}_\Delta}$, which is a bounded self-adjoint operator. Repeating in fact considerations in the proof of Lemma 3.2, we equip the factor space $\widehat{\mathcal{H}}_\Delta := \mathcal{H}_\Delta/\ker A_\Delta$ with the inner product $\{\widehat{f},\widehat{g}\}_\Delta := [A_\Delta f, g]$, $f \in \widehat{f}, g \in \widehat{g}$, which has one negative square. The completion of $\widehat{\mathcal{H}}_\Delta$, which is a Pontryagin space with negative index one, is denoted by $\widetilde{\mathcal{H}}_\Delta$. The operator A_Δ induces bounded self-adjoint operators $\widehat{A}_\Delta$ and $\widetilde{A}_\Delta$ in $\widehat{\mathcal{H}}_\Delta$ and $\widetilde{\mathcal{H}}_\Delta$, respectively, and since in $\widetilde{\mathcal{H}}_\Delta$ all the nonzero spectral points of $\widetilde{A}_\Delta$ are of positive type, there exists a nonzero element $\widetilde{x} \in \ker \widetilde{A}_\Delta$ with $\{\widetilde{x},\widetilde{x}\}_\Delta \le 0$. For the space $\widehat{\mathcal{H}}_\Delta$ this means that there exists a Cauchy sequence $(\widehat{x}_n) \subset \widehat{\mathcal{H}}_\Delta$ which does not converge to zero and is such that $\widehat{A}_\Delta\widehat{x}_n \to 0$ in $\widehat{\mathcal{H}}_\Delta$ and $\lim\{\widehat{x}_n,\widehat{x}_n\} \le 0$ if $n \to \infty$. That $(\widehat{x}_n)$ is a Cauchy sequence means that

$$\{\widehat{x}_n - \widehat{x}_m, \widehat{x}_n - \widehat{x}_m\}_\Delta \longrightarrow 0, \quad \{\widehat{x}_n - \widehat{x}_m, \widehat{y}\}_\Delta \longrightarrow 0$$

for $m,n \to \infty$ and all $\widehat{y} \in \widehat{\mathcal{H}}_\Delta$, or, equivalently,

$$\left[A_\Delta(x_n - x_m), x_n - x_m\right] \longrightarrow 0, \quad \left[A_\Delta(x_n - x_m), y\right] \longrightarrow 0 \tag{3.8}$$

for $m, n \to \infty$ and all $y \in \mathcal{H}_\Delta$, $x_n \in \widehat{x}_n$. By the second relation in (3.8), the elements $A_\Delta x_n$ converge weakly to some $e_0 \in \mathcal{H}_\Delta$, and $\lim_{n\to\infty} [A_\Delta x_n, y_0] \neq 0$ for some $y_0 \in \mathcal{H}_\Delta$ implies $e_0 \neq 0$, and $\lim_{n\to\infty} [A_\Delta^2 x_n, y] = 0$ for all $y \in \mathcal{H}_\Delta$ implies $A_\Delta e_0 = 0$. Finally, $\lim_{n\to\infty} [A_\Delta x_n, x_n] \leq 0$.

Now we replace the sequence (x_n) by the sequence (e_n) of arithmetic means of a subsequence of (x_n) such that $A_\Delta e_n$ converges strongly to e_0. Then the sequence (e_n) has all the properties of the sequence (x_n) mentioned above, and therefore $\|e_n' - e_0\| \to 0$ if $n \to \infty$ and the relations in (3.3) hold. Since $e_0 \in \ker A_\Delta \cap \overline{\operatorname{ran} A_\Delta}$, the element e_0 is neutral.

It remains to be shown that any $\alpha' \in \overline{\mathbb{R}}$, which is different from α in Lemma 3.2, does not satisfy the condition (ii) of Theorem 3.1 with α replaced by α'. By Lemma 3.2 there exists a connected open set $\Delta \in \mathcal{B}(A)$ with $\alpha' \in \Delta$ such that the form $[\cdot,\cdot]_A$ is positive semi-definite on $A \cap (E_A(\Delta)\mathcal{H})^2$. Let $\alpha' \neq 0, \infty$ and assume, in addition, that $0, \infty \notin \overline{\Delta}$. Then for any $f_0 \in \ker(A - \alpha')$, $f_0 \neq 0$, we have $\alpha'[f_0, f_0] > 0$ and hence (ii) with α replaced by α' does not hold.

Now let $\alpha' = 0$ and assume in addition that Δ is bounded. Suppose that there exist an $f_0 \in \ker A$, $f_0 \neq 0$, and a sequence $\left(\begin{pmatrix} f_n \\ f_n' \end{pmatrix}\right) \subset A$ such that $\|f_n' - f_0\| \to 0$ for $n \to \infty$ and the following limit exists and is nonpositive:

$$\lim_{n\to\infty} [f_n', f_n] \leq 0. \tag{3.9}$$

We claim that

$$\lim_{n\to\infty} \left[(I - E_A(\Delta))\, f_n', f_n \right] = 0. \tag{3.10}$$

Indeed, consider $y \in \mathcal{H}$. Then $\widetilde{y} := (I - E_A(\Delta))\, y \in \operatorname{ran} A$, and if $\begin{pmatrix} u \\ \widetilde{y} \end{pmatrix} \in A$ the relation

$$\left[y, (I - E_A(\Delta))\, f_n\right] = \left[\widetilde{y}, f_n\right] = \left[u, f_n'\right], \quad n = 1, 2, \ldots,$$

and the boundedness of the sequence $\left(f_n'\right)$ imply that

$$\sup\left\{ \left\|\left(I - E_A(\Delta)\right) f_n\right\| : n = 1, 2, \ldots \right\} < \infty.$$

From this inequality and

$$\lim_{n\to\infty} \left\|\left(I - E_A(\Delta)\right) f_n'\right\| = \left\|\left(I - E_A(\Delta)\right) f_0\right\| = 0$$

the relation (3.10) follows. Now (3.9) and (3.10) imply

$$\lim_{n\to\infty} [E_A(\Delta) f_n', E_A(\Delta) f_n] = \lim_{n\to\infty} [E_A(\Delta) f_n', f_n] \leq 0,$$

and, since $[\cdot,\cdot]_A$ is positive semi-definite on $A \cap \left(E_A(\Delta)\mathcal{H}\right)^2$, we find

$$\lim_{n\to\infty} [E_A(\Delta) f_n', E_A(\Delta) f_n] = 0. \tag{3.11}$$

If J_Δ is a fundamental symmetry of the Krein space $(E_A(\Delta)\mathcal{H}, [\cdot,\cdot])$ then for $n = 1, 2, \ldots$

$$\begin{aligned}&\left|\left[E_A(\Delta)f'_n, J_\Delta E_A(\Delta)f'_n\right]\right|\\&\quad\le \left|\left[E_A(\Delta)f'_n, E_A(\Delta)f_n\right]\right|^{1/2}\left|\left[AJ_\Delta E_A(\Delta)f'_n, J_\Delta E_A(\Delta)f'_n\right]\right|^{1/2}.\end{aligned}$$

This relation and (3.11) imply that $E_A(\Delta)f'_n \to 0$ and hence $f_0 = 0$, a contradiction. For $\alpha' = \infty$ a similar reasoning applies.. □

Remark 3.3. The proof of Theorem 3.1 shows that this theorem remains true if (3.3) is replaced by $\lim_{n\to\infty}\left[e'_n, e_n\right] \le 0$.

In order to explain Theorem 3.1 we first mention that for a nonnegative self-adjoint relation A with nonempty resolvent set the positive spectral points are of positive type and the negative spectral points are of negative type. In particular, all the eigenvectors corresponding to positive eigenvalues are positive and all eigenvectors corresponding to negative eigenvalues are negative, and there is no non-real spectrum. If A has one negative square, it has either a pair α, $\overline{\alpha}$ of simple non-real complex conjugate eigenvalues, or on the real axis there is a positive eigenvalue α with a nonpositive eigenvector, or a negative eigenvalue α with a nonnegative eigenvector, or at 0 or ∞ there is a neutral eigenvector to which there corresponds an 'approximate associated vector' with a certain sign property, see (3.3).

The following definition of the exceptional eigenvalue will play an important role in this paper.

Definition 3.4. Let A be a self-adjoint relation with $\rho(A) \neq \emptyset$ and one negative square in the Krein space $\mathcal{H}$. The point $\alpha \in \mathbb{C}^+ \cup \overline{\mathbb{R}}$ from Theorem 3.1 is called the *exceptional eigenvalue* of A and it is denoted in the following by $\alpha(A)$.

In the following theorem we give some equivalent descriptions of the exceptional eigenvalue $\alpha(A)$. For the definition of a *subset of* $\overline{\mathbb{R}}$ *of positive type* with respect to an operator function in the class $\mathcal{N}_1(\mathcal{H})$, which will be used in the formulation and the proof of this theorem, we refer the reader to [J2, § 3.1].

Theorem 3.5. *Consider the self-adjoint relation A with one negative square and $\rho(A) \neq \emptyset$ in the Krein space $\mathcal{H}$, and let $\mu \in \rho(A) \setminus \mathbb{R}$, $\alpha \in \mathbb{C}^+ \cup \overline{\mathbb{R}}$. Then each of the following conditions* (i–iv) *is equivalent to the fact that α is the exceptional eigenvalue of A.*

(i) *If A has an eigenvalue in $\mathbb{C}^+$ then this eigenvalue is α. If $\sigma(A)$ is real then α is the unique point in $\overline{\mathbb{R}}$ such that for all open sets $\Delta \in \mathcal{B}(A)$ with $\alpha \in \Delta$ the self-adjoint linear relation $A \cap \left(E_A(\Delta)\mathcal{H}\right)^2$ is not nonnegative in $E_A(\Delta)\mathcal{H}$.*

(ii) *For the rational function r, $r(t) := \left(\psi(t) - \psi(\alpha)\right)\left(\psi(t)^{-1} - \overline{\psi(\alpha)}\right)$, we have*

$$\left[(A-\overline{\mu})^{-1}A(A-\mu)^{-1}r(A)x, x\right] \ge 0, \quad x \in \mathcal{H}.$$

(iii) *α is a pole in $\mathbb{C}^+$ of the operator function $z \mapsto z(A-z)^{-1}$, or $\alpha \in \overline{\mathbb{R}}$ and there exists an $x \in \mathcal{H}$ and a sequence $(z_n) \subset \mathbb{C}^+$ converging to α (in $\overline{\mathbb{C}}$) such that* $\liminf_{n\to\infty} \operatorname{Im}\left\{z_n[(A-z_n)^{-1}x,x]\right\} < 0$.

(iv) *α is a pole in $\mathbb{C}^+$ of the operator function $z \mapsto z(A-z)^{-1}$, or $\alpha \in \overline{\mathbb{R}}$ and α does not belong to an open subset of $\overline{\mathbb{R}}$ of positive type with respect to the operator function $z \mapsto z(A-z)^{-1}$.*

Proof. It was proved in Lemma 3.2 that (i) and (ii) are equivalent to the fact that α is the exceptional eigenvalue of A. For the proof that (i), (iii), and (iv) are equivalent it is sufficient to consider only the case $\alpha \in \overline{\mathbb{R}}$. Therefore in the rest of this proof we assume that $\sigma(A) \subset \mathbb{R}$.

Suppose that (i) holds. Then there exists an $x \in \mathcal{H}$ such that the function

$$G_x: \quad z \mapsto z\big[(A-z)^{-1}x,x\big]$$

belongs to the class $\mathcal{N}_1$. If $\alpha' \in \overline{\mathbb{R}}$ is the generalized pole of nonpositive type of G_x then there exists a sequence $(z_n) \subset \mathbb{C}^+$ which converges to α' (in $\overline{\mathbb{C}}$) and for which

$$\liminf_{n\to\infty} \operatorname{Im} G_x(z_n) < 0. \tag{3.12}$$

Suppose that $\alpha \neq \alpha'$. Let Δ be an open subset of $\overline{\mathbb{R}}$ with $\Delta \in \mathcal{B}(A)$ such that $\alpha \in \Delta$ and $\alpha' \notin \overline{\Delta}$. Then

$$G_x(z) = z\left[E_A(\Delta)(A-z)^{-1}x,x\right] + z\left[E_A(\overline{\mathbb{R}}\setminus\Delta)(A-z)^{-1}x,x\right], \tag{3.13}$$

and by (i) the function $z \mapsto \left[E_A(\overline{\mathbb{R}}\setminus\Delta)(A-z)^{-1}x,x\right]$ belongs to the class $\mathcal{N}_0$. This is in contradiction with (3.12) since the first term on the right-hand side of (3.13) is holomorphic at α'. Therefore $\alpha = \alpha'$ and (iii) holds.

Assume now that (iii) holds. Then for every open $\Delta \in \mathcal{B}(A)$ containing α and for the element $x \in \mathcal{H}$ from (iii) the function G_x belongs to $\mathcal{N}_1$. Therefore $A \cap \left(E_A(\Delta)\mathcal{H}\right)^2$ is not nonnegative. If the point α in (i) would not be unique then A would have more than one negative square, which is not the case. This shows that (i) holds.

By assumption of Theorem 3.5 the function $G: \; z \mapsto z(A-z)^{-1}$ belongs to $\mathcal{N}_1(\mathcal{H})$. Then on account of the additional assumption $\sigma(A) \subset \mathbb{R}$ there exists a unique point $\alpha_0 \in \overline{\mathbb{R}}$ such that $\overline{\mathbb{R}} \setminus \{\alpha_0\}$ is of positive type with respect to G and every open subset Δ_0 of $\overline{\mathbb{R}}$ with $\alpha_0 \in \Delta_0$ is not of positive type with respect to G. With this point α_0 the statement (iv) is equivalent to $\alpha = \alpha_0$. Now [J3, Lemma 3.8, (a)$\Leftrightarrow$(c)] implies that (iii) and (iv) are equivalent. □

Remark 3.6. For $\alpha \in \overline{\mathbb{R}}$ the conditions (i)–(iv) are equivalent to the fact that α is the generalized pole of nonpositive type of the function $G: \quad z \mapsto z(A-z)^{-1}$ in the following sense: For each neighborhood $\mathcal{U}$ of α in $\overline{\mathbb{C}}$ there exists a $\delta_{\mathcal{U}} > 0$ such that for every μ with $\delta_{\mathcal{U}} < \mu < \infty$ there exists a $z \in \mathcal{U} \cap \mathbb{C}^+$ such that $G(z) + i\mu J$ is not injective; here J is an arbitrary fixed fundamental symmetry of $\mathcal{H}$. That this condition is equivalent to (iv) follows from the similar statement [J1,

Proposition 3.5] for generalized Carathéodory functions by means of a fractional linear transformation of the independent variable.

Finally we prove a continuity property of the exceptional eigenvalue.

Theorem 3.7. *Let A and A_n, $n = 1, 2, \ldots$, be self-adjoint relations with one negative square in the Krein space $\mathcal{H}$. Suppose that there exists a point $\mu \in \mathbb{C} \setminus \mathbb{R}$ and an open neighborhood $\mathcal{U}$ of μ such that $\mathcal{U} \subset \rho(A_n)$ for $n = 1, 2, \ldots$, $\mathcal{U} \subset \rho(A)$, and that*

$$\lim_{n\to\infty} (A_n - \mu)^{-1} = (A - \mu)^{-1}, \quad \lim_{n\to\infty} (A_n - \overline{\mu})^{-1} = (A - \overline{\mu})^{-1} \tag{3.14}$$

in the strong operator topology. Then for the exceptional eigenvalues it holds

$$\lim_{n\to\infty} \alpha(A_n) = \alpha(A). \tag{3.15}$$

Proof. We consider the sequence of exceptional eigenvalues $\alpha(A_n)$. Let $\big(\alpha(A_{n_\nu})\big)$ be a subsequence which converges in $\mathbb{C}^+ \cup \overline{\mathbb{R}}$ with respect to the topology of the closed complex plane; denote the limit by α_0. From (3.7) we obtain

$$\Big[\varphi(\psi(A_{n_\nu}))\big(\psi(A_{n_\nu}) - \psi(\alpha(A_{n_\nu}))\big)\big(\psi(A_{n_\nu})^{-1} - \overline{\psi(\alpha(A_{n_\nu}))}\big) f, f\Big] \geq 0, \quad f \in \mathcal{H},$$

where ψ is as in (3.5). If we pass to the limit $\nu \to \infty$ it follows easily from the assumptions (3.14) and $\alpha(A_{n_\nu}) \to \alpha_0$ that

$$\Big[\varphi(\psi(A))\big(\psi(A) - \psi(\alpha_0)\big)\big(\psi(A)^{-1} - \overline{\psi(\alpha_0)}\big) f, f\Big] \geq 0 \text{ for all } f \in \mathcal{H}.$$

Since, on the other hand, the point $\beta = \psi(\alpha)$ in (3.7) is uniquely determined we must have $\alpha_0 = \alpha(A)$. Therefore each convergent subsequence of the sequence $\big(\alpha(A_n)\big)$ has the same limit and (3.15) follows. □

Remark 3.8. It is easy to see that Theorem 3.7 remains true if the assumption (3.14) is replaced by

$$\lim_{n\to\infty} (A_n - \mu)^{-k} = (A - \mu)^{-k}, \quad \lim_{n\to\infty} (A_n - \overline{\mu})^{-k} = (A - \overline{\mu})^{-k} \text{ for } k = 1, 2,$$

with the limits existing only in the weak operator topology.

4. The exceptional eigenvalues of the self-adjoint extensions of a nonnegative closed linear relation of regular defect one

4.1. The family of self-adjoint extensions

We recall some notions and results from [JL3]. Let A_0 be a closed nonnegative linear relation in the Krein space $(\mathcal{H}, [\cdot,\cdot])$. By $r(A_0)$ we denote the set of all points of regular type of A_0, i.e., the set of all $z \in \mathbb{C}$ for which there exists a $c_z > 0$ such that

$$\|f' - zf\| \geq c_z \|f\|, \quad \begin{pmatrix} f \\ f' \end{pmatrix} \in A_0.$$

A_0 is said to be of *regular defect one* if there exist non-real points $z_0, \bar{z}_0 \in r(A_0)$ with $\dim \left(\operatorname{ran}(A_0 - z_0)\right)^{[\perp]} = \dim \left(\operatorname{ran}(A_0 - \bar{z}_0)\right)^{[\perp]} = 1$. By [JL3, Theorem 3.2] the closed nonnegative relations of regular defect one are precisely those closed nonnegative relations of defect one which have at least one self-adjoint extension with nonempty resolvent set; here and in the sequel we consider only self-adjoint extensions without exit, that is within the given Krein space $\mathcal{H}$. The self-adjoint extensions of a closed nonnegative relation of regular defect one are either nonnegative or have one negative square, and their resolvent sets are nonempty with the possible exception of one extension.

In the rest of this paper, A_0 is a closed nonnegative relation in $(\mathcal{H}, [\cdot,\cdot])$ of regular defect one. It is our aim to study the exceptional eigenvalue of the self-adjoint extensions of A_0 which have one negative square.

First we consider the special case where no nonnegative self-adjoint extension A of A_0 with $\rho(A) \neq \emptyset$ exists. Then A_0 has exactly one nonnegative self-adjoint extension $\hat{A}$, this extension has the property $\rho(\hat{A}) = \emptyset$, and all self-adjoint extensions of A_0 except $\hat{A}$ have nonempty resolvent set and one negative square. The relations A_0 with this property were characterized in [JL3, Theorem 8.1]. All self-adjoint extensions A of A_0 different from $\hat{A}$ have the same spectrum σ', the same spectral function E' and the same exceptional eigenvalue α', which is also an eigenvalue of A_0. If $\alpha' \in \mathbb{C} \setminus \mathbb{R}$ then $A \cap \left(E'(\overline{\mathbb{R}})\mathcal{H}\right)^2 \subset A_0$. If $\alpha' \in \overline{\mathbb{R}}$ then $\sigma' \subset \mathbb{R}$ and for every connected closed subset $\Delta \subset \overline{\mathbb{R}} \setminus \{\alpha'\}$ such that $E'(\Delta)$ is defined we have $A \cap \left(E'(\Delta)\mathcal{H}\right)^2 \subset A_0$, see [JL3, Section 4 and Theorem 7.4].

In the rest of this section we exclude this special case, that is we assume that there exists a nonnegative self-adjoint extension A of A_0 with $\rho(A) \neq \emptyset$. We set $R(z) := (A - z)^{-1}$, $z \in \rho(A)$, and denote by E_A the spectral function of A in the Krein space $\mathcal{H}$. We fix a defect element $g \in \left(\operatorname{ran}(A_0 + i)\right)^{[\perp]}$ and define

$$g(z) := \left(I + (z - i)R(z)\right)g, \quad Q(z) := z[g, g] + (z^2 + 1)[R(z)g, g]. \tag{4.1}$$

If Q is constant on $\mathbb{C}^+$ then, according to [JL3, Lemma 7.1] $Q(z) = 0$ for $z \in \mathbb{C}^+$. According to [JL3, Theorem 7.2] the self-adjoint extensions of A_0 coincide with the linear relations $A_{(\gamma)}$, $\gamma \in \overline{\mathbb{R}}$, defined as follows:

$$\begin{aligned} A_{(\infty)} &= A, \\ \left(A_{(\gamma)} - z\right)^{-1} &= R(z) - \frac{[\,\cdot\,, g(\bar{z})]\, g(z)}{\gamma + Q(z)}, \quad \gamma \in \mathbb{R} \setminus \{0\}, \end{aligned} \tag{4.2}$$

for some (and then for all) $z \in \rho(A)$ with $\gamma + Q(z) \neq 0$, and

$$A_{(0)} := \lim_{\gamma \to 0, \gamma \neq 0} A_{(\gamma)},$$

where the limit is to be understood in the sense of the gap metric in $\mathcal{H}^2$. Here $A_{(0)}$ is the only self-adjoint extension of A_0 which can have an empty resolvent set, see [JL3, Theorem 7.2]. If Q is not identically equal to zero then the relation

(4.2) holds also for $\gamma = 0$. For every $\gamma \in \mathbb{R}$ we have

$$\rho(A_{(\gamma)}) \cap \rho(A) = \{z \in \rho(A) : \gamma + Q(z) \neq 0\}. \tag{4.3}$$

According to [JL3, Theorem 8.2] there exists a nonempty open interval (γ_-, γ_+), $-\infty \leq \gamma_- \leq 0 \leq \gamma_+ \leq \infty$, such that

$$A_{(\gamma)} \text{ has one negative square} \iff \gamma \in (\gamma_-, \gamma_+),$$
$$A_{(\gamma)} \text{ is nonnegative} \iff \gamma \in \overline{\mathbb{R}} \setminus (\gamma_-, \gamma_+).$$

Here γ_+ and γ_- can be expressed in terms of A and g (see [JL3, Section 8]): If J is a fundamental symmetry of $\mathcal{H}$, $\|x\|_J := [Jx, x]^{1/2}$, $x \in \mathcal{H}$, and if we denote by T_{op} the operator part of the self-adjoint relation T in the Hilbert space $(\mathcal{H}, [J\,\cdot\,, \cdot\,])$, then

$$\begin{aligned} \gamma_+ &= \begin{cases} \|(JA)_{\mathrm{op}}^{1/2} g\|_J^2 & \text{if } g \in \mathcal{D}((JA)_{\mathrm{op}}^{1/2}), \\ +\infty & \text{otherwise}, \end{cases} \\ \gamma_- &= \begin{cases} -\|(JA^{-1})_{\mathrm{op}}^{1/2} g\|_J^2 & \text{if } g \in \mathcal{D}((JA^{-1})_{\mathrm{op}}^{1/2}), \\ -\infty & \text{otherwise}. \end{cases} \end{aligned} \tag{4.4}$$

If $\gamma \in (\gamma_-, \gamma_+)$ and $\rho(A_{(\gamma)}) \neq \emptyset$, the exceptional eigenvalue of $A_{(\gamma)}$ is denoted by $\alpha(\gamma)$:

$$\alpha(\gamma) := \alpha(A_{(\gamma)}),$$

see Definition 3.4. Recall that these values $\alpha(\gamma)$, $\gamma \in (\gamma_-, \gamma_+)$, belong to $\overline{\mathbb{R}} \cup \mathbb{C}^+$. In the sequel we shall study the function $(\gamma_-, \gamma_+) \ni \gamma \mapsto \alpha(\gamma)$.

4.2. The family of functions $z\big(Q(z) + \gamma\big)$, $\gamma \in \mathbb{R}$

In this subsection we set for $\gamma \in \mathbb{R}$

$$F_\gamma(z) := z(Q(z) + \gamma) = (1 + z^2) z [R(z)g, g] + z^2 [g, g] + \gamma z. \tag{4.5}$$

A simple calculation shows that the function $z \mapsto z[R(z)g, g]$ belongs to $\mathcal{N}_0$ since A is nonnegative. Therefore, the functions F_γ form a family as considered in Section 2:

$$F_\gamma(z) = \gamma z + G_1(z) + z^2 G_2(z), \quad G_1(z) = z\,[R(z)g, g],\ G_2(z) = [g, g] + z\,[R(z)g, g].$$

We express the numbers α_j, β_j, the measures σ_j, $j = 1, 2$, and also γ_0 and γ_∞, which were introduced in Section 2, in terms of A and g. Evidently,

$$\begin{aligned} \alpha_1 &= \operatorname{Re} G_1(i) = -\operatorname{Im}[R(i)g, g] = -[R(i)g, R(i)g], \\ \alpha_2 &= \operatorname{Re} G_2(i) = -[R(i)g, R(i)g] \ + \ [g, g], \\ \beta := \beta_1 \ &= \ \beta_2 = \lim_{\eta \uparrow \infty} \eta^{-1} \operatorname{Im}\{i\eta[R(i\eta)g, g]\} \\ &= \lim_{\eta \uparrow \infty} \operatorname{Re}[R(i\eta)g, g] \ = \ \lim_{\eta \uparrow \infty} \frac{1}{2}\big[(R(i\eta) + R(-i\eta))g, g\big]. \end{aligned}$$

Further, $\sigma_1 = \sigma_2 =: \sigma$, and an application of the Stieltjes-Livšic inversion formula to the relation

$$(1 + z^2)^{-1} z [R(z)g, g] = (1 + z^2)^{-1} \left(\alpha_1 + \beta z + \int_{\mathbb{R}} \frac{1 + zt}{t - z}\, d\sigma(t) \right)$$

yields, for every interval (a,b) with $a,b\neq 0$, the formula

$$\int_{a+0}^{b-0} d\sigma(t) = \big[R(-i)AR(i)E_A((a,b))g,g\big]. \tag{4.6}$$

On the other hand,

$$\begin{aligned}[R(-i)AR(i)g,g] &= \operatorname{Im}\{i\,[R(i)g,g]\}\\ &= \operatorname{Im}\left\{\alpha+\beta i+\int_{\mathbb{R}}\frac{1+it}{t-i}\,d\sigma(t)\right\} = \beta+\int_{\mathbb{R}} d\sigma(t),\end{aligned}$$

and hence, with $\Delta_n := (-n,n)$, $n=1,2,\dots,$

$$\beta = \lim_{n\to\infty}[R(-i)AR(i)(1-E_A(\Delta_n))g,g]. \tag{4.7}$$

Lemma 4.1. *Assume that* $\int_{\mathbb{R}} t^2\,d\sigma(t)<\infty$ *or, equivalently,*

$$\sup\Big\{\big[AE_A(\Delta_n)g,g\big] : n=1,2,\dots\Big\}<\infty. \tag{4.8}$$

Then the sequence $\big(E_A(\Delta_n)\,g\big)$ *converges for* $n\to\infty$ *in* $\mathcal{D}\big((JA)_{\mathrm{op}}^{1/2}\big)$ *with respect to the graph norm of this domain. If* $g'_{(\infty)} := \lim_{n\to\infty} E_A(\Delta_n)g$ *and* $g'_\infty := g-g'_{(\infty)}$*, then* g'_∞ *belongs to the root subspace of* A^{-1} *corresponding to* 0 *and we have*

$$\beta = \big[A^{-1}g'_\infty, g'_\infty\big],\quad \alpha_2-\int_{\mathbb{R}} t\big(1+t^2\big)^{-1}d\sigma(t) = [g'_\infty,g'_\infty].$$

In particular, under the assumption (4.8), $\gamma_\infty<\infty$ *if and only if*

$$\big[A^{-1}g'_\infty,g'_\infty\big] = \big[g'_\infty,g'_\infty\big] = 0,$$

and in this case

$$\gamma_\infty = \lim_{n\to\infty}\big[AE_A(\Delta_n)g,g\big]. \tag{4.9}$$

Proof. On $\operatorname{span}\big\{E_A(\Delta_n\setminus\Delta_1)\mathcal{H} : n=2,\dots\big\}$ the norm $[A\,\cdot,\cdot]^{1/2}$ is equivalent to the graph norm of $(JA)_{\mathrm{op}}^{1/2}$. If $n_1,\ n_2\in\mathbb{N}$, $n_2\geq n_1$, then

$$[A(E_A(\Delta_{n_2})-E_A(\Delta_{n_1}))g,(E_A(\Delta_{n_2})-E_A(\Delta_{n_1}))g] = \int_{\Delta_{n_2}\setminus\Delta_{n_1}}\big(1+t^2\big)d\sigma(t),$$

which implies the first assertion.

For every bounded interval Δ we have

$$E_A(\Delta)g'_\infty = E_A(\Delta)g-\lim_{n\to\infty}E_A(\Delta)E_A(\Delta_n)g = 0.$$

Hence g'_∞ and $A^{-1}g'_\infty$ belong to the root space of A^{-1} at zero, and $[A^{-1}g'_\infty,g'_{(\infty)}] = [g'_\infty,g'_{(\infty)}]=0$. In the subspace $(I-E_A(\Delta_1))\mathcal{H}$ the following identities hold:

$$R(-i)R(i)\Big|_{(I-E_A(\Delta_1))\mathcal{H}} = \big(I-R(-i)R(i)\big)A^{-2}\Big|_{(I-E_A(\Delta_1))\mathcal{H}}, \tag{4.10}$$

$$R(-i)AR(i)\Big|_{(I-E_A(\Delta_1))\mathcal{H}} = \big(A^{-1}-A^{-1}R(-i)R(i)\big)\Big|_{(I-E_A(\Delta_1))\mathcal{H}}. \tag{4.11}$$

Since the operator $A^{-1}|_{(I-E_A(\Delta_1))\mathcal{H}}$ is nonnegative, its root subspace at zero coincides with the kernel of its square. Therefore $A^{-2}g'_\infty = 0$ and, by (4.10) and (4.11),

$$R(-i)R(i)g'_\infty = 0, \quad R(-i)AR(i)g'_\infty = A^{-1}g'_\infty. \tag{4.12}$$

In view of (4.7),

$$\begin{aligned}\beta &= \lim_{n\to\infty} \big[R(-i)AR(i)(I - E_A(\Delta_n))g, g\big] \\ &= \big[R(-i)AR(i)g'_\infty, g'_{(\infty)} + g'_\infty\big] = \big[A^{-1}g'_\infty, g'_{(\infty)} + g'_\infty\big] = \big[A^{-1}g'_\infty, g'_\infty\big].\end{aligned}$$

Moreover,

$$\begin{aligned}\alpha_2 - \int_{\mathbb{R}} t\, d\sigma(t) &= -\big[R(-i)R(i)g, g\big] + [g, g] - \lim_{n\to\infty}\int_{\Delta_n} t\, d\sigma(t) \\ &= \big[(I - R(-i)R(i))g, g\big] - \lim_{n\to\infty}\big[(I - R(-i)R(i))E_A(\Delta_n)g, g\big] \\ &= \lim_{n\to\infty}\big[(I - R(-i)R(i))(I - E_A(\Delta_n))g, g\big] \\ &= \big[(I - R(-i)R(i))g'_\infty, g'_\infty\big] = \big[g'_\infty, g'_\infty\big].\end{aligned}$$

Here the last equality is a consequence of (4.12). Finally, the last assertion of the lemma follows from the definition of γ_∞. □

Remark 4.2. In the terminology of [KK, Section 4], the inequality $\gamma_\infty < \infty$ means that the function $z \longmapsto z[R(z)g, g]$ belongs to the class (R_0). In this case (4.9) and [KK, Theorem S 1.4.2] give also other expressions for γ_∞, e.g.,

$$\gamma_\infty = \sup_{\eta>0}\left\{\eta^2\, \mathrm{Re}\,[R(i\eta)g, g]\right\}.$$

Set $\Gamma_n := \overline{\mathbb{R}} \setminus [-n^{-1}, n^{-1}]$, $n = 1, 2, \ldots$. The following lemma can be proved similarly to Lemma 4.1.

Lemma 4.3. *Assume that* $\int_{\mathbb{R}\setminus\{0\}} t^{-2}\, d\sigma(t) < \infty$ *or, equivalently,*

$$\sup\left\{\big[A^{-1}E_A(\Gamma_n)g, g\big] : n = 1, 2, \ldots\right\} < \infty. \tag{4.13}$$

Then the sequence $\big(E_A(\Gamma_n)g\big)$ *converges for* $n \to \infty$ *in* $\mathcal{D}\big((JA^{-1})^{1/2}_{\mathrm{op}}\big)$ *with respect to the graph norm of this domain. If* $g'_{(0)} := \lim_{n\to\infty} E_A(\Gamma_n)g$ *and* $g'_0 := g - g'_{(0)}$, *then* g'_0 *belongs to the root subspace of* A *at zero,*

$$\big[Ag'_0, g'_{(0)}\big] = \big[g'_0, g'_{(0)}\big] = 0,$$

and we have

$$\sigma(\{0\}) = \big[Ag'_0, g'_0\big], \quad \alpha_1 + \lim_{n\to\infty}\int_{\Gamma_n} t^{-1}\, d\sigma(t) = -\big[g'_0, g'_0\big].$$

In particular, under the assumption (4.13), $\gamma_0 > -\infty$ *if and only if*

$$[Ag'_0, g'_0] = [g'_0, g'_0] = 0,$$

and in this case

$$\gamma_0 = -\lim_{n\to\infty}\big[A^{-1}E_A(\Gamma_n)g, g\big].$$

From the above lemmas, with the definitions of g'_∞ and g'_0 we obtain the following relations:

$$\gamma_\infty = \begin{cases} \lim_{n\to\infty} \left[AE_A(\Delta_n)g, g\right] & \text{if } \sup_n \left[AE_A(\Delta_n)g, g\right] < \infty, \\ & \left[A^{-1}g'_\infty, g'_\infty\right] = \left[g'_\infty, g'_\infty\right] = 0, \\ \infty & \text{otherwise,} \end{cases}$$
$$\gamma_0 = \begin{cases} -\lim_{n\to\infty} \left[A^{-1}E_A(\Gamma_n)g, g\right] & \text{if } \sup_n \left[A^{-1}E_A(\Gamma_n)g, g\right] < \infty, \\ & \left[Ag'_0, g'_0\right] = \left[g'_0, g'_0\right] = 0, \\ -\infty & \text{otherwise.} \end{cases} \tag{4.14}$$

Trivially, $\gamma_0 \le 0 \le \gamma_\infty$, hence

$$\gamma_0 = \gamma_\infty \iff \gamma_0 = \gamma_\infty = 0. \tag{4.15}$$

4.3. The case $\rho(A_{(0)}) = \emptyset$

We shall show that in this case the exceptional eigenvalue $\alpha(\gamma)$, $\gamma \in (\gamma_-, \gamma_+)$, is 0 or ∞. To prove this, we rely on the results of [JL3].

Theorem 4.4. *Assume that there exists a nonnegative self-adjoint extension A of A_0 with $\rho(A) \neq \emptyset$. Then the following conditions are equivalent.*

(i) $\rho(A_{(0)}) = \emptyset$.

(ii) *The element g admits the unique representation $g = g_0 + g_\infty$, $g_0 \in \ker A_0$, $g_\infty \in A_0(0)$*[1]*, and $[g_0, g_0] = [g_\infty, g_\infty] = 0$.*

(iii) $\gamma_0 = \gamma_\infty$.

If these conditions are fulfilled, then

$$\gamma_+ = 0 \iff g_\infty = 0, \qquad \gamma_- = 0 \iff g_0 = 0, \tag{4.16}$$

and

$$\alpha(\gamma) = \begin{cases} 0 & \gamma \in (\gamma_-, 0), \\ \infty & \gamma \in (0, \gamma_+). \end{cases}$$

Proof. If (i) holds then $A_{(0)} \neq A$. Then, by [JL3, Corollary 6.2], $\operatorname{ran}\left(A_0 + i\right)^{[\perp]}$ is spanned by an element of the form $g_0 + g_\infty$ where $g_0 \in \ker A_0$, $g_\infty \in A_0(0)$, and $\left[g_0, g_0\right] = \left[g_\infty, g_\infty\right] = 0$, that is, (ii) holds. Assume now that (ii) holds. Since $\left[g_0, g_\infty\right] = 0$ we have

$$Q(z) := z\left[g_0 + g_\infty, g_0 + g_\infty\right] + \left(z^2 + 1\right)\left[R(z)(g_0 + g_\infty), g_0 + g_\infty\right] = 0$$

for every $z \in \rho(A)$. Then, by [JL3, Lemma 7.1] there exists a self-adjoint extension $\widehat{A}$ of A_0 with $\rho(\widehat{A}) = \emptyset$. By [JL3, Theorem 7.2], $\widehat{A}$ coincides with $A_{(0)}$ and, therefore, (i) holds. Moreover, (ii) implies

$$\begin{aligned} &\sup\left\{\left[AE_A(\Delta_n)g, g\right] : n = 1, 2, \dots\right\} \\ &\qquad = \sup\left\{\left[A^{-1}E_A(\Gamma_n)g, g\right] : n = 1, 2, \dots\right\} = 0, \end{aligned}$$

[1] By [JL3, (3.1)] we have $\ker A_0 \cap A_0(0) = \{0\}$.

and by the definition of g'_∞ and g'_0 in Lemma 4.1 and Lemma 4.3 we have $g'_0 = g_0$ and $g'_\infty = g_\infty$. Then, by (4.14), $\gamma_0 = \gamma_\infty = 0$, that is, (iii) is true.

If (iii) holds, then, by (4.15), $\gamma_0 = \gamma_\infty = 0$. On account of (4.14), this implies for every bounded real interval Δ with $0 \notin \overline{\Delta}$ that $\big[AE_A(\Delta)g, g\big] = 0$ and hence $E_A(\Delta)g = 0$. By the definition of g'_∞ and g'_0 we obtain

$$g'_0 = E_A\big((-1,1)\big)g, \quad g'_\infty = \big(I - E_A\big((-1,1)\big)\big)g.$$

Therefore, $g = g'_0 + g'_\infty$ and, in view of $\big[Ag'_0, g'_0\big] = 0$ (see (4.14)), for every $x \in E_A\big((-1,1)\big)\mathcal{H}$,

$$\big|\big[Ag'_0, x\big]\big| \leq \big[Ag'_0, g'_0\big]^{1/2}\big[Ax, x\big]^{1/2} = 0,$$

which implies $g'_0 \in \ker A$. Similarly, $g'_\infty \in A(0)$. Again by (4.14) we have $\big[g'_0, g'_0\big] = \big[g'_\infty, g'_\infty\big] = 0$. Hence (ii) holds with A_0 replaced by A. This yields $[g(z), g(z)] = 0$ for all $z \in \mathbb{C} \setminus \mathbb{R}$, and, by [JL3, Theorem 6.1], (i) is true.

Assume now that the conditions (i–iii) are satisfied. By (4.4) the relation $\gamma_+ = 0$ is equivalent to $g \in \ker A$. In view of (ii) this is equivalent to $g_\infty \in (\ker A) \cap A(0)$. The latter holds, by the assumption on A, if and only if $g_\infty = 0$ (see [JL3, Theorem 1.3]). Similarly it follows that $\gamma_- = 0$ is equivalent to $g_0 = 0$.

It remains to prove the last assertion about the exceptional eigenvalues. Let $\gamma \in (\gamma_-, 0) \cup (0, \gamma_+)$. Then, by (ii) and [JL3, Theorem 7.2],

$$\big(A_{(\gamma)} - z\big)^{-1} = R(z) - \gamma^{-1}\big[\cdot, iz^{-1}g_0 + g_\infty\big]\big(iz^{-1}g_0 + g_\infty\big),$$

and the following function has one negative square:

$$\begin{aligned} z \mapsto z\big(A_{(\gamma)} - z\big)^{-1} &= zR(z) - \gamma^{-1}z^{-1}\big[\cdot, g_0\big]g_0 \\ &\quad - \gamma^{-1}z\big[\cdot, g_\infty\big]g_\infty - \gamma^{-1}\big(-i\big[\cdot, g_0\big]g_\infty + i\big[\cdot, g_\infty\big]g_0\big). \end{aligned} \tag{4.17}$$

If $\gamma > 0$ then $\gamma_+ > 0$ and, by (4.16), $g_\infty \neq 0$. In this case, the terms on the right-hand side of (4.17) with the exception of the third term are Nevanlinna functions. We choose some $x \in \mathcal{H}$ such that the function $G_x: \; G_x(z) := z\big[(A_{(\gamma)} - z)^{-1}x, x\big]$ belongs to $\mathcal{N}_1$. Then we have $[x, g_\infty] \neq 0$ and the generalized pole of the function $z \mapsto -\gamma^{-1}z|[x, g_\infty]|^2$, which is the point ∞, coincides with the generalized pole of G_x. Therefore there exists a sequence $(z_n) \subset \mathbb{C}^+$ converging to ∞ in $\overline{\mathbb{C}}$ such that $\liminf_{n\to\infty} \operatorname{Im} G_x(z_n) < 0$. Then, on account of Theorem 3.5, $\alpha(\gamma) = \infty$. If $\gamma < 0$ then $\gamma_- < 0$ and, by (4.16), $g_0 \neq 0$. Now all the terms on the right-hand side of (4.17) with the exception of the second term are Nevanlinna functions and, as above, using Theorem 3.5 we find $\alpha(\gamma) = 0$. □

4.4. The generic case

Now we describe the exceptional eigenvalue $\alpha(\gamma)$, $\gamma \in (\gamma_-, \gamma_+)$, if $\rho(A_{(\gamma)}) \neq \emptyset$ for all $\gamma \in \overline{\mathbb{R}}$. To this end we apply the results of Section 2 to the functions F_γ from (4.5). Recall that the function F_γ belongs to the class $\mathcal{N}_1$ if and only if $\gamma \in (-\infty, \gamma_\infty)$, and as in Section 2, for these γ we denote the zero of nonpositive type of F_γ by $\zeta(\gamma)$.

Lemma 4.5. *Suppose that $\rho(A_{(\gamma)}) \neq \emptyset$ for all $\gamma \in \overline{\mathbb{R}}$. Then*

$$-\infty \leq \gamma_- \leq \gamma_0 \leq 0 \leq \gamma_\infty \leq \gamma_+ \leq +\infty, \tag{4.18}$$

and $\alpha(\gamma) = \zeta(\gamma)$ for $\gamma \in (\gamma_0, \gamma_\infty)$. For these γ, if $\alpha(\gamma) \in \mathbb{C}^+$ then $g(\alpha(\gamma))$ (see (4.1)) is an eigenelement of $A_{(\gamma)}$ at the eigenvalue $\alpha(\gamma)$, if $\alpha(\gamma) = \overline{\alpha(\gamma)} \neq 0$ then the limit $g(\alpha(\gamma)) := \lim_{\varepsilon\to 0} g(\alpha(\gamma) + i\varepsilon)$ exists and is an eigenelement of $A_{(\gamma)}$ at $\alpha(\gamma)$ and

$$(\operatorname{sign}\alpha(\gamma))\,[g(\alpha(\gamma)), g(\alpha(\gamma))] \leq 0. \tag{4.19}$$

Proof. By Theorem 4.4 we have $\gamma_0 \neq \gamma_\infty$. The relations (4.14) show that $\gamma_0 \leq 0 \leq \gamma_\infty$ holds. Let $\gamma \in (\gamma_0, \gamma_\infty)$. Then, according to Lemma 2.3 and Theorem 2.4, $\zeta(\gamma) \neq 0, \infty$. We prove Lemma 4.5 in the two cases $\zeta(\gamma) \notin \mathbb{R}$ and $\zeta(\gamma) \in \mathbb{R} \setminus \{0\}$ separately.

If $\zeta(\gamma) \notin \mathbb{R}$ then the second term on the right-hand side of (4.2) has a pole at $\zeta(\gamma)$ and $\zeta(\gamma)$ is an eigenvalue of $A_{(\gamma)}$ in $\mathbb{C}^+$. It follows that $A_{(\gamma)}$ is not nonnegative, i.e., $\gamma \in (\gamma_-, \gamma_+)$ and $\alpha(\gamma) = \zeta(\gamma)$. Since $g(\alpha(\gamma))$ is a defect element of A_0 at the point $\alpha(\gamma)$ it is an eigenelement of the extension $A_{(\gamma)}$ of A_0 to the eigenvalue $\alpha(\gamma)$.

Let $\omega := \zeta(\gamma) \in \mathbb{R} \setminus \{0\}$. Then, by Lemma 2.3, $\int_{\mathbb{R}} |t-\omega|^{-2}\, d\sigma(t) < \infty$. Let Δ be a bounded open interval, such that $\omega \in \Delta$, $0 \notin \bar{\Delta}$. We denote by ι the linear mapping $\phi \mapsto E_A(\Delta)\phi(A)g$ defined for all rational functions ϕ with only non-real poles. Then ι is an isometry from the linear manifold of these elements ϕ of $L^2(\Delta;\sigma)$ into $\big(E_A(\Delta)\mathcal{H}, [R(-i)AR(i)\cdot,\cdot]\big)$. We extend ι to an isometry from $L^2(\Delta;\sigma)$ onto a closed subspace of $\big(E_A(\Delta)\mathcal{H}, [R(-i)AR(i)\cdot,\cdot]\big)$. For any fundamental symmetry J the scalar products $[J\cdot,\cdot]$ and $[R(-i)AR(i)\cdot,\cdot]$ are equivalent on $E_A(\Delta)\mathcal{H}$.

Since the function $(\,\cdot\, - \omega)^{-1}$ belongs to $L^2(\Delta;\sigma)$ and the functions $(\,\cdot\, - \omega - i\epsilon)^{-1}$ converge in $L^2(\Delta;\sigma)$ to $(\,\cdot\, - \omega)^{-1}$ for $\epsilon \to 0$, also

$$\lim_{\epsilon\to 0} R(\omega + i\epsilon)g =: g_\omega \tag{4.20}$$

exists. For arbitrary $z \in \rho(A)$ we have

$$R(z)g_\omega = (z-\omega)^{-1}(R(z)g - g_\omega). \tag{4.21}$$

One verifies without difficulty that for the function F_γ considered in the present section, $\omega \in \mathbb{R} \setminus \{0\}$ and under the assumption (4.20) the relation (2.9) holds if and only if

$$\omega(\omega^2+1)[g_\omega, g_\omega] + (3\omega^2+1)[g_\omega, g] + 2\omega[g,g] + \gamma \leq 0, \tag{4.22}$$

and (2.10) holds if and only if

$$\gamma\omega + \omega^2[g,g] + \omega(1+\omega^2)[g_\omega, g] = 0. \tag{4.23}$$

Therefore the relations (4.22) and (4.23) hold. Inserting (4.23) into (4.22) we obtain

$$\omega(\omega^2+1)[g_\omega, g_\omega] + 2\omega^2[g_\omega, g] + \omega[g,g] \leq 0. \tag{4.24}$$

Using the relations (4.21) and (4.23) it can be verified by an easy calculation that the element

$$g(\omega) = g + (\omega - i)g_\omega \tag{4.25}$$

is not the zero element and an eigenelement of $A_{(\gamma)}$ to the eigenvalue ω. Further,

$$[g(\omega), g(\omega)] = [g,g] + (\omega - i)[g_\omega, g] + (\omega + i)[g, g_\omega] + (\omega^2+1)[g_\omega, g_\omega] \tag{4.26}$$

and, by (4.24),

$$(\operatorname{sign}\omega)[g(\omega), g(\omega)] \le 0. \tag{4.27}$$

Hence $A_{(\gamma)}$ is not nonnegative, i.e., $\gamma \in (\gamma_-, \gamma_+)$, and, by Theorem 3.1, $\omega = \zeta(\gamma)$ coincides with $\alpha(\gamma)$. □

Lemma 4.6. *If $\gamma \in (\gamma_-, \gamma_+)$ and $\alpha(\gamma) \ne 0, \infty$, then $\gamma \in (\gamma_0, \gamma_\infty)$, $\alpha(\gamma) = \zeta(\gamma)$, and we have the direct orthogonal decomposition*

$$\ker\big(A_{(\gamma)} - \alpha(\gamma)\big) = \operatorname{span}\{g(\alpha(\gamma))\} \dot{+} \ker(A - \alpha(\gamma)). \tag{4.28}$$

Proof. If $\alpha(\gamma) \in \mathbb{C}^+$ then, by (4.2), F_γ has a pole at $\alpha(\gamma)$ and therefore $\zeta(\gamma) = \alpha(\gamma)$. Then Theorem 2.4,(1) implies $\gamma \in (\gamma_0, \gamma_\infty)$. In this case (4.28) holds with $\ker(A - \alpha(\gamma)) = \{0\}$.

Assume that $\omega := \alpha(\gamma) \in \mathbb{R} \setminus \{0\}$. Then there exists an $x \in \mathcal{H}$, $x \ne 0$, such that $\begin{pmatrix} x \\ \omega x \end{pmatrix} \in A_{(\gamma)}$ and

$$(\operatorname{sign}\omega)[x,x] \le 0. \tag{4.29}$$

By [JL3, Corollary 7.3], there exists an $x_0 \in \mathcal{H}$, $x_0 \ne 0$, such that

$$\begin{pmatrix} x \\ \omega x \end{pmatrix} = (\gamma + i[g,g])\begin{pmatrix} R(-i)x_0 \\ x_0 - iR(-i)x_0 \end{pmatrix} - [x_0, g]\begin{pmatrix} g \\ ig \end{pmatrix}. \tag{4.30}$$

We have $\gamma + i[g,g] \ne 0$ since otherwise (4.30) would imply $\omega = i$. We also have $[x_0, g] \ne 0$. Indeed, $[x_0, g] = 0$ would imply that $\begin{pmatrix} x \\ \omega x \end{pmatrix} \in A$ and that x is an eigenvector of the bounded nonnegative operator $R(-i)AR(i)$, corresponding to the eigenvalue $\omega(\omega^2+1)^{-1}$, and hence that $(\operatorname{sign}\omega)[x,x] > 0$. This contradicts (4.29).

From (4.30) we get

$$\big(1 - (\omega + i)R(-i)\big)x_0 = Cg \text{ with } C := (i - \omega)[x_0, g](\gamma + i[g,g])^{-1}, \tag{4.31}$$

which implies

$$E_A(\{\omega\})g = 0, \tag{4.32}$$

where E_A denotes again the spectral function of A.

If we set $x_0' := E_A(\{\omega\})x_0$, $\tilde{x}_0 := x_0 - x_0'$, then $[x_0, g] = [\tilde{x}_0, g]$, and we shall show that for

$$\tilde{x} := (\gamma + i[g,g])R(-i)\tilde{x}_0 - [\tilde{x}_0, g]g \tag{4.33}$$

it holds $\begin{pmatrix} \tilde{x} \\ \omega\tilde{x} \end{pmatrix} \in A_{(\gamma)}$ and

$$(\operatorname{sign}\omega)[\tilde{x}, \tilde{x}] \le 0. \tag{4.34}$$

Indeed,

$$\tilde{x} = x - \big(\gamma + i[g,g]\big)R(-i)x_0' = x - \big(\gamma + i[g,g]\big)(\omega+i)^{-1}x_0' \tag{4.35}$$

and

$$\begin{aligned}\begin{pmatrix}\tilde{x}\\ \omega\tilde{x}\end{pmatrix} &= \begin{pmatrix}x\\ \omega x\end{pmatrix} - (\gamma + i[g,g])\begin{pmatrix}(\omega+i)^{-1}x_0'\\ \omega(\omega+i)^{-1}x_0'\end{pmatrix}\\ &= (\gamma + i[g,g])\left\{\begin{pmatrix}R(-i)x_0\\ x_0 - iR(-i)x_0\end{pmatrix} - \begin{pmatrix}R(-i)x_0'\\ x_0' - iR(-i)x_0'\end{pmatrix}\right\}\\ &\quad - [x_0,g]\begin{pmatrix}g\\ ig\end{pmatrix} = (\gamma + i[g,g])\begin{pmatrix}R(-i)\tilde{x}_0\\ \tilde{x}_0 - iR(-i)\tilde{x}_0\end{pmatrix} - [\tilde{x}_0,g]\begin{pmatrix}g\\ ig\end{pmatrix} \in A_{(\gamma)}.\end{aligned} \tag{4.36}$$

To verify (4.34) observe that because of $[\widetilde{x}_0, x_0'] = [g, x_0'] = 0$ we have

$$[\widetilde{x}, x_0'] = 0. \tag{4.37}$$

As A is nonnegative,

$$(\mathrm{sign}\omega)\,[x_0', x_0'] \geq 0. \tag{4.38}$$

Using (4.29), (4.35), (4.37), and (4.38) we find

$$\begin{aligned}0 &\geq (\mathrm{sign}\,\omega)\,[x,x]\\ &= (\mathrm{sign}\,\omega)\,\big[\widetilde{x} + (\gamma + i[g,g])(\omega+i)^{-1}x_0', \widetilde{x} + (\gamma + i[g,g])(\omega+i)^{-1}x_0'\big]\\ &\geq (\mathrm{sign}\,\omega)\,[\widetilde{x},\widetilde{x}].\end{aligned}$$

By (4.31) we have

$$(I - (\omega+i)R(-i))\widetilde{x}_0 = Cg. \tag{4.39}$$

Let Δ be a bounded open interval such that $\omega \in \Delta$ and $0 \notin \bar{\Delta}$. Then, because of (4.6) and (4.39), for every $\varepsilon > 0$,

$$\begin{aligned}&|C|^2 \int_\Delta ((t-\omega)^2 + \epsilon^2)^{-1}\, d\sigma(t)\\ &= |C|^2 \big[R(\omega+i\varepsilon)R(\omega-i\varepsilon)R(-i)AR(i)E_A(\Delta)g, E_A(\Delta)g\big]\\ &= |C|^2 \big[R(\omega+i\varepsilon)(A-\omega)R(\omega-i\varepsilon)(A-\omega)R(-i)^2AR(i)^2E_A(\Delta)\widetilde{x}_0, \widetilde{x}_0\big]\\ &= |C|^2 \int_\Delta (t-\omega)^2((t-\omega)^2+\epsilon^2)^{-1}(1+t^2)^{-2}\, t\, d[E_A(t)\widetilde{x}_0, \widetilde{x}_0],\end{aligned}$$

and therefore,

$$\int_\Delta (t-\omega)^{-2}\, d\sigma(t) < \infty, \tag{4.40}$$

i.e., the function $t \mapsto (t-\omega)^{-1}$ belongs to $L^2(\Delta;\sigma)$. Now we see as in the proof of Lemma 4.5 that $\lim_{\varepsilon\downarrow 0} R(\omega+i\varepsilon)g =: g_\omega$ exists.

Applying the operator $I + (\omega + i + i\varepsilon)R(\omega+i\varepsilon)$ to (4.39) we find

$$\big(I + i\varepsilon R(\omega+i\varepsilon)\big)\widetilde{x}_0 = C\big(g + (\omega+i+i\varepsilon)R(\omega+i\varepsilon)g\big),$$

and for $\varepsilon \downarrow 0$, on account of $E_A(\{\omega\})\widetilde{x}_0 = 0$,

$$\widetilde{x}_0 = C\big(g + (\omega+i)g_\omega\big). \tag{4.41}$$

Taking the inner product with $g = \big(I - E_A(\{\omega\})\big)g$ on both sides of this relation and inserting C from (4.31) we obtain

$$[x_0, g] = (i-\omega)[x_0, g]\big(\gamma + i[g,g]\big)^{-1}\big(g + (\omega+i)g_\omega\big),$$

i.e.,

$$\gamma + \omega[g,g] + (\omega^2+1)[g_\omega, g] = 0. \tag{4.42}$$

The relations (4.33) and (4.41) yield

$$\begin{aligned}\widetilde{x} &= (i-\omega)\big[\widetilde{x}_0, g\big]R(-i)\big(g + (\omega+i)g_\omega\big) - \big[\widetilde{x}_0, g\big]g \\ &= -\big[\widetilde{x}_0, g\big]\left\{g - (i-\omega)\big(R(-i)g + (\omega-i)R(-i)g_\omega\big)\right\}.\end{aligned}$$

This implies in view of

$$(\omega+i)R(-i)g_\omega = g_\omega - R(-i)g$$

(see (4.21)) the relation

$$\widetilde{x} = -\big[\widetilde{x}_0, g\big]\big(g - (i-\omega)g_\omega\big) = -\big[\widetilde{x}_0, g\big]\,g(\omega). \tag{4.43}$$

From (4.34) we obtain

$$\begin{aligned}0 &\geq \omega\big[g - (i-\omega)g_\omega, g - (i-\omega)g_\omega\big] \\ &= \omega(\omega^2+1)\big[g_\omega, g_\omega\big] + 2\omega^2\big[g_\omega, g\big] + \omega[g,g].\end{aligned}$$

Adding (4.42) it follows that

$$0 \geq \omega(\omega^2+1)\big[g_\omega, g_\omega\big] + (3\omega^2+1)\big[g_\omega, g\big] + 2\omega[g,g] + \gamma. \tag{4.44}$$

Now (4.44) and (4.42) imply $\omega = \zeta(\gamma)$, hence $\gamma \in (\gamma_0, \gamma_\infty)$. The first equality in (4.36) and (4.43) give (4.28). □

Now we prove the main result about the curve $\gamma \mapsto \alpha(\gamma)$ of the exceptional eigenvalue. For the definition of the elements g_0', g_∞' see Lemma 4.1 and Lemma 4.3, the measure σ and the number β were defined in Subsection 4.2.

Theorem 4.7. *Suppose that all self-adjoint extensions of A_0 have nonempty resolvent set. Then the following statements hold.*

(i) *The function $\alpha : (\gamma_-, \gamma_+) \to \mathbb{C}^+ \cup \overline{\mathbb{R}}$ is continuous and*

$$\lim_{\gamma \downarrow \gamma_-} \alpha(\gamma) = 0, \qquad \lim_{\gamma \uparrow \gamma_+} \alpha(\gamma) = \infty.$$

(ii) *We have $-\infty \leq \gamma_- \leq \gamma_0 \leq 0 \leq \gamma_\infty \leq \gamma_+ \leq \infty$, $\gamma_0 < \gamma_\infty$, and*

$$\alpha(\gamma) = \begin{cases} 0 & \text{if } \gamma \in (\gamma_-, \gamma_0), \\ \infty & \text{if } \gamma \in (\gamma_\infty, \gamma_+). \end{cases}$$

(iii) *The function $|\alpha| : |\alpha|(\gamma) := |\alpha(\gamma)|$, $\gamma \in (\gamma_0, \gamma_\infty)$, is nondecreasing and $\gamma_1, \gamma_2 \in (\gamma_0, \gamma_\infty)$, $\gamma_1 \neq \gamma_2$ implies $\alpha(\gamma_1) \neq \alpha(\gamma_2)$. If $\sigma \neq 0$ or $\beta \neq 0$, then this function is strictly increasing, if $\sigma = 0$ and $\beta = 0$, then it is strictly increasing if and only if $[g_0', g_0']\,[g_\infty', g_\infty'] \geq 0$.*

(iv) *The function α is real analytic in $\gamma \in (\gamma_0, \gamma_\infty)$ if $\alpha(\gamma) \notin \operatorname{supp}\sigma$ with possible exception of branch points of order ≤ 3 for which $\alpha(\gamma) \in \mathbb{R}$.*

(v) $\sigma\left(\{\alpha(\gamma) : \gamma \in (\gamma_0, \gamma_\infty)\} \cap \mathbb{R}\right) = 0$.

(vi) *For every component I of the open set $\mathbb{R} \setminus \left(\operatorname{supp}\sigma \cup \{0\}\right)$ the set $I \cap \{\alpha(\gamma) : \gamma \in (\gamma_0, \gamma_\infty)\}$ is empty or connected.*

The statement (v) means that $\alpha(\gamma)$ cannot move within the support of σ, (vi) means that every bounded component I as in the statement is 'visited' by the exceptional eigenvalue at most once.

Proof of Theorem 4.7. The first assertion of (i) follows from (4.2) and Theorem 3.7. By Theorem 4.4 we have $\gamma_0 < \gamma_\infty$. Then, according to Lemma 4.5, $\alpha(\gamma) = \zeta(\gamma)$ for $\gamma \in (\gamma_0, \gamma_\infty)$, and Theorem 2.4 implies

$$\lim_{\gamma \downarrow \gamma_0} \alpha(\gamma) = 0, \qquad \lim_{\gamma \uparrow \gamma_\infty} \alpha(\gamma) = \infty. \tag{4.45}$$

If $\gamma \in (\gamma_-, \gamma_0) \cup (\gamma_\infty, \gamma_+)$ then by Lemma 4.6 either $\alpha(\gamma) = 0$ or $\alpha(\gamma) = \infty$. Hence, on account of the continuity of $\gamma \mapsto \alpha(\gamma)$ and (4.45), $\alpha(\gamma) = 0$ for $\gamma \in (\gamma_-, \gamma_0)$ and $\alpha(\gamma) = \infty$ for $\gamma \in (\gamma_\infty, \gamma_+)$. This proves the assertions (i) and (ii). Lemma 4.6 and Theorem 2.4 imply the first assertion of (iii) and that the function $\gamma \mapsto |\alpha(\gamma)|$ is nondecreasing.

The criterion on strict monotonicity follows from Theorem 2.4 in view of the fact that we have

$$\alpha_1 = -\left[g_0', g_0'\right], \qquad \alpha_2 = -\left[g_0', g_0'\right] + [g, g] = \left[g_\infty', g_\infty'\right].$$

This proves assertion (iii). The assertions (iv), (v), and (vi) are consequences of Theorem 2.4 and Theorem 2.5. □

The following two examples show that intervals of constancy of the functions $\gamma \mapsto \alpha(\gamma)$ and $\gamma \mapsto |\alpha(\gamma)|$ may occur, even in finite-dimensional situations.

Example 4.8. In the Krein space $\left(\mathbb{C}^2, \left(\begin{pmatrix} 0 & 1 \\ 1 & 0 \end{pmatrix} \cdot, \cdot\right)\right)$ we consider the nonnegative operator $T_a := \begin{pmatrix} 0 & 0 \\ a & 0 \end{pmatrix}$, $a > 0$. Define $\mathcal{H}_0 := \mathcal{H}_\infty := \mathbb{C}^2$, $\mathcal{H} := \mathcal{H}_0 \times \mathcal{H}_\infty$ and the nonnegative relation $A = T_{a_0} \times T_{a_\infty}^{-1}$ for some $a_0,\ a_\infty > 0$. Let $g = (b_1, b_2, c_1, c_2)^T$. With the help of Lemmas 4.1 and 4.3 we can easily compute the quantities γ_+, γ_∞, γ_-, γ_0:

$$\gamma_+ = \begin{cases} \infty & \text{if } c_1 \neq 0 \\ a_0|b_1|^2 + a_\infty^{-1}|c_2|^2 & \text{if } c_1 = 0 \end{cases}, \qquad \gamma_\infty = \begin{cases} \infty & \text{if } c_1 \neq 0 \\ a_0|b_1|^2 & \text{if } c_1 = 0 \end{cases},$$

$$\gamma_- = \begin{cases} -\infty & \text{if } b_1 \neq 0 \\ -a_\infty|c_1|^2 - a_0^{-1}|b_2|^2 & \text{if } b_1 = 0 \end{cases}, \qquad \gamma_0 = \begin{cases} -\infty & \text{if } b_1 \neq 0 \\ -a_\infty|c_1|^2 & \text{if } b_1 = 0 \end{cases}.$$

Example 4.9. Let $(\mathcal{H}, [\cdot,\cdot])$ be the orthogonal direct sum of the Krein subspaces $(\mathcal{H}', [\cdot,\cdot])$ and $(\mathcal{H}'', [\cdot,\cdot])$. Then $A := \left\{\binom{x'}{y''} : x' \in \mathcal{H}',\ y'' \in \mathcal{H}''\right\}$ is a nonnegative self-adjoint relation. We choose some $g = g' + g''$ with $g' \in \mathcal{H}'$, $g'' \in \mathcal{H}''$ and $[g', g'][g'', g''] < 0$. Then $(\gamma_0, \gamma_\infty) \ni \gamma \mapsto |\alpha(\gamma)|$ is not strictly increasing.

In the following theorem we give some criteria for the coincidence of the numbers γ_∞ and γ_+ as well as of γ_o and γ_-. Recall that $\Delta_n = (-n, n)$, $\Gamma_n = \overline{\mathbb{R}} \setminus [-n^{-1}, n^{-1}]$.

Theorem 4.10.

(1) *Each of the following conditions implies* $\gamma_\infty = \gamma_+$*:*
   (i) $\lim_{n\to\infty} [AE(\Delta_n)g, g] = \infty$.
   (ii) $A(0)$ *is trivial or a definite subspace.*

(2) *Each of the following conditions implies* $\gamma_0 = \gamma_-$*:*
   (i) $\lim_{n\to\infty} [A^{-1}E(\Gamma_n)g, g] = \infty$.
   (ii) $\ker A$ *is trivial or a definite subspace.*

Proof. We prove assertion (1), assertion (2) can be proved by a similar reasoning. Condition (i) implies $\gamma_\infty = \infty$ (see (4.14)), and therefore $\gamma_\infty = \gamma_+ = \infty$. Assume that (ii) holds. If $\lim_{n\to\infty} [AE(\Delta_n)g, g] < \infty$, then, by Lemma 4.1, $(E(\Delta_n)g)$ converges in $\mathcal{D}((JA)_{op}^{1/2})$ to an element $g'_{(\infty)}$ with respect to the graph norm, and $g'_\infty := g - g'_{(\infty)}$ belongs to the root space of A^{-1} corresponding to 0. Then, by (ii), we have either $[g'_\infty, g'_\infty] \neq 0$ or $g'_\infty = 0$. In the first case Lemma 4.1 implies $\gamma_\infty = \infty$. In the second case we have $g = g'_{(\infty)} \in \mathcal{D}\big((JA)_{\rm op}^{1/2}\big)$, and Lemma 4.1 and (4.4) give

$$\gamma_\infty = \lim_{n\to\infty} \big[AE(\Delta_n)g, g\big] = \lim_{n\to\infty} \big\|(JA)_{\rm op}^{1/2} E(\Delta_n)g\big\|_J^2 = \gamma_+. \qquad \square$$

4.5. The root subspace at a real nonzero exceptional eigenvalue

We suppose as in Subsection 4.4 that all self-adjoint extensions of A_0 have nonempty resolvent set. Let $\gamma \in (\gamma_0, \gamma_\infty)$ be such that $\omega := \alpha(\gamma) \in \mathbb{R} \setminus \{0\}$. We show that modulo $\ker(A - \omega)$ the elements of the Jordan chain of $A_{(\gamma)}$ at ω can be expressed as limits of the defect elements $g(\omega + i\varepsilon)$, $\varepsilon \downarrow 0$.

First we recall that by the Lemmas 4.5 and 4.6 with $g(\omega) = \lim_{\varepsilon\downarrow 0} g(\omega + i\varepsilon)$ we have

$$\ker(A_{(\gamma)} - \omega) = \operatorname{span}\{g(\omega)\} \dot{+} \ker(A - \omega)$$

and $(\operatorname{sign}\omega)[g(\omega), g(\omega)] \le 0$. Moreover, $g(\omega) = g + (\omega - i)g_\omega$, where $g_\omega := \lim_{\varepsilon\downarrow 0} R(\omega + i\varepsilon)g$ (see (4.20)). If $E_{(\gamma)}$ denotes the spectral function of $A_{(\gamma)}$ and Δ is a bounded open interval such that $\omega \in \Delta$, $0 \notin \overline{\Delta}$, then $\big(E_{(\gamma)}(\Delta)\mathcal{H}, [A_{(\gamma)}\cdot, \cdot]\big)$ is a Pontryagin space with negative index 1, and $A_{(\gamma)}$ is self-adjoint in this space. According to [JL2], the length of any Jordan chain of $A_{(\gamma)}$ at ω is ≤ 3,

$$\dim \ker(A_{(\gamma)} - \omega)^{k+1}) / \ker(A_{(\gamma)} - \omega)^k \le 1, \quad k = 1, 2,$$

and exactly one of the following cases holds:

(e) $\ker(A_{(\gamma)} - \omega) = \ker(A_{(\gamma)} - \omega)^2$ and ω is not a singular critical point of $A_{(\gamma)}$.

($\mathrm{p_{1s}}$) $\ker(A_{(\gamma)} - \omega) = \ker(A_{(\gamma)} - \omega)^2$ and ω is a singular critical point of $A_{(\gamma)}$.

($\mathrm{p_{2r}}$) $\ker(A_{(\gamma)} - \omega) \neq \ker(A_{(\gamma)} - \omega)^2 = \ker(A_{(\gamma)} - \omega)^3$ and ω is a regular critical point of $A_{(\gamma)}$.

(p_{2s}) $\ker(A_{(\gamma)} - \omega) \neq \ker(A_{(\gamma)} - \omega)^2 = \ker(A_{(\gamma)} - \omega)^3$ and ω is a singular critical point of $A_{(\gamma)}$.

(p_3) $\ker(A_{(\gamma)} - \omega) \neq \ker(A_{(\gamma)} - \omega)^2 \neq \ker(A_{(\gamma)} - \omega)^3$.

In the case (p_3) ω is a regular critical point of $A_{(\gamma)}$.

Theorem 4.11. *Assume that all self-adjoint extensions of A_0 have nonempty resolvent set and let $\gamma \in (\gamma_0, \gamma_\infty)$ be such that $\omega := \alpha(\gamma) \in \mathbb{R} \setminus \{0\}$. Then*

$$\left(A_{(\gamma)} - \omega\right) g(\omega) = 0,$$

and the following equivalences hold:

$$\begin{aligned}
(\mathrm{e}) &\iff (\operatorname{sign}\omega)[g(\omega), g(\omega)] < 0.\\
(\mathrm{p_{1s}}) &\iff [g(\omega), g(\omega)] = 0,\ \lim_{\varepsilon\downarrow 0} R(\omega + i\varepsilon)g(\omega + i\varepsilon) \ \text{ does not exist.}\\
(\mathrm{p_{2r}}) &\iff [g(\omega), g(\omega)] = 0,\ f_1 := \lim_{\varepsilon\downarrow 0} R(\omega + i\varepsilon)g(\omega + i\varepsilon) \ \text{ exists, } \ [g(\omega), f_1] \neq 0.\\
(\mathrm{p_{2s}}) &\iff [g(\omega), g(\omega)] = 0,\ f_1 := \lim_{\varepsilon\downarrow 0} R(\omega + i\varepsilon)g(\omega + i\varepsilon) \ \text{ exists, } [g(\omega), f_1] = 0,\\
&\qquad \lim_{\varepsilon\downarrow 0} R(\omega + i\varepsilon)^2 g(\omega + i\varepsilon) \ \text{ does not exist.}\\
(\mathrm{p_3}) &\iff [g(\omega), g(\omega)] = 0,\ f_1 := \lim_{\varepsilon\downarrow 0} R(\omega + i\varepsilon)g(\omega + i\varepsilon) \ \text{ exists, } \ [g(\omega), f_1] = 0,\\
&\qquad f_2 := \lim_{\varepsilon\downarrow 0} R(\omega + i\varepsilon)^2 g(\omega + i\varepsilon) \ \text{ exists.}
\end{aligned}$$

Moreover, in the cases (p_{2r}) *and* (p_{2s}), *the elements $g(\omega), f_1$ form a Jordan chain of $A_{(\gamma)}$ at ω and*

$$\ker\left((A_{(\gamma)} - \omega)^2\right) = \operatorname{span}\{g(\omega), f_1\} \,[\dot{+}]\, \ker(A - \omega);$$

in the case (p_3) *the elements $g(\omega), f_1, f_2$ form a Jordan chain of $A_{(\gamma)}$ at ω and*

$$\ker\left((A_{(\gamma)} - \omega)^3\right) = \operatorname{span}\{g(\omega), f_1, f_2\} \,[\dot{+}]\, \ker(A - \omega).$$

Proof. First we mention that a real eigenvalue of a self-adjoint relation B with non-empty resolvent set in a Pontryagin space is a regular critical point of B if and only if the corresponding root subspace is non-degenerated, see [KL1]. The statement of the theorem is now a consequence of the following two claims and of Lemma 4.6.

Claim 1: If $[g(\omega), g(\omega)] = 0$ then there exists a Jordan chain of $A_{(\gamma)}$ at ω of length ≥ 2 if and only if

$$f_1 := \lim_{\varepsilon\downarrow 0} R(\omega + i\varepsilon)g(\omega + i\varepsilon) \tag{4.46}$$

exists; in this case the elements $g(\omega), f_1$ form a Jordan chain of $A_{(\gamma)}$ at ω, and this Jordan chain is orthogonal to $\ker(A - \omega)$.

To prove this, assume first that $f_1 = \lim_{\varepsilon\downarrow 0} R(\omega+i\varepsilon)g(\omega+i\varepsilon)$ exists. If $z, \zeta \in \mathbb{C}\setminus\mathbb{R}$, $z\neq\zeta$, we find from (4.2) by a straightforward calculation that

$$\begin{aligned}&\big((A_{(\gamma)}-\zeta)^{-1}-(z-\zeta)^{-1}\big)R(z)g(z)\\&=-(z-\zeta)^{-2}g(z)-\frac{[g(z),g(\overline{z})]-(\gamma+Q(z))(z-\zeta)^{-1})}{(z-\zeta)(\gamma+Q(\zeta))}g(\zeta).\end{aligned}$$

If we set $z=\omega+i\varepsilon$, then in view of

$$\lim_{\varepsilon\to 0}\,[g(\omega+i\varepsilon),g(\omega-i\varepsilon)]=[g(\omega),g(\omega)]=0$$

and

$$\lim_{\varepsilon\to 0}\big(\gamma+Q(\omega+i\varepsilon)\big)=\gamma+\omega[g,g]+\big(\omega^2+1\big)[g_\omega,g]=0$$

(see (4.23)), we find

$$\big((A_{(\gamma)}-\zeta)^{-1}-(\omega-\zeta)^{-1}\big)f_1=-(\omega-\zeta)^{-2}g(\omega). \tag{4.47}$$

This relation is equivalent to

$$\begin{pmatrix}f_1\\ g(\omega)\end{pmatrix}\in A_{(\gamma)}-\omega. \tag{4.48}$$

By (4.32), for $z\neq\overline{z}$ we have

$$E(\{\omega\})R(z)g(z)=R(z)\big(I+(z-i)R(z)\big)E(\{\omega\})g=0,$$

therefore $f_1[\perp]\ker(A-\omega)$ and $\operatorname{span}\{g(\omega),f_1\}[\perp]\ker(A-\omega)$.

Conversely, assume that there exist elements $x'\in\mathcal{H}$, $y\in\ker\big(A_{(\gamma)}-\omega\big)$, $y\neq 0$, such that $\begin{pmatrix}x'\\ y\end{pmatrix}\in A_{(\gamma)}-\omega$. This implies $[y,y]=0$, and by Lemma 4.6 and $[g(\omega),g(\omega)]=0$ we find $y=\alpha g(\omega)$ for some number $\alpha\neq 0$. Hence

$$\begin{pmatrix}x\\ g(\omega)\end{pmatrix}\in A_{(\gamma)}-\omega\quad\text{for}\quad x:=\alpha^{-1}x'. \tag{4.49}$$

If $\zeta\neq\overline{\zeta}$ then

$$\begin{pmatrix}x\\ -(\omega-\zeta)^{-2}g(\omega)\end{pmatrix}\in(A_{(\gamma)}-\zeta)^{-1}-(\omega-\zeta)^{-1},$$

or

$$\begin{aligned}&\big(R(\zeta)-(\omega-\zeta)^{-1}\big)x-(\gamma+Q(\zeta))^{-1}[x,g(\overline{\zeta})]g(\zeta)\\&=-(\omega-\zeta)^{-2}g(\omega)=-(\omega-\zeta)^{-2}g-(\omega-\zeta)^{-2}(\omega-i)g_\omega\end{aligned} \tag{4.50}$$

(see (4.20), (4.25)). Let Δ be a real bounded open interval such that $\omega\in\Delta$, $0\notin\overline{\Delta}$, and denote the spectral function of A again by E_A. Evidently,

$$\begin{aligned}&\big(R(\zeta)-(\omega-\zeta)^{-1}\big)E_A(\Delta)x\\&=-(\omega-\zeta)^{-1}E_A(\Delta)(A-\omega)R(\zeta)E_A(\Delta)x\in\operatorname{ran}\Big((A-\omega)\big|_{E_A(\Delta)\mathcal{H}}\Big).\end{aligned} \tag{4.51}$$

For $\zeta \neq \overline{\zeta}$ we have

$$\begin{aligned} E_A(\Delta)g(\zeta) &= E_A(\Delta)\big(I + (\zeta - i)R(\zeta)\big)g \\ &= E_A(\Delta)\big(I + (\zeta - i)R(\zeta)\big)(A - \omega)E_A(\Delta)g_\omega \in \operatorname{ran}\Big((A - \omega)\big|_{E_A(\Delta)\mathcal{H}}\Big). \end{aligned} \tag{4.52}$$

We apply $E_A(\Delta)$ to the left- and the right-hand side of (4.50). Then (4.51) and (4.52) yield

$$E_A(\Delta)g_\omega \in \operatorname{ran}\Big((A - \omega)\big|_{E_A(\Delta)\mathcal{H}}\Big),$$

and that the set $\big\{R(\omega + i\varepsilon)E_A(\Delta)g_\omega : \varepsilon \in \mathbb{R} \setminus \{0\}\big\}$ is bounded in $E_A(\Delta)\mathcal{H}$. Making use of the isometry ι defined in the proof of Lemma 4.5 we conclude that the integrals

$$\int_\Delta |t - \omega|^{-2}\big((t - \omega)^2 + \varepsilon^2\big)^{-1}\, d\sigma(t), \quad \varepsilon > 0,$$

are uniformly bounded. Therefore the function $t \mapsto (t - \omega)^{-4}$ is σ-integrable. This implies that the functions $\Delta \ni t \mapsto (t - i)(t - \omega - i\varepsilon)^{-2}$ converge in $L^2(\Delta; \sigma)$ for $\varepsilon \downarrow 0$, which is equivalent to the fact that the limit (4.46) exists. Then, by (4.48) and (4.49), $x - f_1 \in \ker\big(A_{(\gamma)} - \omega\big)$, which on account of Lemma 4.6 gives

$$\begin{aligned} \ker\big(A_{(\gamma)} - \omega\big)^2 &= \operatorname{span}\{f_1\} + \ker\big(A_{(\gamma)} - \omega\big) \\ &= \operatorname{span}\{g(\omega), f_1\}[\dot{+}]\ker(A - \omega). \end{aligned}$$

Claim 2: If $[g(\omega), g(\omega)] = 0$, $f_1 := \lim_{\varepsilon\downarrow 0} R(\omega + i\varepsilon)g(\omega + i\varepsilon)$ exists and $[f_0, f_1] = 0$ then there exists a Jordan chain of $A_{(\gamma)}$ at ω of length 3 if and only if $f_2 := \lim_{\varepsilon\downarrow 0} R(\omega + i\varepsilon)^2 g(\omega + i\varepsilon)$ exists; in this case the elements $g(\omega), f_1, f_2$ form a Jordan chain of $A_{(\gamma)}$ at ω, and this chain is orthogonal to $\ker(A - \omega)$.

The proof of this claim is analogous to the proof of the Claim 1; e.g., the relation (4.47) is to be replaced by

$$\big((A_{(\gamma)} - \zeta)^{-1} - (\omega - \zeta)^{-1}\big) f_2 = -(\omega - \zeta)^{-2} f_1 + (\omega - \zeta)^{-3} g(\omega). \qquad \square$$

Remark 4.12. If the assumptions of Theorem 4.11 are fulfilled, it follows from the proof of Theorem 4.11 that $\lim_{\varepsilon\downarrow 0} R(\omega+i\varepsilon)g(\omega+i\varepsilon)$ exists if and only if the function $t \mapsto (t - \omega)^{-4}$ is σ-integrable. Analogously, the limits $\lim_{\varepsilon\downarrow 0} R(\omega + i\varepsilon)g(\omega + i\varepsilon)$ and $\lim_{\varepsilon\downarrow 0} R(\omega + i\varepsilon)^2 g(\omega + i\varepsilon)$ exist if and only if the function $t \mapsto (t - \omega)^{-6}$ is σ-integrable.

If, in addition to the assumptions of Theorem 4.11, $\omega \notin \operatorname{supp}\sigma$, then only the cases (e), ($\mathrm{p}_{2\mathrm{r}}$) and (p_3) can occur. With the reasoning at the end of Section 2, the behavior of the curve $\gamma \mapsto \alpha(\gamma)$ in a neighborhood of ω can be characterized by the structure of the root subspace of $A_{(\gamma)}$ at ω.

Proposition 4.13. *Let the assumptions of Theorem* 4.11 *be fulfilled with γ replaced by some $\gamma_1 \in (\gamma_0, \gamma_\infty)$ such that $\omega := \alpha(\gamma_1) \notin \operatorname{supp}\sigma$. If by* (rr), (cr), (rc), *and* (cc) *we denote the properties of a curve in the neighborhood of some real point* *as introduced at the end of Section* 2 *(with* ζ_1 *replaced by* ω*), then the following equivalences hold:*

$$\begin{aligned}
&\text{(e)} &&\Longleftrightarrow \quad \omega \text{ is of type (rr)},\\
(\mathrm{p}_{2\mathrm{r}}) \text{ and } &[f_1, g(\omega)] < 0 &&\Longleftrightarrow \quad \omega \text{ is of type (cr)},\\
(\mathrm{p}_{2\mathrm{r}}) \text{ and } &[f_1, g(\omega)] > 0 &&\Longleftrightarrow \quad \omega \text{ is of type (rc)},\\
&(\mathrm{p}_3) &&\Longleftrightarrow \quad \omega \text{ is of type (cc)}.
\end{aligned}$$

4.6. Rank one perturbations of a nonnegative self-adjoint relation

In the preceding considerations the starting point was a nonnegative relation A_0 of regular defect one. After excluding some special cases a nonnegative self-adjoint extension A of A_0 with $\rho(A) \neq \emptyset$ was fixed, and this self-adjoint extension played an essential role. This viewpoint may be changed by starting from a nonnegative self-adjoint relation A with $\rho(A) \neq \emptyset$ and a rank one perturbation B of A, either in resolvent sense or in the sense of generalized sums.

In the first case, let A be a nonnegative self-adjoint relation A with $\rho(A) \neq \emptyset$ and let B be a self-adjoint relation with $\rho(B) \neq \emptyset, B \neq A$, such that for some (and hence for all) $\lambda \in \rho(A) \cap \rho(B)$ the operator $(A-\lambda)^{-1} - (B-\lambda)^{-1}$ is of rank one, say $(A-\lambda)^{-1} - (B-\lambda)^{-1} = [\,\cdot\,, h]\, h'$ with some $h, h' \in \mathcal{H}$. Setting $g := \big(I + (i - \overline{\lambda})(A-i)^{-1}\big)\, h$, then A and B are self-adjoint extensions of the relation

$$A_0 := \left\{ \begin{pmatrix} x \\ y \end{pmatrix} : \begin{pmatrix} x \\ y \end{pmatrix} \in A, [y - \lambda x, h] = 0 \right\} = \left\{ \begin{pmatrix} x \\ y \end{pmatrix} : \begin{pmatrix} x \\ y \end{pmatrix} \in A, [y - ix, g] = 0 \right\}.$$

This relation A_0 is nonnegative and of regular defect one. Hence the relation B belongs to the family of operators $A_{(\gamma)}$ as constructed above with the help of A_0 and γ, and the results of this paper can be used in order to study the spectral properties of the relation B.

To define rank one perturbations B of a nonnegative self-adjoint operator A with $\rho(A) \neq \emptyset$ in the sense of generalized sums we recall some definitions from [JL1]. Denote by J a fundamental symmetry of the Krein space $\mathcal{H}$ and by $\mathcal{H}_{\frac{1}{2}}(A)$ the linear space $\mathcal{D}(JA)^{1/2}$ provided with the graph norm

$$\|x\|_{\frac{1}{2}} = \sqrt{\|x\|^2 + \|(JA)^{1/2}x\|^2},\ x \in \mathcal{D}(JA)^{1/2}.$$

Define the negative norm

$$\|x\|_{-\frac{1}{2}} := \sup\left\{ |[x,y]| : y \in \mathcal{H}_{\frac{1}{2}}(A), \|y\|_{\frac{1}{2}} \leq 1 \right\},$$

and let $\mathcal{H}_{-\frac{1}{2}}(A)$ be the completion of $\mathcal{H}$ with respect to this norm. Then $[\,\cdot\,,\,\cdot\,]$ can be extended by continuity to $\mathcal{H}_{\frac{1}{2}}(A) \times \mathcal{H}_{-\frac{1}{2}}(A)$. The operator A can be extended to a bounded operator $\widetilde{A}$ from $\mathcal{H}_{\frac{1}{2}}(A)$ into $\mathcal{H}_{-\frac{1}{2}}(A)$, and $(A-z)^{-1}$, $z \in \rho(A)$, can be extended by continuity to an isomorphism $\widetilde{R}(z)$ of $\mathcal{H}_{-\frac{1}{2}}(A)$ onto $\mathcal{H}_{\frac{1}{2}}(A)$. If $\widetilde{f} \in \mathcal{H}_{-\frac{1}{2}}(A)$ and $\alpha \in \mathbb{R}$ we define the operator $A \uplus \alpha\,[\,\cdot\,, \widetilde{f}]\widetilde{f}$ as the restriction of $\widetilde{A} + \alpha\,[\,\cdot\,, \widetilde{f}]\widetilde{f} \in \mathcal{L}\Big(\mathcal{H}_{\frac{1}{2}}(A), \mathcal{H}_{-\frac{1}{2}}(A)\Big)$ to those $x \in \mathcal{H}_{\frac{1}{2}}(A)$ for which $\widetilde{A}x +$

$\alpha\,[x,\widetilde{f}]\widetilde{f} \in \mathcal{H}$. The operator $A \uplus \alpha\,[\,\cdot\,,\widetilde{f}]\widetilde{f}$ is self-adjoint and its resolvent set is not empty.

Now we consider for some $\widetilde{f} \in \mathcal{H}_{-\frac{1}{2}}(A)$ the family of self-adjoint operators

$$A_{[\alpha]} := A \uplus \alpha\,[\,\cdot\,,\widetilde{f}]\,\widetilde{f}, \qquad \alpha \in \mathbb{R}.$$

In particular, $\widetilde{f}$ may belong to $\mathcal{H}$. If $g := \widetilde{R}(i)\widetilde{f}$, then $[x,\widetilde{f}] = [(A+i)x, g]$, $x \in \mathcal{D}(A)$, and we set

$$A_0 := A|_{\{x\in\mathcal{D}(A):\ [(A+i)x,g]=0\}}.$$

Starting from A and the defect element g, we define the self-adjoint extensions $A_{(\gamma)}$ of A_0 as above. Then $\gamma_+ = [\widetilde{R}(-i)\widetilde{A}\widetilde{R}(i)\widetilde{f},\widetilde{f}] < \infty$, and Theorem 4.10 implies that in this case $\gamma_\infty = \gamma_+$. It was shown in [JL1, 2.2] that

$$A_{[\alpha]} = A_{(\gamma)}, \quad \text{where} \quad \gamma = \alpha^{-1} + \gamma_+,\ \alpha \in \mathbb{R}.$$

For an expression of $\left(\gamma_- - \gamma_+\right)^{-1}$ in terms of $\widetilde{f}$ see [JL1, Proposition 2.1].

References

[ADo] Aronszajn, N., Donoghue, W.F.: On exponential representations of analytic functions in the upper half-plane with positive imaginary part, J. d'Analyse Math. **5** (1956-57), 321–388.

[DaL1] Daho, K., Langer, H.: Matrix functions of the class N_κ, Math. Nachr. **120** (1985), 275–294.

[DaL2] Daho, K., Langer, H.: Sturm–Liouville operators with an indefinite weight function: the periodic case, Radovi Matematički **2** (1986), 165–188.

[DHdS] Derkach, V., Hassi, S., de Snoo, H.S.V.: Rank one perturbations in a Pontryagin space with one negative square, J. Funct. Anal. **188** (2002), 317–349.

[DdS1] Dijksma, A., de Snoo, H.S.V.: Symmetric and selfadjoint relations in Krein spaces I, Operator Theory: Advances and Applications, vol. **24** (1987), Birkhäuser Verlag Basel, 145–166.

[DdS2] Dijksma, A., de Snoo, H.S.V.: Symmetric and selfadjoint relations in Krein spaces II, Ann. Acad. Sci. Fenn., Ser. A. I. Mathematica **12** (1987), 199–216.

[DLShZ] Dijksma, A., Langer, H., Shondin, Yu., Zeinstra, C.: Selfadjoint differential operators with inner singularities and Pontryagin spaces, Operator Theory: Adv. Appl., vol. **118**, Birkhäuser Verlag, Basel, 2000, 105–175.

[DLSh] Dijksma, A., Luger, A., Shondin, Yu.: Minimal models for $\mathcal{N}_\kappa^\infty$-functions, Operator Theory: Adv. Appl., vol. **163**, Birkhäuser Verlag, Basel, 2006, 97–134.

[J1] Jonas, P.: A class of operator-valued meromorphic functions on the unit disc, Ann. Acad. Sci. Fenn. Ser. A. I: Mathematica **17** (1992), 257–284.

[J2] Jonas, P.: Operator representations of definitizable functions, Ann. Acad. Sci. Fenn. Ser. A. I: Mathematica **25** (2000), 41–72.

[J3] Jonas, P.: On locally definite operators in Krein spaces, in: Spectral Theory and Its Applications, Bucharest, Theta 2003, 95–127.

[JL1] Jonas, P., Langer, H.: Some questions in the perturbation theory of J-nonnegative operators in Krein spaces, Math. Nachr. **114** (1983), 205–226.

[JL2] Jonas, P., Langer, H.: A model for π-self-adjoint operators and a special linear pencil, Integral Equations Operator Theory **8** (1985), 13–35.

[JL3] Jonas, P., Langer, H.: Selfadjoint extensions of a closed linear relation of defect one in a Krein space, Operator Theory: Advances and Applications, vol. **80** (1995), Birkhäuser Verlag Basel, 176–205.

[KK] Kac, I.S., Krein, M.G.: R-functions - analytic functions mapping the upper half-plane into itself, Supplement I of the Russian translation of the book by F.V. Atkinson, Discrete and continuous boundary problems, Moscow 1968. English translation: Amer. Math. Soc. Transl. (2) vol. **103** (1974), 1–18.

[KL1] Krein, M.G., Langer, H.: On the spectral function of a selfadjoint operator in a space with an indefinite metric, Dokl. Akad. Nauk SSSR **152** (1963), 39–42 (Russian)

[KL] Krein, M.G., Langer, H.: Some propositions on analytic matrix functions related to the theory of operators in the space Π_κ, Acta Sci. Math. **43** (1981), 181–205.

[L1] Langer, H.: Spektraltheorie linearer Operatoren in J-Räumen und einige Anwendungen auf die Schar $L(\lambda) = \lambda^2 I + \lambda B + C$, Habilitationsschrift Technische Universität Dresden, 1965.

[L2] Langer, H.: Characterization of generalized zeros of negative type of functions of the class N_κ, Operator Theory: Advances and Applications, vol. **17** (1986), Birkhäuser Verlag Basel, 202–212.

P. Jonas
Institut für Mathematik
Technische Universität Berlin
Straße des 17. Juni 135
D-10623 Berlin, Germany
e-mail: jonas@math.tu-berlin.de

H. Langer
Institut für Analysis und Computational Mathematics
Technische Universität Wien
Wiedner Hauptstrasse 8–10
A-1040 Vienna, Austria
e-mail: hlanger@mail.zserv.tuwien.ac.at

Operator Theory:
Advances and Applications, Vol. 175, 159–168

Canonical Differential Equations of Hilbert-Schmidt Type

Michael Kaltenbäck and Harald Woracek

Abstract. A canonical system of differential equations, or Hamiltonian system, is a system of order two of the form $Jy'(x) = -zH(x)y(x)$, $x \in \mathbb{R}^+$. We characterize the property that the selfadjoint operators associated to a canonical system have resolvents of Hilbert-Schmidt type in terms of the Hamiltonian H as well as in terms of the associated Titchmarsh-Weyl coefficient.

Mathematics Subject Classification (2000). Primary 47B25, 47E05, 34B20; Secondary 34A55, 34L05, 47A57.

Keywords. Canonical differential equation, Hilbert-Schmidt.

1. Introduction

A canonical system of differential equations is an equation of the form

$$Jy'(x) = -zH(x)y(x), \ x \in \mathbb{R}^+, \tag{1.1}$$

where $y(x)$ is a $\mathbb{C}^2$-valued function on $\mathbb{R}^+$,

$$J = \begin{pmatrix} 0 & -1 \\ 1 & 0 \end{pmatrix},$$

and $H(x)$ is a $\mathbb{R}^{2\times 2}$-valued function on $\mathbb{R}^+$ such that $H(x) \geq 0$ and $H|_{(0,x]} \in L^1$ for all $x \in \mathbb{R}^+$. The function $H(x)$ is called the Hamiltonian corresponding to the canonical differential equation (1.1). We always assume that moreover $\operatorname{tr}(H(x)) = 1$, $x \in \mathbb{R}^+$.

Canonical differential equations are usually studied with operator theoretic methods. A very good and detailed account on the operator model associated with the equation (1.1) can be found in [5]. Let us recall the basic notions: Denote by $L^2(H, \mathbb{R}^+)$ the Hilbert space of all measurable $\mathbb{C}^2$-valued functions $f(x)$ on $\mathbb{R}^+$ such that

$$\|f\|^2 = \int_0^{+\infty} f(x)^* H(x) f(x) dx < +\infty.$$

One considers the closed subspace $L^2_s(H,\mathbb{R}^+)$ of $L^2(H,\mathbb{R}^+)$ consisting of all $f \in L^2(H,\mathbb{R}^+)$ such that $H(x)f(x)$ is constant on H-indivisible intervals. Thereby an interval $I \subseteq \mathbb{R}^+$ is called H-indivisible if for some $\varphi \in \mathbb{R}$

$$H(x) = \xi_\varphi \xi_\varphi^T, \ x \in I \text{ a.e.}, \ \xi_\varphi = \begin{pmatrix} \cos\varphi \\ \sin\varphi \end{pmatrix}.$$

Moreover one considers the linear relation

$$T_{\max,s} = \{(f;g) \in L^2_s(H,\mathbb{R}^+)^2 : f \text{ is loc. abs. cont., } Jf' = -Hg\}, \tag{1.2}$$

and its restriction

$$T_{\min,s} = \{(f;g) \in T_{\max,s} : f(0+) = 0\}, \tag{1.3}$$

which turns out to be a symmetric operator such that

$$T^*_{\min,s} = T_{\max,s}.$$

All the selfadjoint extensions of $T_{\min,s}$ are given by

$$A(\nu) = \{(f;g) \in T_{\max,s} : \sin\nu f_1(0+) = \cos\nu f_2(0+)\}, \ \nu \in \mathbb{R}, \tag{1.4}$$

and thereby we have $A(\nu_1) = A(\nu_2)$ if and only if $\nu_1 - \nu_2 \in \pi\mathbb{Z}$. For a detailed discussion of linear relations in Hilbert spaces see [2].

A fundamental notion in the theory of canonical systems is the Titchmarsh-Weyl coefficient associated with the equation (1.1). Let $W(x,z)=(w_{ij}(x,z))_{i,j=1,2}$, $x \in \mathbb{R}^+$, $z \in \mathbb{C}$, be the 2×2-matrix-valued solution of the initial value problem

$$\frac{dW(x,z)}{dx}J = zW(x,z)H(x), \ x > 0, \ W(0+,z) = I,$$

and define

$$q_H(z) = \lim_{x\to+\infty} \frac{w_{11}(x,z)\tau + w_{12}(x,z)}{w_{21}(x,z)\tau + w_{22}(x,z)}. \tag{1.5}$$

This limit exists for $z \in \mathbb{C}\setminus\mathbb{R}$ and does not depend on $\tau \in \mathbb{R}$. The function q_H is called the Titchmarsh-Weyl coefficient of (1.1). It belongs to the Nevanlinna class $\mathcal{N}_0$, i.e., is holomorphic, satisfies $q_H(\bar z) = \overline{q_H(z)}$, $z \in \mathbb{C}\setminus\mathbb{R}$, and has the property that the kernel

$$L_{q_H}(w,z) = \frac{q_H(z) - \overline{q_H(w)}}{z - \bar w}$$

is positive semidefinite.

The inverse spectral theorem, a deep result due to L. de Branges (see [1]), shows that (1.5) sets up a bijective correspondence between the set of all trace normed Hamiltonians and the Nevanlinna class $\mathcal{N}_0$.

In a generalization to the indefinite setting one allows the Titchmarsh-Weyl coefficient q to belong to the generalized Nevanlinna class $\mathcal{N}_{<\infty}$, i.e., allows the kernel L_q to have a finite number of negative squares, and asks for an equation of similar type as (1.1), or for an operator model similar to $L^2_s(H,\mathbb{R}^+)$, $T_{\max,s}$, such that again there is a bijective correspondence between $\mathcal{N}_{<\infty}$ and the respective indefinite Hamiltonians.

It turns out that an indefinite Hamiltonian is composed out of a finite number $H_1, \dots, H_n$ of positive Hamiltonians and a finite number of real parameters. Thereby the occurring Hamiltonians and parameters are subject to certain conditions. One among them is that selfadjoint extensions of the symmetry $T_{\min,s}$ in $L^2_s(H_i, \mathbb{R}^+)$ have resolvents of Hilbert-Schmidt type.

It is the aim of this note to characterize this property explicitly in terms of the Hamiltonian as well as in terms of the spectral measure of the associated Titchmarsh-Weyl coefficient. In §2 we prove our main result Theorem 2.4 which gives a necessary and sufficient integrability condition on H in order that selfadjoint extensions have resolvents of Hilbert-Schmidt type. To this end we reduce the problem to a specific boundary condition and apply classical criteria for integral operators which can be found, e.g., in [3]. It is the subject of §3 to reformulate these conditions in terms of the spectral measure of the Titchmarsh-Weyl coefficient q_H, see Theorem 3.1. Thereby we use the theory of integral representations and operator models for Nevanlinna functions as developed, e.g., in [4]. In general our exposition relies on the material presented in [5] and the classical theory of integral operators, otherwise is fairly elementary.

2. Integral operators of Hilbert-Schmidt type

In the subsequent lemma we are going to use a well-known criterion on whether an integral operator is of Hilbert-Schmidt type. Concerning this lemma note that the square root of $H(t)$ exists because $H(t)$ is assumed to be positive semidefinite.

Lemma 2.1. *Let $H(t) = (h_{ij}(t))_{i,j=1,2}$ be a trace normed Hamiltonian on $\mathbb{R}^+$ such that*

$$\int_0^{+\infty} h_{11}(t)dt < +\infty. \tag{2.1}$$

Consider the space $L^2(\mathbb{R}^+)^2$ of all $\mathbb{C}^2$-valued functions on $\mathbb{R}^+$ which are square integrable with respect to the Lebesgue measure, and consider the kernel

$$K(x,y) = H(x)^{\frac{1}{2}} \begin{pmatrix} 0 & -\chi_{\{y<x\}} \\ -\chi_{\{y>x\}} & 0 \end{pmatrix} H(y)^{\frac{1}{2}}. \tag{2.2}$$

On $L^2(\mathbb{R}^+)^2$ define the integral operator

$$(Cg)(x) = \int_0^{+\infty} K(x,y)g(y)dy \tag{2.3}$$

with $g \in \operatorname{dom}(C)$ if the integral exists for almost every x and $Cg \in L^2(\mathbb{R}^+)^2$. Moreover, we set

$$M(x) = (m_{ij}(x))_{i,j=1,2} = \int_0^x H(t)dt.$$

Then C is a continuous, everywhere defined operator which belongs to the Hilbert-Schmidt class if and only if

$$2\int_0^{+\infty} m_{22}(t)h_{11}(t)dt < +\infty. \tag{2.4}$$

In this case the Hilbert-Schmidt norm of C coincides with the square root of this number.

Proof. It is well known (see for example [3]) that C is a continuous, everywhere defined operator which belongs to the Hilbert-Schmidt class if and only if

$$\int_0^{+\infty}\int_0^{+\infty} \operatorname{tr}(K(x,y)^*K(x,y))dxdy < +\infty.$$

In this case $\|C\|_2^2$ coincides with the value of integral. We calculate

$$\begin{aligned}
&K(x,y)^*K(x,y)\\
&= H(y)^{\frac{1}{2}}\begin{pmatrix} 0 & -\chi_{\{y>x\}} \\ -\chi_{\{y<x\}} & 0 \end{pmatrix} H(x) \begin{pmatrix} 0 & -\chi_{\{y<x\}} \\ -\chi_{\{y>x\}} & 0 \end{pmatrix} H(y)^{\frac{1}{2}}\\
&= H(y)^{\frac{1}{2}}\left(\chi_{\{y>x\}}\begin{pmatrix} 0 & 1 \\ 0 & 0 \end{pmatrix} H(x)\begin{pmatrix} 0 & 0 \\ 1 & 0 \end{pmatrix} + \chi_{\{y<x\}}\begin{pmatrix} 0 & 0 \\ 1 & 0 \end{pmatrix} H(x)\begin{pmatrix} 0 & 1 \\ 0 & 0 \end{pmatrix}\right) H(y)^{\frac{1}{2}}\\
&= H(y)^{\frac{1}{2}}\left(\chi_{\{y>x\}}h_{22}(x)\begin{pmatrix} 1 & 0 \\ 0 & 0 \end{pmatrix} + \chi_{\{y<x\}}h_{11}(x)\begin{pmatrix} 0 & 0 \\ 0 & 1 \end{pmatrix}\right) H(y)^{\frac{1}{2}}.
\end{aligned} \tag{2.5}$$

Set $(k_{ij}(y))_{i,j=1,2} = H(y)^{\frac{1}{2}}$, and note that $k_{12} = k_{21}$. We then rewrite (2.5) as

$$\begin{aligned}
&\chi_{\{y>x\}}h_{22}(x)\begin{pmatrix} k_{11}(y)k_{11}(y) & k_{11}(y)k_{12}(y) \\ k_{11}(y)k_{12}(y) & k_{12}(y)k_{12}(y) \end{pmatrix}\\
&+\chi_{\{y<x\}}h_{11}(x)\begin{pmatrix} k_{12}(y)k_{12}(y) & k_{12}(y)k_{22}(y) \\ k_{12}(y)k_{22}(y) & k_{22}(y)k_{22}(y) \end{pmatrix}.
\end{aligned}$$

As $k_{11}^2(y) + k_{12}^2(y) = h_{11}(y)$ and $k_{12}^2(y) + k_{22}^2(y) = h_{22}(y)$ we obtain

$$\operatorname{tr}(K(x,y)^*K(x,y)) = \chi_{\{y>x\}}h_{22}(x)h_{11}(y) + \chi_{\{y<x\}}h_{11}(x)h_{22}(y),$$

and by Fubini's theorem

$$\begin{aligned}
&\int_0^{+\infty}\int_0^{+\infty} \operatorname{tr}(K(x,y)^*K(x,y))dxdy\\
&= \int_0^{+\infty} h_{11}(y)\int_0^y h_{22}(x)dxdy + \int_0^{+\infty} h_{22}(y)\int_y^{+\infty} h_{11}(x)dxdy\\
&= 2\int_0^{+\infty} h_{11}(y)\int_0^y h_{22}(x)dxdy = 2\int_0^{+\infty} m_{22}(y)h_{11}(y)dy.
\end{aligned}$$

□

Lemma 2.2. *Let $H(t) = (h_{ij}(t))_{i,j=1,2}$ be a trace normed Hamiltonian on $\mathbb{R}^+$ such that (2.1) holds. Moreover, let B be the operator defined in $L_s^2(H,\mathbb{R}^+)$ by*

$$(Bf)(x) = \int_0^x JH(t)f(t)dt - \begin{pmatrix} 0 & 0 \\ 0 & 1 \end{pmatrix}\int_0^{+\infty} JH(t)f(t)dt,$$

with $f \in \operatorname{dom}(B)$ *if* $f \in L_s^2(H,\mathbb{R}^+)$ *such that* $(Bf)(x)$ *exists for almost every* $x \in \mathbb{R}^+$ *and* $Bf \in L_s^2(H,\mathbb{R}^+)$.

Then B *is a continuous, everywhere defined operator which belongs to the Hilbert-Schmidt class if and only if* (2.4) *holds true.*

Proof. The integral operator B can be written as

$$(Bf)(x) = \int_0^{+\infty} k(x,y)f(y)dy,$$

where $k(x,y)$ is the kernel

$$k(x,y) = \begin{pmatrix} \chi_{\{y<x\}} & 0 \\ 0 & -\chi_{\{y>x\}} \end{pmatrix} JH(y).$$

Now we consider the mapping ϕ from $L_s^2(H,\mathbb{R}^+)$ into $L^2(\mathbb{R}^+)^2$ given by

$$\phi(f)(t) = H(t)^{\frac{1}{2}} f(t).$$

It is straightforward to show that ϕ is an isometry. Let $K(x,y)$ be as in (2.2) and the operator C defined as in (2.3). By Lemma 2.1 the operator C is a continuous, everywhere defined operator which belongs to the Hilbert-Schmidt class if and only if (2.4) holds true.

We have $C\phi = \phi B$. In fact, $\phi B \subseteq C\phi$ is obvious from the definitions of the respective kernels. For the other inclusion let $f \in \operatorname{dom}(C\phi)$. From (2.2) we see that $C(\phi(f))(x)$ is of the form $H(x)^{\frac{1}{2}} g(x)$ for $g \in L^2(H,\mathbb{R}^+)$ with

$$\begin{aligned} g(x) &= \int_0^{+\infty} k(x,y)f(y)dy \\ &= \int_0^x JH(t)f(t)dt - \begin{pmatrix} 0 & 0 \\ 0 & 1 \end{pmatrix} \int_0^{+\infty} JH(t)f(t)dt. \end{aligned}$$

It remains to show that $g \in L_s^2(H,\mathbb{R}^+)$. So let (a,b) be an indivisible interval of type $\varphi \in \mathbb{R}$ and calculate for $x \in (a,b)$

$$\frac{d}{dx}\xi_\varphi^T g(x) = \xi_\varphi^T JH(x)f(x) = \xi_\varphi^T J\xi_\varphi \xi_\varphi^T f(x) = 0.$$

Hence $g \in L_s^2(H,\mathbb{R}^+)$, and we proved that $C\phi = \phi B$.

If $g \in L^2(\mathbb{R}^+)^2 \ominus \phi(L_s^2(H,\mathbb{R}^+))$, then

$$(Cg)(x) = H(x)^{\frac{1}{2}} \left(\int_0^x JH(y)^{\frac{1}{2}} g(y)dy - \begin{pmatrix} 0 \\ 1 \end{pmatrix} \int_0^{+\infty} \begin{pmatrix} 1 \\ 0 \end{pmatrix}^T H(y)^{\frac{1}{2}} g(y)dy \right).$$

The second integral vanishes because

$$H(y)^{\frac{1}{2}} \begin{pmatrix} 1 \\ 0 \end{pmatrix} \in \phi(L_s^2(H,\mathbb{R}^+)).$$

If x is contained in an indivisible interval (a,b) of type $\varphi \in \mathbb{R}$, we assume that (a,b) is maximal, i.e., (a,b) is not contained in a larger indivisible interval. For $a < y < x$ we have

$$H(x)^{\frac{1}{2}} J H(y)^{\frac{1}{2}} = \xi_\varphi \xi_\varphi^T J \xi_\varphi \xi_\varphi^T = 0.$$

Therefore,

$$(Cg)(x) = H(x)^{\frac{1}{2}} \int_0^a J H(y)^{\frac{1}{2}} g(y) dy.$$

Since the columns of the matrix function $\chi_{\{y \leq a\}} H(y)^{\frac{1}{2}} J^T$ belong to $\phi(L_s^2(H,\mathbb{R}^+))$, we see that $Cg = 0$.

We conclude that $\operatorname{dom}(C) = \phi(\operatorname{dom}(B)) \oplus \phi(L_s^2(H,\mathbb{R}^+))^\perp$ and $C = \phi B \phi^{-1} P$, where P is the orthogonal projection onto $\phi(L_s^2(H,\mathbb{R}^+))$. Thus C is a Hilbert-Schmidt operator if and only if B is. □

The previous lemma is of importance if we consider the operator theoretical background of canonical differential equations. Using the notation from the introduction we obtain the subsequent result.

Corollary 2.3. *Let* $H(t) = (h_{ij}(t))_{i,j=1,2}$ *be a trace normed Hamiltonian on* $\mathbb{R}^+$ *such that* (2.1) *holds true. Then the selfadjoint extensions* $A(\nu)$, $\nu \in (-\frac{\pi}{2}, \frac{\pi}{2}]$ *of* $T_{\min,s}$ *have a resolvent*

$$(A(\nu) - z)^{-1}, \ z \in \rho(A(\nu))$$

which is of Hilbert-Schmidt type if and only if (2.4) *holds.*

Proof. Because of the resolvent identity and since the product of a bounded operator and a Hilbert-Schmidt operator is of Hilbert-Schmidt type, the selfadjoint extension $A(\nu)$ of $T_{\min,s}$ has a resolvent which is of Hilbert-Schmidt type if and only if $(A(\nu) - z)^{-1}$ is of Hilbert-Schmidt type for one $z \in \rho(A(\nu))$.

Moreover, by Krein's formula (see Proposition 4.4 in [5]) the resolvents of different selfadjoint extensions are one-dimensional perturbations of each other. Thus, the resolvents of $A(\nu)$, $\nu \in (-\frac{\pi}{2}, \frac{\pi}{2}]$ being of Hilbert-Schmidt type is equivalent to the fact that

$$\left(A\left(\frac{\pi}{2}\right) - z \right)^{-1} \tag{2.6}$$

is a Hilbert-Schmidt operator for some $z \in \rho(A(\frac{\pi}{2}))$.

Note that by our assumption (2.1)

$$\operatorname{span}\left\{ \begin{pmatrix} 1 \\ 0 \end{pmatrix} \right\} = \ker T_{\max,s}, \tag{2.7}$$

and hence $\ker A(\frac{\pi}{2}) = \{0\}$. This means that $A(\frac{\pi}{2})^{-1}$ is a densely defined selfadjoint operator on $L_s^2(H,\mathbb{R}^+)$. Moreover, as the elements $f \in \operatorname{dom} A(\frac{\pi}{2})$ vanish at 0 in the upper entry we see that $B \subseteq A(\frac{\pi}{2})^{-1}$, where B is defined as in Lemma 2.2.

If (2.4) is satisfied, then B is an everywhere defined, bounded operator belonging to the Hilbert-Schmidt class. We conclude $B = A(\frac{\pi}{2})^{-1}$, and $0 \in \rho(A(\frac{\pi}{2}))$.

For the converse direction assume that (2.6) is of Hilbert-Schmidt type. We are going to show that $A(\frac{\pi}{2})^{-1}$ is an everywhere defined, bounded operator belonging to the Hilbert-Schmidt class. If $z = 0$, we are done. Otherwise, besides zero the spectrum of (2.6) consists of eigenvalues. Since $\ker A(\frac{\pi}{2}) = \{0\}$, the number $-z^{-1}$ cannot be an eigenvalue of (2.6). Because of

$$A\left(\frac{\pi}{2}\right)^{-1} = \frac{1}{z}\left(A\left(\frac{\pi}{2}\right) - z\right)^{-1}\left(\frac{1}{z} + \left(A\left(\frac{\pi}{2}\right) - z\right)^{-1}\right)^{-1}, \tag{2.8}$$

$A(\frac{\pi}{2})^{-1}$ is an everywhere defined, bounded operator belonging to the Hilbert-Schmidt class.

If $g \in L^2_s(H, \mathbb{R}^+)$ and $f = A(\frac{\pi}{2})^{-1}g$, then $f' = JHg$. It follows from Lemma 7.8, [5] and from (2.7) that $f(x)$ tends to zero in the lower component for $x \to +\infty$. Moreover, $f(0+)$ is zero in the upper component. Thus Bg exists and coincides with $f = A(\frac{\pi}{2})^{-1}g$, and we proved that $B = A(\frac{\pi}{2})^{-1}$. By Lemma 2.2 condition (2.4) holds true. □

The above assertions yield a criterion for a canonical differential equation to have a compact resolvent of Hilbert-Schmidt type. Put

$$\xi_\varphi := \begin{pmatrix} \cos\varphi \\ \sin\varphi \end{pmatrix}.$$

Theorem 2.4. *Let $H(t) = (h_{ij}(t))_{i,j=1,2}$ be a trace normed Hamiltonian on $\mathbb{R}^+$. Then the selfadjoint extensions $A(\nu)$, $\nu \in (-\frac{\pi}{2}, \frac{\pi}{2}]$ of $T_{\min,s}$ have a resolvent*

$$(A(\nu) - z)^{-1},\ z \in \rho(A(\nu)),$$

which is of Hilbert-Schmidt type if and only if there exists a real φ such that

$$\int_0^{+\infty} \xi_\varphi^T H(t)\xi_\varphi dt < +\infty, \tag{2.9}$$

and

$$\int_0^{+\infty} \xi_{\varphi+\frac{\pi}{2}}^T M(t)\xi_{\varphi+\frac{\pi}{2}}\ \xi_\varphi^T H(t)\xi_\varphi dt < +\infty. \tag{2.10}$$

Here $M(x)$ is defined by

$$M(x) = (m_{ij}(x))_{i,j=1,2} = \int_0^x H(t)dt.$$

Proof. First we consider a transformation of the Hamiltonian $H(t)$. For $\mu \in \mathbb{R}$ set

$$N_\mu = \begin{pmatrix} \sin\mu & -\cos\mu \\ \cos\mu & \sin\mu \end{pmatrix},$$

and $H_\mu(t) = N_\mu^* H(t) N_\mu$ (see [5]). It is straightforward to verify that $\psi : f \mapsto N_\mu^* f$ is an isomorphism from $L^2_s(H, \mathbb{R}^+)$ onto $L^2_s(H_\mu, \mathbb{R}^+)$.

If we denote by $T^{\mu}_{\min,s}$ ($T^{\mu}_{\max,s}$) the respective differential operators on the space $L^2_s(H_\mu,\mathbb{R}^+)$ and by $A^\mu(\nu)$ the corresponding selfadjoint extensions of $T^{\mu}_{\min,s}$, then

$$\psi T_{\min,s} = T^{\mu}_{\min,s}\psi,\ \psi T_{\max,s} = T^{\mu}_{\max,s}\psi,$$

and $\psi A(\nu) = A^\mu(\nu+\mu-\frac{\pi}{2})\psi$. Thus, the $A(\nu)$'s have resolvents of Hilbert-Schmidt type if and only if the $A^\mu(\nu)$'s do so.

Condition (2.9) for the given $H(t)$ is equivalent to condition (2.1) when $h_{11}(t)$ now denotes the left upper entry of $H_\mu(t)$ with $\mu = \frac{\pi}{2} - \varphi$. Similarly condition (2.10) for $H(t)$ is equivalent to condition (2.4) when $h_{11}(t), h_{22}(t)$ $(m_{11}(t), m_{22}(t))$ now denote the diagonal entries of $H_\mu(t)$ (the primitive of $H_\mu(t)$) with $\mu = \frac{\pi}{2} - \varphi$.

If there exists a real φ such that (2.9) and (2.10) hold, then by Corollary 2.3 the resolvents of the $A^\mu(\nu)$'s are Hilbert-Schmidt type for $\mu = \frac{\pi}{2} - \varphi$. As we mentioned above this is equivalent to the $A(\nu)$'s having resolvents of Hilbert-Schmidt type.

Conversely, assume that $A(\nu)$, $\nu \in (-\frac{\pi}{2}, \frac{\pi}{2}]$, have resolvents of Hilbert-Schmidt type. If there exists a $\varphi \in \mathbb{R}$ such that $\xi_\varphi \in L^2_s(H,\mathbb{R}^+)$, then we set $\mu = \frac{\pi}{2} - \varphi$ and see that (2.9) holds. We just mentioned that this is the same as saying that (2.1) holds true when $h_{11}(t)$ denotes the left upper entry of $H_\mu(t)$. Since also the $A^\mu(\nu)$'s have resolvents of Hilbert-Schmidt type, Corollary 2.3 shows that (2.4) holds for $H_\mu(t)$ which in turn yields (2.10).

It remains to exclude the case that none of the constant functions ξ_φ belong to $L^2_s(H,\mathbb{R}^+)$. In fact, if we were in this situation, then all of the selfadjoint relations $A(\nu)$ would have a trivial kernel because the scalar multiples of ξ_φ are the only possible candidates for elements of $L^2_s(H,\mathbb{R}^+)$ to belong to $\ker A(\varphi)$.

Thus all $A(\nu)^{-1}$ are operators, and by (2.8) with $A(\frac{\pi}{2})$ replaced by $A(\nu)$ obtain that the operators $A(\nu)^{-1}$ are everywhere defined, bounded and of Hilbert-Schmidt type. Hence the closed one-dimensional restriction $(T_{\min,s})^{-1}$ is a bounded, symmetric operator with a closed domain of co-dimension one. It is easy to check that

$$D = (T_{\min,s})^{-1} \oplus (\{0\} \times \operatorname{dom}((T_{\min,s})^{-1})^\perp)$$

is a selfadjoint extension of $(T_{\min,s})^{-1}$. Its inverse would then be a selfadjoint extension of $T_{\min,s}$, and would, therefore, coincide with $A(\nu)$ for some ν. But this contradicts the fact that D^{-1} has a non-trivial kernel. □

3. Titchmarsh-Weyl coefficients and the Hilbert-Schmidt property

Recall that a Nevanlinna function has a unique integral representation

$$a + bz + \int_{\mathbb{R}} \left(\frac{1}{t-z} - \frac{t}{t^2+1} \right) d\sigma(t), \tag{3.1}$$

where $a, b \in \mathbb{R}$, $b \geq 0$ and σ is a non-negative Borel measure on $\mathbb{R}$ such that

$$\int_{\mathbb{R}} \frac{1}{t^2+1} d\sigma(t) < +\infty.$$

Theorem 3.1. *Let $H(t) = (h_{ij}(t))_{i,j=1,2}$ be a trace normed Hamiltonian on $\mathbb{R}^+$, and let $q_H(z)$ be the Titchmarsh-Weyl coefficient of the canonical differential equation* (1.1). *Assume that* (3.1) *is the integral representation of $q_H(z)$.*

Then $H(x)$ satisfies (2.9) *and* (2.10) *if and only of σ is a discrete measure*

$$\sigma = \sum_n \alpha_n \delta_{t_n}, \tag{3.2}$$

such that

$$\sum_n \frac{1}{t_n^2} < +\infty. \tag{3.3}$$

Proof. Corollary 4.2, [5] shows that $T_{\min,s}$ is a completely non-selfadjoint symmetric relation with defect index $(1,1)$, and by Theorem 4.3, [5] the Titchmarsh-Weyl coefficient $q_H(z)$ is the Q-function of $A(\frac{\pi}{2})$ and $T_{\min,s}$.

Thus according to Theorem 2.5, [4] the triplet $(L^2_s(H,\mathbb{R}^+), T_{\min,s}, A(\frac{\pi}{2}))$ is unitarily equivalent to $(\mathcal{H}, S, A)$, where

$$\begin{aligned}
\mathcal{H} &= L^2(\sigma),\\
A &= \{(\varphi(t); t\varphi(t)) : \varphi(t), t\varphi(t) \in L^2(\sigma)\},\\
S &= \{(\varphi(t); t\varphi(t)) : \varphi(t), t\varphi(t) \in L^2(\sigma),\ \int_{\mathbb{R}} \varphi(t)d\sigma(t) = 0\},
\end{aligned}$$

in the case that $b = 0$ in the representation (3.1) of $q_H(z)$ and

$$\begin{aligned}
\mathcal{H} &= L^2(\sigma) \oplus \mathbb{C},\\
A &= \left\{\left(\begin{pmatrix}\varphi(t)\\0\end{pmatrix}; \begin{pmatrix}t\varphi(t)\\c\end{pmatrix}\right) : \varphi(t), t\varphi(t) \in L^2(\sigma),\ c \in \mathbb{C}\right\},\\
S &= \left\{\left(\begin{pmatrix}\varphi(t)\\0\end{pmatrix}; \begin{pmatrix}t\varphi(t)\\c\end{pmatrix}\right) : \varphi(t), t\varphi(t) \in L^2(\sigma),\ bc + \int_{\mathbb{R}} \varphi(t)d\sigma(t) = 0\right\},
\end{aligned}$$

otherwise. We conclude that $A(\frac{\pi}{2} - z)^{-1}$ is of Hilbert-Schmidt type if and only if the operator

$$\varphi(t) \mapsto \frac{\varphi(t)}{t - z},$$

on $L^2(\sigma)$ is of Hilbert-Schmidt type. This in turn is equivalent to the fact that σ is a discrete measure (3.2) such that (3.3) holds. □

References

[1] L. de Branges, *Hilbert spaces of entire functions* Prentice-Hall, London 1968.

[2] A. Dijksma, H. de Snoo, *Self-adjoint extensions of symmetric subspaces* Pacific J. Math. **54** (1974), 71–100.

[3] I. Gohberg, S. Goldberg, M.A. Kaashoek, *Classes of linear operators. Vol. I* Operator Theory: Advances and Applications, 49, Birkhäuser Verlag, Basel, 1990.

[4] S. Hassi, H. Langer, H. de Snoo, *Selfadjoint extensions for a class of symmetric operators with defect numbers* $(1,1)$ Topics in operator theory, operator algebras and applications (Timişoara, 1994), 115–145, Rom. Acad., Bucharest, 1995.

[5] S. Hassi, H. de Snoo, H. Winkler, *Boundary-value problems for two-dimensional canonical systems* Integral Equations Operator Theory **36** (2000), no. 4, 445–479.

Michael Kaltenbäck and Harald Woracek
Institut für Analysis und Scientific Computing
Technische Universität Wien
Wiedner Hauptstraße 8-10
A 1040 Wien, Austria
e-mail: michael.kaltenbaeck@tuwien.ac.at
e-mail: harald.woracek@tuwien.ac.at

Operator Theory:
Advances and Applications, Vol. 175, 169–191

Spectral Analysis of Differential Operators with Indefinite Weights and a Local Point Interaction

Ilia Karabash and Aleksey Kostenko

Abstract. We consider quasi-self-adjoint extensions of the symmetric operator $A = -(\operatorname{sgn} x)\frac{d^2}{dx^2}$, $\operatorname{dom}(A) = \{f \in W_2^2(\mathbb{R}) : f(0) = f'(0) = 0\}$, in the Hilbert space $L^2(\mathbb{R})$. The main result is a criterion of similarity to a normal operator for operators of this class. The spectra and resolvents of these extensions are described. As an application we describe the main spectral properties of the operators $(\operatorname{sgn} x)\left(-\frac{d^2}{dx^2} + c\delta\right)$ and $(\operatorname{sgn} x)\left(-\frac{d^2}{dx^2} + c\delta'\right)$.

Mathematics Subject Classification (2000). Primary 47A45; Secondary 47B50.

Keywords. Symmetric operator, quasi-self-adjoint extensions, similarity problem, boundary triplets, Weyl functions.

Introduction

Consider the symmetric operator A in the Hilbert space $L^2(\mathbb{R})$ defined by

$$\operatorname{dom}(A) = \{f \in W_2^2(\mathbb{R}) : f(0) = f'(0) = 0\},$$
$$(Af)(x) = -(\operatorname{sgn} x)f''(x) \qquad \text{for} \qquad f \in \operatorname{dom} A. \tag{0.1}$$

The object of investigation is the similarity of quasi-self-adjoint extensions of A (see [1]) to a normal operator. Let us recall that two operators T_1 and T_2 in a Hilbert space $\mathfrak{H}$ are called similar if there exists a bounded operator C with bounded inverse C^{-1} such that $T_1 = C^{-1}T_2C$.

Spectral problems

$$(Ly)(x) = \lambda r(x)y(x), \tag{0.2}$$

where L is an elliptic operator and the function $r(x)$ change sign, occur in certain physical models (see [4] and references therein). The question whether the system of eigenfunctions of the problem (0.2) forms a Riesz basis was studied in [3], [4],

[19], [32] [33], [34] (see also references in [34]). If the operator $\frac{1}{r}L$ has a nonempty continuous spectrum, then the corresponding problem is the similarity of $\frac{1}{r}L$ to a self-adjoint (normal) operator.

In [9], [6], [15], [7], [8], [16] the Krein–Langer spectral theory of definitizable operators (see [28]) was applied to similarity problems for quasi J-nonnegative operators (see [15]) of the form $\frac{1}{r}L$. In particular, B. Ćurgus and B. Najman [7] showed that the operator

$$\tilde{A} = -(\operatorname{sgn} x)\frac{d^2}{dx^2}, \qquad \operatorname{dom}(\tilde{A}) = W_2^2(\mathbb{R}), \tag{0.3}$$

is similar to a self-adjoint one.

This result was proved by another method in [21]; the method is based on the Naboko–Malamud criterion of similarity to a self-adjoint operator [31], [29] (see also [5]). One more proof is presented in [20]. In the recent papers [22], [13], [14], [24] the Naboko-Malamud criterion was applied to different J-self-adjoint differential operators.

Differential operators with an indefinite weight are of interest from one more point of view. The characteristic function $W(\cdot)$ of the operator $\frac{1}{r}L$ as well as the corresponding J-form $J - W^*JW$ is unbounded in $\mathbb{C}_+$. Therefore known sufficient conditions of similarity to a self-adjoint operator cannot be applied here (see [30], [20] and bibliography therein).

In the present paper we describe quasi-self-adjoint extensions A_B of the symmetric operator A in terms of boundary triplets (see [18], [11]). In Sections 3–4 we formulate a criterion of similarity of A_B to a normal (self-adjoint) operator. In order to illustrate these results in Section 5 we obtain simple similarity criteria for operators with local point interactions at zero

$$\tilde{A}_1 := \operatorname{sgn} x\left(-\frac{d^2}{dx^2} + c_1\delta\right), \; c_1 \in \mathbb{C}, \quad \tilde{A}_2 := \operatorname{sgn} x\left(-\frac{d^2}{dx^2} + c_2\delta'\right), \; c_2 \in \mathbb{C}.$$

(See definitions of the operators $\tilde{A}_1$, $\tilde{A}_2$ in [2] and also in Section 5 of the present paper.)

The results of the paper were announced in [23].

Notation: By $\mathfrak{H}, \mathcal{H}$ we denote separable Hilbert spaces. The set of all bounded linear operators from $\mathfrak{H}$ to $\mathcal{H}$ is denoted by $[\mathfrak{H}, \mathcal{H}]$ or $[\mathfrak{H}]$ if $\mathfrak{H} = \mathcal{H}$. $\mathcal{C}(\mathfrak{H})$ stands for the set of closed densely defined operators in $\mathfrak{H}$. Let T be a linear operator in a Hilbert space $\mathfrak{H}$. In what follows $\operatorname{dom}(T)$, $\ker(T)$, $\operatorname{ran}(T)$ are the domain, kernel, range of T, respectively. We denote by $\sigma(T)$, $\sigma_r(T)$, $\sigma_c(T)$ the point, residual and continuous spectra of T. By $\sigma_p(T)$ the set of eigenvalues of T is indicated. We denote the resolvent set by $\rho(T)$; $R_T(\lambda) := (T - \lambda I)^{-1}$, $\lambda \in \rho(T)$, is the resolvent of T. Recall that $\sigma_r(T) = \{\lambda \in \sigma(T) \setminus \sigma_p(T) : \overline{\operatorname{ran}(T - \lambda I)} \neq \mathfrak{H}\}$, $\sigma_c(T) = \sigma(T) \setminus (\sigma_p(T) \bigcup \sigma_r(T))$.

We set $\mathbb{C}_\pm := \{\lambda \in \mathbb{C} : \pm \operatorname{Im} \lambda > 0\}$, $\mathbb{R}_+ := (0, +\infty)$, $\mathbb{R}_- := (-\infty, 0)$. By $\chi_{\mathcal{I}}(t)$ we denote the characteristic function of the interval $\mathcal{I}$, i.e., $\chi_{\mathcal{I}}(t) = 1$ for $t \in \mathcal{I}$, $\chi_{\mathcal{I}}(t) = 0$ for $t \notin \mathcal{I}$. Finally, we set $\chi_\pm(t) := \chi_{\mathbb{R}_\pm}(t)$.

1. Preliminaries

1.1. A similarity criterion

Our approach is based on the concept of boundary triplets (see [18], [11]) and the resolvent similarity criterion obtained by S.N. Naboko [31] and M.M. Malamud [29] (in [5] this criterion was obtained under an additional assumption).

Theorem 1.1 ([29, 31]). *A closed operator T in a Hilbert space $\mathfrak{H}$ is similar to a self-adjoint one if and only if $\sigma(A) \subset \mathbb{R}$ and for all $f \in \mathfrak{H}$ the inequalities*

$$\begin{aligned} \sup_{\varepsilon>0} \int_{-\infty}^{+\infty} \varepsilon \left\| R_T (\mu + i\varepsilon) f \right\|^2 d\mu \leq C \left\| f \right\|^2, \\ \sup_{\varepsilon>0} \int_{-\infty}^{+\infty} \varepsilon \left\| R_{T^*} (\mu + i\varepsilon) f \right\|^2 d\mu \leq C^* \left\| f \right\|^2, \end{aligned} \tag{1.1}$$

are valid with constants C and C^* independent of f.

1.2. Linear relations

Definition 1.1.

(i) *A closed linear relation Θ in $\mathcal{H}$ is a closed subspace Θ of $\mathcal{H} \oplus \mathcal{H}$.*

(ii) *The closed linear relation Θ is symmetric if for all $\{f_1, g_1\}, \{f_2, g_2\} \in \Theta$ the condition*

$$(g_1, f_2) - (f_1, g_2) = 0, \tag{1.2}$$

is satisfied.

(iii) *The closed linear relation Θ is self-adjoint if it is maximal symmetric, i.e., Θ is symmetric and there does not exist a closed symmetric relation $\tilde{\Theta}$ such that Θ is properly contained in $\tilde{\Theta}$.*

Let us illustrate closed linear relations by simple examples.

Example 1.1.

(i) *Let B be a closed operator in $\mathcal{H}$, not necessarily bounded. Then the graph $G(B)$ of B is a closed relation in $\mathcal{H}$. Moreover, if $B = B^*$ is a self-adjoint operator, then $G(B)$ is a self-adjoint relation in $\mathcal{H}$.*

(ii) *The subspaces $\Theta_0 := \{0\} \times \mathcal{H}$, $\Theta_1 := \mathcal{H} \times \{0\}$ of $\mathcal{H} \times \mathcal{H}$ are self-adjoint relations in $\mathcal{H}$. Obviously, Θ_0 is not the graph of any operator.*

1.3. Boundary triplets

Let $A \in \mathcal{C}(\mathfrak{H})$ be a closed symmetric operator with equal deficiency indices $n_+(A) = n_-(A)$ ($n_\pm(T) := \dim \mathfrak{N}_{\pm i}$ and by $\mathfrak{N}_\lambda := \ker(T^* - \lambda)$ the deficiency subspaces of A are indicated). Without loss of generality we can assume that A is simple. This means that A has no self-adjoint parts.

Definition 1.2 ([1]).

(i) *A closed extension* $\tilde{A}$ *of* A *is called a proper extension if* $A \subset \tilde{A} \subset A^*$. *The set of all proper extensions is denoted by* Ext_A.

(ii) *A proper extension* $\tilde{A}$ *is called a quasi-self-adjoint if*

$$\dim(\operatorname{dom}(\tilde{A})/\operatorname{dom}(A)) = n_\pm(A). \tag{1.3}$$

We recall the definition of a boundary triplet which may be considered as an abstract version of the second Green formula.

Definition 1.3 ([18]). *A triplet* $\Pi = \{\mathcal{H}, \Gamma_0, \Gamma_1\}$ *consisting of an auxiliary Hilbert space* $\mathcal{H}$ *and linear mappings*

$$\Gamma_j : \operatorname{dom}(A^*) \longrightarrow \mathcal{H}, \qquad j \in \{0,1\}, \tag{1.4}$$

is called a boundary triplet for the adjoint operator A^* *of* A *if the following two conditions are satisfied:*

(i) *The second Green formula*

$$(A^*f, g) - (f, A^*g) = (\Gamma_1 f, \Gamma_0 g)_{\mathcal{H}} - (\Gamma_0 f, \Gamma_1 g)_{\mathcal{H}}, \quad f, g \in \operatorname{dom}(A^*), \tag{1.5}$$

takes place and

(ii) *the mapping*

$$\Gamma : \operatorname{dom}(A^*) \longrightarrow \mathcal{H} \oplus \mathcal{H}, \quad \Gamma f := \{\Gamma_0 f, \Gamma_1 f\}, \tag{1.6}$$

is surjective.

The above definition allows one to describe the set Ext_A in the following way (see [10, 11]).

Proposition 1.1 ([10, 11]). *Let* $\Pi = \{\mathcal{H}, \Gamma_0, \Gamma_1\}$ *be a boundary triplet for* A^*. *Then the mapping* Γ *establishes a bijective correspondence* $\tilde{A} \to \Theta := \Gamma(\operatorname{dom}(\tilde{A}))$ *between the set* Ext_A *and the set of closed linear relations in* $\mathcal{H}$.

By Proposition 1.1 the following definition is natural.

Definition 1.4. *Let* $\Pi = \{\mathcal{H}, \Gamma_0, \Gamma_1\}$ *be a boundary triplet for the operator* A^*.

(i) *Denote* $A_\Theta = \tilde{A}$ *if* $\Theta = \Gamma(\operatorname{dom}(\tilde{A}))$, *that is*

$$A_\Theta := A^* | D_\Theta, \quad \text{where} \quad D_\Theta := \{f \in \operatorname{dom}(A^*) : \{\Gamma_0 f, \Gamma_1 f\} \in \Theta\}. \tag{1.7}$$

(ii) *If* $\Theta = G(B)$ *is the graph of* $B \in \mathcal{C}(\mathcal{H})$, *then* $\operatorname{dom}(A_\Theta)$ *is determined by the equation* $\operatorname{dom}(A_B) = D_B := D_\Theta = \ker(\Gamma_1 - B\Gamma_0)$. *We set* $A_B := A_\Theta$.

Let us make the following remarks.

Remark 1.1.

1) *The deficiency indices* $n_\pm(A)$ *are equal to the dimension of* $\mathcal{H}$, *i.e.,* $\dim(\mathcal{H}) = n_\pm(A)$.

2) *There exist two self-adjoint extensions* $A_j := A^* | \ker(\Gamma_j)$ *which are naturally associated to a boundary triplet. According to Definition 1.4* $A_j = A_{\Theta_j}, j \in \{0,1\}$, *where* $\Theta_0 = \{0\} \times \mathcal{H}$, $\Theta_1 = \mathcal{H} \times \{0\}$. *Conversely, if* A_0 *is a self-adjoint*

extension of A*, then there exists a boundary triplet* $\Pi = \{\mathcal{H}, \Gamma_0, \Gamma_1\}$ *such that* $A_0 = A^*|\ker(\Gamma_0)$.

3) Θ *is the graph of an operator* $B \in \mathcal{C}(\mathcal{H})$ *iff* $\tilde{A}$ *and* A_0 *are disjoint, i.e.,* $\mathrm{dom}(\tilde{A}) \cap \mathrm{dom}(A_0) = \mathrm{dom}(A)$.
4) $\Theta = G(B)$ *with* $B \in [\mathcal{H}]$ *iff* $\tilde{A}$ *and* A_0 *are transversal, i.e.,* $\tilde{A}$ *and* A_0 *are disjoint and* $\mathrm{dom}(\tilde{A}) + \mathrm{dom}(A_0) = \mathrm{dom}(A^*)$.

Definition 1.5 ([12]). *The proper extension* $\tilde{A} \in \mathrm{Ext}_A$ *is called almost solvable if there exists a boundary triplet* $\Pi = \{\mathcal{H}, \Gamma_0, \Gamma_1\}$ *and an operator* $B \in [\mathcal{H}]$ *such that*

$$\mathrm{dom}(\tilde{A}) = \mathrm{dom}(A_B) := \ker(\Gamma_1 - B\Gamma_0). \tag{1.8}$$

The set of almost solvable extensions is denoted by $\mathcal{A}s_A$. Note that the class $\mathcal{A}s_A$ is sufficiently wide. Proper extensions having two regular points $\lambda_1, \lambda_2 \in \mathbb{C}$ such that $\mathrm{Im}\,\lambda_1 \cdot \mathrm{Im}\,\lambda_2 < 0$ belong to $\mathcal{A}s_A$. All quasi-self-adjoint extensions are in $\mathcal{A}s_A$ if $n_\pm(A) < \infty$.

1.4. Weyl functions

It is well known that Weyl functions are an important tool in the direct and inverse spectral theory of singular Sturm–Liouville operators. In [10, 11] the concept of Weyl function was generalized to an arbitrary symmetric operator A with infinite deficiency indices $n_+(A) = n_-(A)$. In this subsection we recall basic facts about Weyl functions.

Definition 1.6 ([10, 11]). *Let* $\Pi = \{\mathcal{H}, \Gamma_0, \Gamma_1\}$ *be a boundary triplet for the operator* A^**. The Weyl function of* A *corresponding to the boundary triplet* $\{\mathcal{H}, \Gamma_0, \Gamma_1\}$ *is a unique mapping*

$$M(\cdot) : \rho(A_0) \longrightarrow [\mathcal{H}] \tag{1.9}$$

satisfying

$$\Gamma_1 f_\lambda = M(\lambda)\Gamma_0 f_\lambda \qquad \text{for all} \quad f_\lambda \in \mathfrak{N}_\lambda, \quad \lambda \in \rho(A_0), \tag{1.10}$$

where $\mathfrak{N}_\lambda := \ker(A^* - \lambda I)$.

It is well known (see [10, 11]) that the above implicit definition of the Weyl function is correct and $M(\cdot)$ is an R-function obeying $0 \in \rho(\mathrm{Im}(M(i)))$. The Weyl function immediately provides some information about the "spectral properties" of proper extensions. We confine ourselves to the case of almost solvable extensions of the symmetric operator A.

Proposition 1.2 ([11, 12]). *Suppose that* $\Pi = \{\mathcal{H}, \Gamma_0, \Gamma_1\}$ *is a boundary triplet for* A^*, $M(\cdot)$ *is the corresponding Weyl function,* $\lambda \in \rho(A_0)$ *and* $B \in [\mathcal{H}]$*. Then:*

1) $\lambda \in \rho(A_B)$ *if and only if* $0 \in \rho(B - M(\lambda))$;
2) $\lambda \in \sigma_i(A_B)$ *if and only if* $0 \in \sigma_i(B - M(\lambda))$, $i \in \{p, r, c\}$.

1.5. γ-fields

With each boundary triplet we can associate a so-called γ-field.

Definition 1.7 ([11]). *Let $\Pi = \{\mathcal{H}, \Gamma_0, \Gamma_1\}$ be a boundary triplet for A^*. The γ-field $\gamma(\cdot)$ corresponding to Π is defined by*

$$\gamma(\lambda) := (\Gamma_0|\mathfrak{N}_\lambda)^{-1} : \mathcal{H} \longrightarrow \mathfrak{N}_\lambda, \quad \lambda \in \rho(A_0). \tag{1.11}$$

One can easily check that

$$\gamma(\lambda) = (A_0 - \lambda_0)(A_0 - \lambda)^{-1}\gamma(\lambda_0), \quad \lambda, \lambda_0 \in \rho(A_0), \tag{1.12}$$

and consequently $\gamma(\cdot)$ is a γ-field in the sense of [26]. It is shown in [11] that the γ-field $\gamma(\cdot)$ and the Weyl function $M(\cdot)$ are related by

$$M(\lambda) - M(\lambda_0)^* = (\lambda - \bar{\lambda}_0)\gamma(\lambda_0)^*\gamma(\lambda), \quad \lambda, \lambda_0 \in \rho(A_0). \tag{1.13}$$

The relation (1.13) means the $M(\cdot)$ is a Q-function in the sense of [26].

The following version of the Krein-Naimark formula for canonical resolvents (see for instance [26]) is based on the notion of boundary triplets.

Theorem 1.2 ([10, 11]). *Let $\tilde{A}$ be an almost solvable extension of A ($\tilde{A} \in \mathcal{A}s_A$), i.e., $\tilde{A} = A_B$ with $B \in [\mathcal{H}]$ for some boundary triplet $\Pi = \{\mathcal{H}, \Gamma_0, \Gamma_1\}$. Then*

$$(A_B - \lambda)^{-1} = (A_0 - \lambda)^{-1} + \gamma(\lambda)(B - M(\lambda))^{-1}\gamma^*(\overline{\lambda}), \qquad \lambda \in \rho(A_B). \tag{1.14}$$

Here $M(\cdot)$ and $\gamma(\cdot)$ are the Weyl function and γ-field corresponding to the triplet Π.

2. Extensions of the minimal operator

2.1. Boundary conditions

Consider the operator A of the form (0.1). It is obvious that A is a closed simple symmetric operator with deficiency indices $n_\pm(A) = 2$.

We denote by $\sqrt{z}$ the branch of the multifunction on the complex plane $\mathbb{C}$ with the cut along $\mathbb{R}_-$, singled out by the condition $\sqrt{-1 + i0} = i$.

Theorem 2.1.

(i) *The adjoint operator A^* has the form*

$$A^* = -(\operatorname{sgn} x)\frac{d^2}{dx^2}, \quad \operatorname{dom}(A^*) = W_2^2(\mathbb{R} \setminus \{0\}) := W_2^2(\mathbb{R}_-) \oplus W_2^2(\mathbb{R}_+). \tag{2.1}$$

(ii) *Let mappings $\Gamma_j : W_2^2(\mathbb{R} \setminus \{0\}) \to \mathbb{C}^2$, $j = \{0, 1\}$, be given by*

$$\Gamma_0 f = \begin{pmatrix} f(+0) \\ f'(-0) \end{pmatrix}, \quad \Gamma_1 f = \begin{pmatrix} f'(+0) \\ -f(-0) \end{pmatrix}. \tag{2.2}$$

Then $\Pi = \{\mathbb{C}^2, \Gamma_0, \Gamma_1\}$ is a boundary triplet for A^*.

(iii) *The corresponding Weyl function $M(\cdot)$ is*

$$M(\lambda) := \begin{pmatrix} -\sqrt{-\lambda} & 0 \\ 0 & -1/\sqrt{\lambda} \end{pmatrix}, \qquad \lambda \in \mathbb{C} \setminus \mathbb{R}. \tag{2.3}$$

(iv) *The corresponding* γ*-field* $\gamma(\lambda) : \mathbb{C}^2 \to \mathfrak{N}_\lambda$ *is*

$$\gamma(\lambda)\begin{pmatrix} c_+ \\ c_- \end{pmatrix} := c_+ \cdot e^{-\sqrt{-\lambda}x}\chi_+(x) + \frac{c_-}{\sqrt{\lambda}} \cdot e^{\sqrt{\lambda}x}\chi_-(x), \qquad c_\pm \in \mathbb{C}. \tag{2.4}$$

Proof. The first statement is obvious. Moreover, we have

$$\begin{aligned}(A^*f,\ g) - (f,\ A^*g) = f'(+0)\overline{g(+0)} + f'(-0)\overline{g(-0)} - \\ - f(+0)\overline{g'(+0)} - f(-0)\overline{g'(-0)}, \quad f,\ g \in \operatorname{dom}(A^*).\end{aligned} \tag{2.5}$$

Hence (ii) follows from Definition 1.3.

Note that

$$\mathfrak{N}_\lambda = \{f_\lambda(x) := c_+ \cdot e^{-\sqrt{-\lambda}x}\chi_+(x) + c_- \cdot e^{\sqrt{\lambda}x}\chi_-(x) : c_\pm \in \mathbb{C}\}. \tag{2.6}$$

Combining (2.2) and (2.6), one gets

$$\Gamma_0 f_\lambda = \begin{pmatrix} c_+ \\ c_-\sqrt{\lambda} \end{pmatrix}, \qquad \Gamma_1 f_\lambda = \begin{pmatrix} -c_+\sqrt{-\lambda} \\ -c_- \end{pmatrix}. \tag{2.7}$$

By Definitions 1.6 and 1.7, we easily obtain (2.3) and (2.4). □

Let us introduce the following boundary conditions at zero

$$\left\{\begin{array}{l} a_{11}f(-0) + a_{12}f'(-0) + a_{13}f(+0) + a_{14}f'(+0) = 0 \\ a_{21}f(-0) + a_{22}f'(-0) + a_{23}f(+0) + a_{24}f'(+0) = 0 \end{array}\right., \qquad a_{ij} \in \mathbb{C}. \tag{2.8}$$

By Definition 1.2, a quasi-self-adjoint extension $\tilde{A}$ of the operator A has the form

$$\begin{aligned}\tilde{A} = A_{(a_{ij})} = A^* | \operatorname{dom}(A_{(a_{ij})}), \\ \operatorname{dom}(A_{(a_{ij})}) = \{f \in W_2^2(\mathbb{R} \setminus \{0\}) :\ f \text{ satisfies conditions (2.8)}\},\end{aligned} \tag{2.9}$$

with the matrix

$$(a_{ij}) := \begin{pmatrix} a_{11} & a_{12} & a_{13} & a_{14} \\ a_{21} & a_{22} & a_{23} & a_{24} \end{pmatrix}$$

such that

$$\operatorname{rank}(a_{ij}) = n_\pm(A) = 2. \tag{2.10}$$

Consider three cases.

1) Let

$$\begin{pmatrix} a_{11} & a_{12} \\ a_{21} & a_{22} \end{pmatrix} = \begin{pmatrix} 0 & 0 \\ 0 & 0 \end{pmatrix} \quad \text{or} \quad \begin{pmatrix} a_{13} & a_{14} \\ a_{23} & a_{24} \end{pmatrix} = \begin{pmatrix} 0 & 0 \\ 0 & 0 \end{pmatrix}.$$

Then $\tilde{A} = A_- \oplus A_+$, where $A_\pm$ is an operator in $L^2(\mathbb{R}_\pm)$. By condition (2.10), we see that one of the operators A_-, A_+ is a symmetric with deficiency indices (1,1) and another one is an adjoint to a symmetric operator with deficiency indices (1,1). Hence $\mathbb{C} \setminus \mathbb{R} \subset \sigma_p(\tilde{A})$ and $\tilde{A}$ is not similar to a normal operator.

2) Suppose that $\begin{pmatrix} a_{11} & a_{12} \\ a_{21} & a_{22} \end{pmatrix}$ and $\begin{pmatrix} a_{13} & a_{14} \\ a_{23} & a_{24} \end{pmatrix}$ have a zero column. Since $\operatorname{rank}(a_{ij}) = 2$, it follows that $\tilde{A} = A_- \oplus A_+$, where A_+ and A_- are self-adjoint operators in $L^2(\mathbb{R}_+)$ and $L^2(\mathbb{R}_-)$, respectively. Thus $\tilde{A} = \tilde{A}^*$.

3) Suppose that there are three nonzero columns in (a_{ij}). In this case one of the determinants

$$\begin{aligned} \Delta_1 &= \begin{vmatrix} a_{11} & a_{14} \\ a_{21} & a_{24} \end{vmatrix}, \quad \Delta_2 = \begin{vmatrix} a_{12} & a_{13} \\ a_{22} & a_{23} \end{vmatrix}, \\ \Delta_3 &= \begin{vmatrix} a_{12} & a_{14} \\ a_{22} & a_{24} \end{vmatrix}, \quad \Delta_4 = \begin{vmatrix} a_{11} & a_{13} \\ a_{21} & a_{23} \end{vmatrix} \end{aligned} \tag{2.11}$$

does not vanish.

Evidently, only the case (3) is of interest to us.

2.2. The case $\Delta_1 \neq 0$

Let $\Delta_1 \neq 0$. (The cases $\Delta_2 \neq 0$, $\Delta_3 \neq 0$, and $\Delta_4 \neq 0$ will be considered in Section 4.) Then conditions (2.8) take the form

$$\begin{cases} f'(+0) = b_{11} f(+0) + b_{12} f'(-0) \\ -f(-0) = b_{21} f(+0) + b_{22} f'(-0). \end{cases} \tag{2.12}$$

Hence, by Definition 1.5, $A_{(a_{ij})} = A_B = A^* | \ker(\Gamma_1 - B\Gamma_0)$. Here

$$B = \begin{pmatrix} b_{11} & b_{12} \\ b_{21} & b_{22} \end{pmatrix} \in \mathbb{C}^{2\times 2}$$

and the boundary triplet $\Pi = \{\mathbb{C}^2, \Gamma_0, \Gamma_1\}$ is of the form (2.2).

In what follows A_B stands for the operator

$$A_B := -\operatorname{sgn} x \frac{d^2}{dx^2}, \quad \operatorname{dom}(A_B) = \{f \in W_2^2(\mathbb{R} \setminus \{0\}) : f \text{ satisfies } (2.12)\}. \tag{2.13}$$

For $B \in \mathbb{C}^{2\times 2}$ and the Weyl function $M(\cdot)$ of the form (2.3) we define the function $\varphi_B(\cdot) : \mathbb{C} \setminus \mathbb{R} \to \mathbb{C}$ by

$$\varphi_B(\lambda) := \det(B - M(\lambda)), \qquad \lambda \in \mathbb{C} \setminus \mathbb{R}. \tag{2.14}$$

Lemma 2.1. *Suppose A_B is the operator of the form* (2.13) *and* $|b_{12}| + |b_{21}| \neq 0$*; then:*

(i) $\sigma_c(A_B) = \mathbb{R}$;
(ii) $\sigma_r(A_B) = \emptyset$;
(iii) $\sigma_p(A_B) = \{\lambda \in \mathbb{C} \setminus \mathbb{R} : \varphi_B(\lambda) = 0\} = \{\lambda \in \mathbb{C}_+ : \varphi_B(\lambda) = 0\} \cup \{\lambda \in \mathbb{C}_- : \varphi_{B^*}(\overline{\lambda}) = 0\}$.

Proof. Simple calculations show that there are no eigenvalues on the real axis and $\sigma_c(A_B) = \mathbb{R}$. The second and the third statements evidently follow from Proposition 1.2. □

Lemma 2.2. *Let $\Pi = \{\mathcal{H}, \Gamma_0, \Gamma_1\}$ be a boundary triplet of the form (2.2). Then*

$$\gamma^*(\overline{\lambda})f = \begin{pmatrix} \int\limits_0^{+\infty} f(t)e^{-\sqrt{-\lambda}t}dt \\ \frac{1}{\sqrt{\lambda}} \int\limits_{-\infty}^{0} f(t)e^{\sqrt{\lambda}t}dt \end{pmatrix}, \qquad f \in L^2(\mathbb{R}). \tag{2.15}$$

Proof. By Theorem 2.1.(iv), $\gamma^*(\cdot)$ is a map from $L^2(\mathbb{R})$ to $\mathbb{C}^2$. To find $\gamma^*(\cdot)$ we use the equation

$$(\gamma(\lambda)c, f)_{L^2(\mathbb{R})} = (c, \gamma^*(\lambda)f)_{\mathbb{C}^2}, \qquad c = \begin{pmatrix} c_1 \\ c_2 \end{pmatrix} \in \mathbb{C}^2, \quad f \in L^2(\mathbb{R}). \tag{2.16}$$

Combining (2.4) and (2.16), one gets

$$c_1 \cdot \int\limits_0^{+\infty} \overline{f(t)}e^{-\sqrt{-\lambda}t}dt + \frac{c_2}{\sqrt{\lambda}} \cdot \int\limits_{-\infty}^{0} \overline{f(t)}e^{\sqrt{\lambda}t}dt = c_1 \cdot \overline{(\gamma^*(\lambda)f)_1} + c_2 \cdot \overline{(\gamma^*(\lambda)f)_2}. \tag{2.17}$$

Hence (2.15) immediately follows from (2.17). □

Let us denote

$$y_+(f,\lambda) := \int\limits_0^{+\infty} f(t)e^{-\sqrt{-\lambda}t}dt, \qquad y_-(f,\lambda) := \int\limits_{-\infty}^{0} f(t)e^{\sqrt{\lambda}t}dt, \qquad \lambda \in \mathbb{C} \setminus \mathbb{R}. \tag{2.18}$$

Lemma 2.3. *Let the operator A_B be of the form (2.13) and $A_0 = A^*|\ker\Gamma_0$. Then*

$$\begin{aligned} &\left((A_B - \lambda)^{-1}f - (A_0 - \lambda)^{-1}f\right)(x) \\ &\quad = \frac{e^{-\sqrt{-\lambda}x} \cdot \chi_+(x)}{\varphi_B(\lambda)} \left(\left(b_{22} + \frac{1}{\sqrt{\lambda}}\right) y_+(f,\lambda) - \frac{b_{12}}{\sqrt{\lambda}} \cdot y_-(f,\lambda) \right) \\ &\qquad + \frac{e^{\sqrt{\lambda}x} \cdot \chi_-(x)}{\sqrt{\lambda} \cdot \varphi_B(\lambda)} \left(\frac{b_{11} + \sqrt{-\lambda}}{\sqrt{\lambda}} \cdot y_-(f,\lambda) - b_{21} \cdot y_+(f,\lambda) \right), \\ &\qquad\qquad f \in L^2(\mathbb{R}), \qquad \lambda \in \rho(A_B). \end{aligned} \tag{2.19}$$

Here $\varphi_B(\cdot)$ and $y_\pm(f,\cdot)$ are given by (2.14) and (2.18), respectively.

Proof. By (2.3), for $\lambda \in \rho(A_B)$ (see Lemma 2.1) one obtains

$$\begin{aligned} (B - M(\lambda))^{-1} &= \begin{pmatrix} b_{11} + \sqrt{-\lambda} & b_{12} \\ b_{21} & b_{22} + 1/\sqrt{\lambda} \end{pmatrix}^{-1} \\ &= \frac{1}{\varphi_B(\lambda)} \begin{pmatrix} b_{22} + 1/\sqrt{\lambda} & -b_{12} \\ -b_{21} & b_{11} + \sqrt{-\lambda} \end{pmatrix}. \end{aligned} \tag{2.20}$$

Combining (2.4), (2.15), (2.20) with formula (1.14), we get (2.19). □

3. Similarity to a normal operator

3.1. The main result

For each $B \in \mathbb{C}^{2\times 2}$ let us define the function $\varphi_B^+ : \overline{\mathbb{C}_+} \to \overline{\mathbb{C}}$ in the following way. We set

$$\varphi_B^+(\lambda) := \varphi_B(\lambda) = \det(B - M(\lambda)) \qquad \text{for} \quad \lambda \in \mathbb{C}_+, \tag{3.1}$$

and for $x \in \overline{\mathbb{R}}$ by $\varphi_B^+(x)$ we denote the boundary values of $\varphi_B(\lambda)$ in $\mathbb{C}_+$,

$$\varphi_B^+(x) := \lim_{\substack{z\to x\\ z\in\mathbb{C}_+}} \det(B - M(z)), \qquad x \in \mathbb{R}\cup\{\infty\}. \tag{3.2}$$

Note that the function φ_B^+ is analytic on $\mathbb{C}_+$ and continuous on $\overline{\mathbb{C}_+}\setminus\{0\}$.

The following similarity criterion is the main result of the paper.

Theorem 3.1 (Main Theorem). *Assume that $\Delta_1 \neq 0$ and the operator A_B is defined by (2.13). Let φ_B^+ and $\varphi_{B^*}^+$ be the functions defined in (3.1)–(3.2) and $|b_{12}|+|b_{21}| \neq 0$. Then A_B is similar to a normal operator if and only if the following conditions hold:*

(i) *φ_B^+ and $\varphi_{B^*}^+$ have no zeroes in the set $\mathbb{R}\cup\{\infty\}$;*
(ii) *φ_B^+ and $\varphi_{B^*}^+$ have no zeroes of the second order in $\mathbb{C}_+$.*

Remark 3.1. *Suppose that $|b_{12}| + |b_{21}| = 0$. Then the operator A_B has the form*

$$A_B = A_- \oplus A_+,$$

where the operators $A_\pm : L^2(\mathbb{R}_\pm) \to L^2(\mathbb{R}_\pm)$ are given by

$$A_\pm := \mp\frac{d^2}{dx^2}, \qquad \operatorname{dom}(A_\pm) = \{f \in W_2^2(\mathbb{R}_\pm) : f(\pm 0) + b_\pm \cdot f'(\pm 0) = 0\}. \tag{3.3}$$

Here $b_+ := -1/b_{11}$, $b_- := b_{22}$. Operators $A_\pm$ are well studied.

Remark 3.2. *The function φ_B^+ has a simple form. Indeed, by (2.14) and (2.3), we have*

$$\varphi_B^+(\lambda) = -ib_{22}\sqrt{\lambda} + (b_{11}b_{22} - b_{12}b_{21}) - i + \frac{b_{11}}{\sqrt{\lambda}}, \qquad \lambda \in \overline{\mathbb{C}_+}\setminus\{0\}. \tag{3.4}$$

So the conditions of Theorem 3.1 can be easily checked (see Section 5). Let us remark that the function φ_B^+ has at most two zeroes (a zero of multiplicity k is counted as k zeroes).

A criterion of similarity to a self-adjoint operator immediately follows from Theorem 3.1.

Theorem 3.2. *Let $|b_{12}|+|b_{21}| \neq 0$. Then the operator A_B is similar to a self-adjoint one iff the functions φ_B^+ and $\varphi_{B^*}^+$ do not vanish in $\overline{\mathbb{C}_+}$.*

Proof. By Lemma 2.1.(iii), $\sigma(A_B) = \mathbb{R}$ iff the functions φ_B^+ and $\varphi_{B^*}^+$ have no zeroes in $\mathbb{C}_+$. Combining this fact with Theorem 3.1, we get Theorem 3.2. □

To prove the main theorem we recall the following Lemma.

Lemma 3.1. *If an operator T is similar to a normal one, then the inequality*

$$\|(T-\lambda I)^{-1}\| \le \frac{C}{\operatorname{dist}(\lambda,\ \sigma(T))} \tag{3.5}$$

holds with some constant $C > 0$.

3.2. Some estimates

We start with the following lemma.

Lemma 3.2. *Let $|b_{12}| + |b_{21}| \neq 0$. Suppose there exists $\lambda_0 \in \mathbb{R} \cup \{\infty\}$ such that $\varphi_B^+(\lambda_0) = 0$ or $\varphi_{B^*}^+(\lambda_0) = 0$. Then the operator A_B of the form (2.13) is not similar to a normal operator.*

Proof. Without loss of generality suppose that $b_{21} \neq 0$ and $\varphi_B^+(\lambda_0) = 0$ for some $\lambda_0 \in \mathbb{R}$.

It is obvious that for $\lambda \in \mathbb{C} \setminus \mathbb{R}$

$$\begin{aligned}\sup_{\|f\|\le 1} |y_+(f,\lambda)|^2 &= \sup_{\|f\|\le 1} \left| \int_{\mathbb{R}_+} f(t) e^{-\sqrt{-\lambda}t} dt \right|^2 \\ &= \|e^{-\sqrt{-\lambda}x}\chi_+(x)\|_{L^2}^2 = \frac{1}{|2\operatorname{Re}\sqrt{-\lambda}|} = \frac{1}{|2\operatorname{Im}\sqrt{\lambda}|}.\end{aligned} \tag{3.6}$$

Further, we set $f_+(\cdot) := f(\cdot)\chi_+(\cdot),\ f \in L^2(\mathbb{R})$. By (2.19), we have

$$\begin{aligned}&\left\|(A_B-\lambda I)^{-1} f_+ - (A_0-\lambda I)^{-1} f_+\right\|_{L^2}^2 \\ &= \left\| \frac{e^{-\sqrt{-\lambda}x}\cdot \chi_+(x)}{\varphi_B^+(\lambda)} \left(b_{22} + \frac{1}{\sqrt{\lambda}}\right) y_+(f_+,\lambda) \right\|_{L_2}^2 \\ &\quad + |b_{21}| \cdot \left\| \frac{e^{\sqrt{\lambda}x}\cdot \chi_-(x)}{\sqrt{\lambda}\cdot \varphi_B^+(\lambda)} \cdot y_+(f_+,\lambda) \right\|_{L_2}^2, \qquad \lambda \in \rho(A_B) \cap \mathbb{C}_+.\end{aligned} \tag{3.7}$$

Combining (3.6) and (3.7), one obtains

$$\begin{aligned}&\left\|(A_B-\lambda I)^{-1} - (A_0-\lambda I)^{-1}\right\|_{L^2}^2 \\ &\ge \left| \left(b_{22} + \frac{1}{\sqrt{\lambda}}\right) \frac{1}{2\varphi_B^+(\lambda)\cdot \operatorname{Im}\sqrt{\lambda}} \right|^2 + \left| \frac{b_{21}}{4\sqrt{\lambda}\cdot \varphi_B^+(\lambda)} \right|^2 \cdot \frac{1}{|\operatorname{Re}\sqrt{\lambda}\cdot \operatorname{Im}\sqrt{\lambda}|}.\end{aligned} \tag{3.8}$$

Now if we recall Lemma 2.1, we obtain that $\operatorname{dist}(\lambda, \sigma(A_B)) = |\operatorname{Im}\lambda|$ in some neighborhood of λ_0. Therefore, for sufficiently small ε and $\lambda = \lambda_0 + i\varepsilon$

$$\begin{aligned}&\operatorname{dist}(\lambda, \sigma(A_B))^2 \cdot \left\|(A_B-\lambda I)^{-1} - (A_0-\lambda I)^{-1}\right\|_{L^2}^2 \\ &\ge \left| \frac{b_{21}\cdot \operatorname{Im}\lambda}{4\sqrt{\lambda}\cdot \varphi_B^+(\lambda)} \right|^2 \cdot \frac{1}{|\operatorname{Re}\sqrt{\lambda}\cdot \operatorname{Im}\sqrt{\lambda}|} \ge C_1 \cdot \frac{1}{|\operatorname{Im}\lambda|}.\end{aligned} \tag{3.9}$$

Then the left part of inequality (3.9) is unbounded in the neighborhood of λ_0. Note that A_0 is a self-adjoint operator. Hence inequality (3.5) is valid for A_0. Therefore, the function

$$\operatorname{dist}(\lambda, \sigma(A_B)) \cdot \left\|(A_B - \lambda I)^{-1}\right\|_{L^2} \tag{3.10}$$

is unbounded in the neighborhood of λ_0. By Lemma 3.1, the operator A_B is not similar to a normal operator.

If $\varphi_B^+(\infty) = 0$, then formula (3.4) implies $\varphi_B^+(\lambda) = b_{11}/\sqrt{\lambda}$ for $\lambda \in \mathbb{C}_+$. Hence for ε large enough

$$\begin{aligned} &\operatorname{dist}(i\varepsilon, \sigma(A_B))^2 \cdot \left\|(A_B - i\varepsilon I)^{-1} - (A_0 - i\varepsilon I)^{-1}\right\|_{L^2}^2 \\ &\quad \geq \frac{|b_{21}|}{4|b_{11}|} \cdot \frac{\varepsilon^2}{|\operatorname{Re}\sqrt{i\varepsilon} \cdot \operatorname{Im}\sqrt{i\varepsilon}|} = C_2 \varepsilon\ . \end{aligned} \tag{3.11}$$

Since the right part of (3.11) is unbounded in $\mathbb{C}_+$, we see that A_B is not similar to a normal operator. □

Lemma 3.3. *Let $|b_{12}| + |b_{21}| \neq 0$. Suppose that the function φ_B^+ has a zero of algebraic multiplicity 2 in $\mathbb{C}_+$. Then A_B is not similar to a normal operator.*

Proof. Let $b_{21} \neq 0$ (the case $b_{12} \neq 0$ can be considered in the same way). Suppose $\lambda_0 \in \mathbb{C}_+$ is a zero of multiplicity 2 of $\varphi_B^+(\cdot)$. By (3.8), we have for $\lambda \in \rho(A_B) \cap \mathbb{C}_+$

$$\left\|(A_B - \lambda I)^{-1} - (A_0 - \lambda I)^{-1}\right\|_{L^2} \geq \left|\frac{b_{21}}{4\sqrt{\lambda} \cdot \varphi_B^+(\lambda)}\right| \cdot \frac{1}{|\operatorname{Re}\sqrt{\lambda} \cdot \operatorname{Im}\sqrt{\lambda}|^{1/2}}. \tag{3.12}$$

Since $\lambda_0 \in \rho(A_0)$, we see that (3.12) implies

$$\left\|(A_B - \lambda I)^{-1}\right\|_{L^2} \geq \frac{C_{\lambda_0}}{|\varphi_B^+(\lambda)|}, \qquad C_{\lambda_0} = const > 0, \tag{3.13}$$

in some neighborhood of λ_0. Therefore λ_0 is a pole of multiplicity 2 of the resolvent $(A_B - \lambda I)^{-1}$. Consequently, the operator A_B is not similar to a normal one. □

We also need the following estimates.

Lemma 3.4. *Let $\lambda = \mu + i\varepsilon$, $(\varepsilon > 0)$. Let $y_\pm(f, \lambda)$ be of the form (2.18). Then the following inequalities*

$$\varepsilon \int_{-\infty}^{+\infty} \left\| \frac{y_\pm(f,\lambda)}{\sqrt{\lambda}} \cdot e^{-\sqrt{-\lambda}x} \chi_+(x) \right\|_{L^2}^2 d\mu \leq 2\pi \cdot C_1 \|f\|_{L^2}^2, \tag{3.14}$$

$$\varepsilon \int_{-\infty}^{+\infty} \left\| \frac{y_\pm(f,\lambda)}{\sqrt{\lambda}} \cdot e^{\sqrt{\lambda}x} \chi_-(x) \right\|_{L^2}^2 d\mu \leq 2\pi \cdot C_2 \|f\|_{L^2}^2 \tag{3.15}$$

are valid for all $f \in L^2(\mathbb{R})$ with constants C_1, C_2 independent of ε and f.

Proof. Let us prove the inequality (3.14) for $y_-(f,\lambda)$.

Put $f_-(\cdot) := f(\cdot)\chi_-(\cdot)$, $f \in L^2(\mathbb{R})$. Denote by $F(z)$ the Fourier transform of f_-,

$$F(z) := \frac{1}{\sqrt{2\pi}} \int_{-\infty}^{+\infty} f_-(t) e^{-izt} dt, \qquad z \in \mathbb{C}_+. \tag{3.16}$$

Note that $F(\cdot) \in H^2(\mathbb{C}_+)$ and $\|F\|_{H^2} = \|f_-\|_{L^2} \le \|f\|_{L^2}$. Further, we obtain

$$\begin{aligned}
&\int_{-\infty}^{+\infty} \left\| \frac{y_-(f,\lambda)}{\sqrt{\lambda}} \cdot e^{-\sqrt{-\lambda}x} \chi_+(x) \right\|_{L^2}^2 d\mu \\
&\quad = 2\pi \int_{-\infty}^{+\infty} \frac{1}{|\lambda|} |F(i\sqrt{\lambda})|^2 \|e^{-\sqrt{-\lambda}x}\|_{L^2}^2 d\mu \\
&\quad = 2\pi \int_{-\infty}^{+\infty} \frac{1}{2|\sqrt{\lambda}\,\mathrm{Im}\,\sqrt{\lambda}|} \left|F(i\sqrt{\lambda})\right|^2 \frac{d\mu}{2\sqrt{|\lambda|}} \\
&\quad \le 2\pi \int_{i\varepsilon-\infty}^{i\varepsilon+\infty} \frac{1}{2|\,\mathrm{Im}\,\sqrt{\lambda}\,\mathrm{Re}\,\sqrt{\lambda}|} \left|F(i\sqrt{\lambda})\right|^2 |d\sqrt{\lambda}| \\
&\quad = 2\pi \int_{i\varepsilon-\infty}^{i\varepsilon+\infty} \frac{1}{\mathrm{Im}\,\lambda} \left|F(i\sqrt{\lambda})\right|^2 |d\sqrt{\lambda}| \\
&\quad = \frac{2\pi}{\varepsilon} \int_{i\varepsilon-\infty}^{i\varepsilon+\infty} \left|F(i\sqrt{\lambda})\right|^2 |d\sqrt{\lambda}|.
\end{aligned} \tag{3.17}$$

Hence we find the estimate

$$\varepsilon \int_{-\infty}^{+\infty} \left\| \frac{y_-(f,\lambda)}{\sqrt{\lambda}} \cdot e^{-\sqrt{-\lambda}x} \chi_+(x) \right\|_{L^2}^2 d\mu \le 2\pi \int_{i\varepsilon-\infty}^{i\varepsilon+\infty} \left|F(i\sqrt{\lambda})\right|^2 |d\sqrt{\lambda}|. \tag{3.18}$$

Finally, let us remark that $|d\sqrt{\lambda}|$ is the Carleson measure for all $\varepsilon > 0$ (see [17]). It is easy to see that the Carleson norms of these measures are uniformly bounded. Then, by the Carleson embedding theorem (see [17]), there exists $C_1 > 0$ such that for all $\varepsilon > 0$ the inequality

$$\int_{i\varepsilon-\infty}^{i\varepsilon+\infty} \left|F(i\sqrt{\lambda})\right|^2 |d\sqrt{\lambda}| \le C_1 \|F\|_{H^2}^2 = C_1 \|f_-\|_{L^2} \le C_1 \|f\|_{L^2} \tag{3.19}$$

holds. Combining (3.18) and (3.19), one gets (3.14).

Other inequalities can be obtained analogously. □

4. Other boundary conditions

4.1. The case $\Delta_2 \neq 0$

The following theorem is an obvious corollary of Theorem 2.1.

Theorem 4.1.

(i) *The triplet* $\Pi_2 = \{\mathbb{C}^2, \Gamma_0, \Gamma_1\}$, *where* $\Gamma_j : W_2^2(\mathbb{R} \setminus \{0\}) \to \mathbb{C}^2, j \in \{0,1\}$,

$$\Gamma_0 f = \begin{pmatrix} -f'(+0) \\ f(-0) \end{pmatrix}, \qquad \Gamma_1 f = \begin{pmatrix} f(+0) \\ f'(-0) \end{pmatrix}, \tag{4.1}$$

is a boundary triplet for A^*.

(ii) *The corresponding Weyl function* $M(\cdot)$ *is*

$$M(\lambda) = M_2(\lambda) := \begin{pmatrix} 1/\sqrt{-\lambda} & 0 \\ 0 & \sqrt{\lambda} \end{pmatrix}, \qquad \lambda \in \mathbb{C} \setminus \mathbb{R}. \tag{4.2}$$

(iii) *The corresponding* γ*-field* $\gamma(\lambda) : \mathbb{C}^2 \to \mathfrak{N}_\lambda$ *is*

$$\gamma(\lambda) \begin{pmatrix} c_+ \\ c_- \end{pmatrix} := c_+ \frac{1}{\sqrt{-\lambda}} \cdot e^{-\sqrt{-\lambda}x} \chi_+(x) + c_- \cdot e^{\sqrt{\lambda}x} \chi_-(x), \qquad c_\pm \in \mathbb{C}. \tag{4.3}$$

If $\Delta_2 \neq 0$ then the boundary conditions (2.8) take the form

$$\begin{cases} f(+0) = -b_{11} f'(+0) + b_{12} f(-0) \\ f'(-0) = -b_{21} f'(+0) + b_{22} f(-0) \end{cases}. \tag{4.4}$$

Further, by Definition 1.5, we get

$$A_{(a_{ij})} = A_B = A^* \upharpoonright \ker(\Gamma_1 - B\Gamma_0), \qquad B = \begin{pmatrix} b_{11} & b_{12} \\ b_{21} & b_{22} \end{pmatrix} \in \mathbb{C}^{2\times 2}. \tag{4.5}$$

As before, denote by $\varphi_B(\cdot)$ the function as in (2.14) with $M(\cdot)$ as in (4.2).

Combining Proposition 1.2 and Theorem 1.2, one obtains

Lemma 4.1. *If* A_B *is the operator* (4.5) *and* $|b_{12}| + |b_{21}| \neq 0$ *then:*

(i) $\sigma_c(A_B) = \mathbb{R}$, $\quad \sigma_r(A_B) = \emptyset$;

(ii) $\sigma_p(A_B) = \{\lambda \in \mathbb{C} \setminus \mathbb{R} : \varphi_B(\lambda) = 0\} = \{\lambda \in \mathbb{C}_+ : \varphi_B(\lambda) = 0\} \cup \{\lambda \in \mathbb{C}_- : \varphi_{B^*}(\overline{\lambda}) = 0\}$.

(iii) *The Krein formula has the form*

$$\begin{aligned} &\left((A_B - \lambda)^{-1} f - (A_0 - \lambda)^{-1} f\right)(x) \\ &= \frac{e^{-\sqrt{-\lambda}x} \cdot \chi_+(x)}{\varphi_B(\lambda) \cdot \sqrt{-\lambda}} \left(\frac{b_{22} - \sqrt{\lambda}}{\sqrt{-\lambda}} \cdot y_+(f,\lambda) - b_{12} \cdot y_-(f,\lambda) \right) \\ &\quad + \frac{e^{\sqrt{\lambda}x} \cdot \chi_-(x)}{\varphi_B(\lambda)} \left(-\frac{b_{21}}{\sqrt{-\lambda}} \cdot y_+(f,\lambda) + \left(b_{11} - \frac{1}{\sqrt{-\lambda}} \right) \cdot y_-(f,\lambda) \right), \\ &\hspace{20em} f \in L^2(\mathbb{R}), \end{aligned} \tag{4.6}$$

where $A_0 = A^* \upharpoonright \ker \Gamma_0$, $\lambda \in \rho(A_B) \cap \rho(A_0)$, *and* $y_\pm(f,\lambda)$ *are defined by* (2.18).

4.2. The case $\Delta_3 \neq 0$

Let $\Delta_3 \neq 0$. In this case we write the boundary conditions (2.8) in the following form

$$\begin{cases} f'(+0) = b_{11}f(+0) + b_{12}f(-0) \\ f'(-0) = b_{21}f(+0) + b_{22}f(-0). \end{cases} \tag{4.7}$$

Let $B = \begin{pmatrix} b_{11} & b_{12} \\ b_{21} & b_{22} \end{pmatrix}$ and

$$A_B := A^*|\operatorname{dom}(A_B), \quad \operatorname{dom}(A_B) = \{f \in W_2^2(\mathbb{R} \setminus \{0\}) : f \text{ satisfies } (4.7)\}. \tag{4.8}$$

Theorem 4.2.

(i) *The triplet* $\Pi_3 = \{\mathbb{C}^2, \Gamma_0, \Gamma_1\}$, *where* $\Gamma_j : W_2^2(\mathbb{R} \setminus \{0\}) \to \mathbb{C}^2, j \in \{0,1\}$,

$$\Gamma_0 f = \begin{pmatrix} f(+0) \\ f(-0) \end{pmatrix}, \qquad \Gamma_1 f = \begin{pmatrix} f'(+0) \\ f'(-0) \end{pmatrix}, \tag{4.9}$$

is a boundary triplet for A^*.

(ii) *The corresponding Weyl function* $M(\cdot)$ *is*

$$M(\lambda) = M_3(\lambda) := \begin{pmatrix} -\sqrt{-\lambda} & 0 \\ 0 & \sqrt{\lambda} \end{pmatrix}, \qquad \lambda \in \mathbb{C} \setminus \mathbb{R}. \tag{4.10}$$

(iii) *The corresponding* γ*-field* $\gamma(\lambda) : \mathbb{C}^2 \to \mathfrak{N}_\lambda$ *is*

$$\gamma(\lambda) \begin{pmatrix} c_+ \\ c_- \end{pmatrix} := c_+ \cdot e^{-\sqrt{-\lambda}x} \chi_+(x) + c_- \cdot e^{\sqrt{\lambda}x} \chi_-(x), \qquad c_\pm \in \mathbb{C}. \tag{4.11}$$

Therefore, A_B of the form (4.8) is an almost solvable extension of A and $A_B = A^*|\ker(\Gamma_1 - B\Gamma_0)$, where Γ_i, $i \in \{0,1\}$, are defined by (4.9).

Lemma 4.2. *Let the function* $\varphi_B(\cdot)$ *be of the form* (2.14) *with* $M(\cdot)$ *defined by* (4.10)*; let the operator* A_B *is given by* (4.8) *and* $|b_{12}| + |b_{21}| \neq 0$. *Then:*

(i) $\sigma_c(A_B) = \mathbb{R}, \quad \sigma_r(A_B) = \emptyset$;

(ii) $\sigma_p(A_B) = \{\lambda \in \mathbb{C} \setminus \mathbb{R} : \varphi_B(\lambda) = 0\} = \{\lambda \in \mathbb{C}_+ : \varphi_B(\lambda) = 0\} \cup \{\lambda \in \mathbb{C}_- : \varphi_{B^*}(\overline{\lambda}) = 0\}$.

(iii) *The Krein formula has the form*

$$\begin{aligned} &\left((A_B - \lambda)^{-1} f - (A_0 - \lambda)^{-1} f\right)(x) \\ &= \frac{e^{-\sqrt{-\lambda}x} \cdot \chi_+(x)}{\varphi_B(\lambda)} \left((b_{22} - \sqrt{\lambda}) \cdot y_+(f,\lambda) - b_{12} \cdot y_-(f,\lambda)\right) \\ &\quad + \frac{e^{\sqrt{\lambda}x} \cdot \chi_-(x)}{\varphi_B(\lambda)} \left(-b_{21} \cdot y_+(f,\lambda) + (b_{11} + \sqrt{-\lambda}) \cdot y_-(f,\lambda)\right), \\ &\qquad\qquad f \in L^2(\mathbb{R}), \end{aligned} \tag{4.12}$$

where $A_0 = A^*|\ker\Gamma_0$, $\lambda \in \rho(A_B) \cap \rho(A_0)$, *and* $y_\pm(f,\lambda)$ *are given by* (2.18).

4.3. The case $\Delta_4 \neq 0$

If $\Delta_4 \neq 0$, then boundary conditions (2.8) take the form

$$\begin{cases} -f(+0) = b_{11} f'(+0) + b_{12} f'(-0) \\ -f(-0) = b_{21} f'(+0) + b_{22} f'(-0). \end{cases} \tag{4.13}$$

For $B = \begin{pmatrix} b_{11} & b_{12} \\ b_{21} & b_{22} \end{pmatrix}$ we set

$$A_B := A^* \upharpoonright \operatorname{dom}(A_B),$$
$$\operatorname{dom}(A_B) = \{ f \in W_2^2(\mathbb{R} \setminus \{0\}) : f \text{ satisfies (4.13)} \}. \tag{4.14}$$

Theorem 4.3.

(i) *The triplet* $\Pi_4 = \{\mathbb{C}^2, \Gamma_0, \Gamma_1\}$, *where* $\Gamma_j : W_2^2(\mathbb{R} \setminus \{0\}) \to \mathbb{C}^2, j \in \{0,1\}$,

$$\Gamma_0 f = \begin{pmatrix} f'(+0) \\ f'(-0) \end{pmatrix}, \qquad \Gamma_1 f = \begin{pmatrix} -f(+0) \\ -f(-0) \end{pmatrix}, \tag{4.15}$$

is a boundary triplet for A^*.

(ii) *The corresponding Weyl function* $M(\cdot)$ *is*

$$M(\lambda) = M_4(\lambda) := \begin{pmatrix} 1/\sqrt{-\lambda} & 0 \\ 0 & -1/\sqrt{\lambda} \end{pmatrix}, \qquad \lambda \in \mathbb{C} \setminus \mathbb{R}. \tag{4.16}$$

(iii) *The corresponding* γ*-field* $\gamma(\lambda) : \mathbb{C}^2 \to \mathfrak{N}_\lambda$ *is*

$$\gamma(\lambda) \begin{pmatrix} c_+ \\ c_- \end{pmatrix} := -\frac{c_+}{\sqrt{-\lambda}} \cdot e^{-\sqrt{-\lambda}x} \chi_+(x) + \frac{c_-}{\sqrt{\lambda}} \cdot e^{\sqrt{\lambda}x} \chi_-(x), \qquad c_\pm \in \mathbb{C}. \tag{4.17}$$

Therefore, A_B as in (4.14) is an almost solvable extension of A and $A_B = A^* \upharpoonright \ker(\Gamma_1 - B\Gamma_0)$, where Γ_j, $j \in \{0,1\}$ are given by (4.15).

Lemma 4.3. *Suppose the function* $\varphi_B(\cdot)$ *is given by* (2.14) *with* $M(\cdot)$ *as in* (4.16). *If the operator* A_B *is given by* (4.14) *and* $|b_{12}| + |b_{21}| \neq 0$ *then:*

(i) $\sigma_c(A_B) = \mathbb{R}, \quad \sigma_r(A_B) = \emptyset$;

(ii) $\sigma_p(A_B) = \{\lambda \in \mathbb{C} \setminus \mathbb{R} : \varphi_B(\lambda) = 0\} = \{\lambda \in \mathbb{C}_+ : \varphi_B(\lambda) = 0\} \cup \{\lambda \in \mathbb{C}_- : \varphi_{B^*}(\overline{\lambda}) = 0\}$.

(iii) *The Krein formula has the form*

$$\begin{aligned} &\left((A_B - \lambda)^{-1} f - (A_0 - \lambda)^{-1} f\right)(x) \\ &= \frac{e^{-\sqrt{-\lambda}x} \cdot \chi_+(x)}{\varphi_B(\lambda) \cdot \sqrt{-\lambda}} \left((b_{22} + 1/\sqrt{\lambda}) \cdot \frac{y_+(f,\lambda)}{\sqrt{-\lambda}} + b_{12} \cdot \frac{y_-(f,\lambda)}{\sqrt{\lambda}} \right) \\ &\quad + \frac{e^{\sqrt{\lambda}x} \cdot \chi_-(x)}{\varphi_B(\lambda) \cdot \sqrt{\lambda}} \left(b_{21} \cdot \frac{y_+(f,\lambda)}{\sqrt{-\lambda}} + (b_{11} - 1/\sqrt{-\lambda}) \cdot \frac{y_-(f,\lambda)}{\sqrt{\lambda}} \right), \\ &\qquad f \in L^2(\mathbb{R}), \end{aligned} \tag{4.18}$$

where $A_0 = A^* \upharpoonright \ker \Gamma_0$, $\lambda \in \rho(A_B) \cap \rho(A_0)$, *and* $y_\pm(f,\lambda)$ *are given by* (2.18).

4.4. The similarity criterion

Arguing as above, we see that for the cases considered in Subsections 4.1–4.3 analogues of Theorems 3.1 and 3.2 hold.

Theorem 4.4. *Theorems* 3.1 *and* 3.2 *are valid for the extensions* A_B *with boundary conditions* (4.4), (4.7) *or* (4.13) *if the function* $M(\cdot)$ *is replaced by* (4.2), (4.10) *or* (4.16), *respectively.*

5. On similarity of $(\operatorname{sgn} x)\left(-\frac{d^2}{dx^2}+c\delta\right)$ and $(\operatorname{sgn} x)\left(-\frac{d^2}{dx^2}+c\delta'\right)$ to normal operators

5.1. Let us illustrate the previous results by several examples. We start with the operator $\tilde{A} = -(\operatorname{sgn} x)\frac{d^2}{dx^2}$, see (0.3). It is obvious that $\tilde{A} = A_B$, where A_B is an almost solvable extension of the form (2.13) with $B = \begin{pmatrix} 0 & 1 \\ -1 & 0 \end{pmatrix}$. It follows from (3.4) that in this case

$$\varphi_B^+(\lambda) = \varphi_{B^*}^+(\lambda) \equiv 1 - i, \qquad \lambda \in \overline{\mathbb{C}_+}.$$

Hence, by Theorem 3.2, one obtains the result of [7].

Theorem 5.1 ([7]). *The operator* $\tilde{A} = -(\operatorname{sgn} x)\frac{d^2}{dx^2}$ *is similar to a self-adjoint operator.*

5.2. Let δ be the Dirac delta. Let us introduce the following differential expression

$$-\frac{d^2}{dx^2} + c\delta, \qquad c \in \mathbb{C} \setminus \{0\}. \tag{5.1}$$

Here δ formally represents a contact interaction at zero if $c \in \mathbb{R}$.

In $L^2(\mathbb{R})$, expression (5.1) generates the differential operator $L_{c\delta}$ defined by

$$\operatorname{dom}(L_{c\delta}) = \{f \in W_2^2(\mathbb{R} \setminus \{0\}) : f(+0) = f(-0), f'(+0) - f'(-0) = cf(-0)\},$$

$$L_{c\delta} := -\frac{d^2}{dx^2} \tag{5.2}$$

(see for example [2, 35]). We put $A_{c\delta} := JL_{c\delta}$, where $(Jf)(x) = (\operatorname{sgn} x)f(x)$, i.e., the operator $A_{c\delta}$ is defined in $L^2(\mathbb{R})$ by the differential expression

$$(\operatorname{sgn} x)\left(-\frac{d^2}{dx^2} + c\delta\right), \qquad c \in \mathbb{C} \setminus \{0\}. \tag{5.3}$$

It is clear that the operator $A_{c\delta}$ is an extension of the operator A of the form (0.1). Moreover, $A_{c\delta} = A_B$ where A_B is given by (2.13) and

$$B = B_c := \begin{pmatrix} c & 1 \\ -1 & 0 \end{pmatrix}. \tag{5.4}$$

Theorem 5.2. *Let $c \neq 0$.*

(i) *The operator $A_{c\delta}$ is similar to a normal one if and only if* $\operatorname{Re} c \neq -|\operatorname{Im} c|$.

(ii) *$A_{c\delta}$ is similar to a self-adjoint operator if and only if* $\operatorname{Re} c > -|\operatorname{Im} c|$.

Proof. Let B be given by (5.4). By (3.4), one gets

$$\varphi_B^+(\lambda) = 1 - i + \frac{c}{\sqrt{\lambda}}, \qquad \varphi_{B^*}^+(\lambda) = 1 - i + \frac{\overline{c}}{\sqrt{\lambda}}, \qquad \lambda \in \overline{\mathbb{C}_+} \setminus \{0\}. \tag{5.5}$$

Note that $\varphi_B^+(\cdot)$ or $\varphi_{B^*}^+(\cdot)$ have a real zero iff $\operatorname{Re}(c) = -|\operatorname{Im}(c)|$. Furthermore, $\varphi_B^+(\cdot)$ and $\varphi_{B^*}^+(\cdot)$ do not vanish in $\overline{\mathbb{C}_+}$ iff $\operatorname{Re}(c) > -|\operatorname{Im} c|$. Hence the statements of Theorem 5.2 obviously follow from Theorems 3.1 and 3.2. □

5.3. Let us consider the extension $A_{\widetilde{B}}$ of the form (2.13) with

$$\widetilde{B} = \widetilde{B}_c = \begin{pmatrix} 0 & 1 \\ -1 & c \end{pmatrix}, \quad c \neq 0.$$

This is a so-called "operator with δ'-interaction" (see [2]). The formal differential expression corresponding to $A_{\widetilde{B}}$ is

$$(\operatorname{sgn} x)\left(-\frac{d^2}{dx^2} + c\delta'\right), \qquad c \in \mathbb{C} \setminus \{0\}. \tag{5.6}$$

Therefore we will denote the operator $A_{\widetilde{B}}$ by $A_{c\delta'}$.

Theorem 5.3. *Let $c \neq 0$.*

(i) *The operator $A_{c\delta'}$ is similar to a normal one if and only if* $\operatorname{Re} c \neq -|\operatorname{Im} c|$.

(ii) *$A_{c\delta'}$ is similar to a self-adjoint operator if and only if* $\operatorname{Re} c > -|\operatorname{Im} c|$.

Proof. So, by (3.4), we have

$$\varphi_{c\delta'}^+(\lambda) = 1 - i - ic\sqrt{\lambda}, \qquad \varphi_{c\delta'}^+(\lambda) = 1 - i - i\overline{c}\sqrt{\lambda}, \qquad \lambda \in \overline{\mathbb{C}_+}. \tag{5.7}$$

These functions have a real zero iff $\operatorname{Re} c = -|\operatorname{Im} c|$, have no zeros in $\mathbb{C}_+$ iff $\operatorname{Re} c > -|\operatorname{Im} c|$. Hence the statements of Theorem 5.3 follow from Theorems 3.1 and 3.2. □

Remark 5.1. *Let $\widehat{L}$ be a self-adjoint extension of the symmetric operator*

$$(Lf)(x) := -f''(x), \quad \operatorname{dom}(L) = \{f \in W_2^2(\mathbb{R}) : f(0) = f'(0) = 0\}. \tag{5.8}$$

Assume that the boundary conditions at 0 associated with the extension $\widehat{L}$ are nonseparate, i.e., the operator $\widehat{L}$ does not admit the following decomposition $\widehat{L}_+ \oplus \widehat{L}_-$, where $\widehat{L}_\pm := \widehat{L} \upharpoonright (\operatorname{dom}(\widehat{L}) \cap L^2(\mathbb{R}_\pm))$. Then $\widehat{L}$ can be considered as an operator with a singular interaction (see [27] for the details). Using the arguments of this section, one can describe the main spectral properties of the corresponding J-self-adjoint operator $\widehat{A} := J\widehat{L}$.

Acknowledgment

The authors are deeply grateful to professor M.M. Malamud for statement of the problem and his constant attention to this work. The authors thanks a referee for a careful reading of the manuscript and a number of remarks, which helped to improve it. Also, it is a pleasure for the first author to acknowledge Technische Universität Berlin for support of his participation in 4th Workshops "Operator Theory in Krein Spaces".

References

[1] N.I. Akhiezer, I.M. Glazman, *Theory of linear operators in Hilbert space.* Dover Publications, Inc., New York, 1993.

[2] S. Albeverio, F. Gesztesy, R. Hoegh–Krohn, H. Holden, *Solvable models in quantum mechanics.* Springer, New York, 1988.

[3] R. Beals, *An abstract treatment of some forward-backward problems of transport and scattering.* J. Funct. Anal. **34** (1979), 1–20.

[4] R. Beals, *Indefinite Sturm-Liouville problems and half-range completeness.* J. Differential Equation **56** (1985), 391–407.

[5] J.A. Casteren, *Operators similar to unitary or selfadjoint ones.* Pacific J. Math. **104** (1983), 241–255.

[6] B. Ćurgus, B. Najman, *A Krein space approach to elliptic eigenvalue problems with indefinite weights.* Differential and Integral Equations **7** (1994), 1241–1252.

[7] B. Ćurgus, B. Najman, *The operator* $-(\operatorname{sgn} x)\frac{d^2}{dx^2}$ *is similar to selfadjoint operator in* $L^2(\mathbb{R})$. Proc. Amer. Math. Soc. **123** (1995), 1125–1128.

[8] B. Ćurgus, B. Najman, *Positive differential operator in Krein space* $L^2(\mathbb{R}^n)$. Oper. Theory Adv. Appl., Birkhäuser, Basel, **106** (1998), 113–130.

[9] B. Ćurgus, H. Langer, *A Krein space approach to symmetric ordinary differential operators with an indefinite weight function.* J. Differential Equations **79** (1989), 31–61.

[10] V.A. Derkach, M.M. Malamud, *On the Weyl function and Hermitian operators with gaps.* Dokl. AN SSSR. **293** (1987), 1041–1046 [in Russian].

[11] V.A. Derkach, M.M. Malamud, *Generalized resolvents and the boundary value problems for Hermitian operators with gaps.* J. Funct. Anal. **95** (1991), 1–95.

[12] V.A. Derkach, M.M. Malamud, *Characteristic functions of almost solvable extensions of Hermitian operators.* Ukrainian Math. J. **44** (1992), 435–459.

[13] M.M. Faddeev, R.G. Shterenberg, *On similarity of some singular operators to selfadjoint ones.* Zapiski Nauchnykh Seminarov POMI **270** (2000), 336–349 [in Russian]; *English translation*: J. Math. Sciences **115** (2003), no. 2, 2279–2286.

[14] M.M. Faddeev, R.G. Shterenberg, *On similarity of differential operators to a selfadjoint one.* Math. Notes **72** (2002), 292–303.

[15] A. Fleige, *A spectral theory of indefinite Krein-Feller differential operators.* Mathematical Research, **98**, Berlin: Akademie Verlag, 1996.

[16] A. Fleige, B. Najman, *Nonsingularity of critical points of some differential and difference operators.* Oper. Theory Adv. Appl., Birkhäuser, Basel, **106** (1998), 147–155.

[17] J.B. Garnett, *Bounded analytic functions.* Academic Press, 1981.

[18] V.I. Gorbachuk, M.L. Gorbachuk, *Boundary value problems for operator differential equations.* Mathematics and Its Applications, Soviet Series **48**, Dordrecht eds., Kluwer Academic Publishers, 1991.

[19] H.G. Kaper, M.K. Kwong, C.G. Lekkerkerker, A. Zettl, *Full- and partial-range eigenfunction expansions for Sturm-Liouville problems with indefinite weights.* Proc. Roy. Soc. Edinburgh **A. 98** (1984), 69–88.

[20] V.V. Kapustin, *Non-self-adjoint extensions of symmetric operators.* Zapiski Nauchnykh Seminarov POMI **282** (2001), 92–105 [in Russian]; *English translation*: J. Math. Sciences **120** (2004), no. 5, 1696–1703.

[21] I.M. Karabash, *The operator* $-(\operatorname{sgn} x)\frac{d^2}{dx^2}$ *is similar to a self–adjoint operator in* $L^2(\mathbb{R})$. Spectral and Evolution Problems, **8**, Proc. of the VIII Crimean Autumn Math. School–Symposium, Simferopol, 1998, 23–26.

[22] I.M. Karabash, *J-self-adjoint ordinary differential operators similar to self-adjoint operators.* Methods Funct. Anal. Topology **6**, no. 2, (2000), 22–49.

[23] I.M. Karabash, A.S. Kostenko, *On similarity of the operators type* $(\operatorname{sgn} x)(-d^2/dx^2 + c\delta)$ *to a normal or to a self-adjoint one.* Math. Notes **74** (2003), 134–139.

[24] I.M. Karabash, M.M. Malamud, *The similarity of a J-self-adjoint Sturm-Liouville operator with finite-gap potential to a self-adjoint operator.* Doklady Mathematics **69**, no. 2, (2004), 195–199 [in Russian].

[25] T. Kato, *Perturbation theory for linear operators.* Springer-Verlag, Berlin-Heidelberg-New York, 1966.

[26] M.G. Krein, H. Langer, *On defect subspaces and generalized resolvents of Hermitian operator acting in Pontrjagyn space.* Funktsional. Anal. i Prilozhen. **5**, no. 3, (1971), 54-69 [in Russian]; *English translation:* Functional Anal. Appl. **5** (1971).

[27] P. Kurasov, *Distribution theory for discontinuous test functions and differential operators with generalized coefficients.* J. Math. Anal. Appl. **201** (1996), 297–323.

[28] H. Langer, *Spectral functions of definitizable operators in Krein space.* Lecture Notes in Mathematics **948**, 1982, 1–46.

[29] M.M. Malamud, *A criterion for similarity of a closed operator to a self-adjoint one.* Ukrainian Math. J. **37** (1985) 49–56.

[30] M.M. Malamud, *Similarity of a triangular operator to a diagonal operator.* Zapiski Nauchnykh Seminarov POMI **270** (2000), 201–241 [in Russian]; *English translation:* J. Math. Sciences **115** (2000), 2199–2222.

[31] S.N. Naboko, *Conditions for similarity to unitary and selfadjoint operators.* Funktsional. Anal. i Prilozhen. **18**, no. 1, (1984), 16–27 [in Russian].

[32] S.G. Pyatkov, *Properties of eigenfunctions of certain spectral problem with some applications.* Nekotorie prilozhenia funktsionalnogo analiza k zadacham matematicheskoi fiziki, Novosibirsk: IM SO AN SSSR, 1986, 65–84 [in Russian].

[33] S.G. Pyatkov, *Some properties of eigenfunctions of linear pencils.* Sibirskii Mat. Zh. **30**, no. 4 (1989), 111–124 [in Russian]; *English translation:* Siberian Math. J. **30** (1989), 587–597.

[34] S.G. Pyatkov, *Indefinite elliptic spectral problems.* Sibirskii Mat. Zh. **39**, no. 2 (1998), 409–426 [in Russian]; *English translation:* Siberian Math. J. **30** (1989), 358–372.

[35] M. Reed, B. Simon, *Methods of modern mathematical physics. II.* Academic Press, New York-San Francisco-London, 1975.

Ilia Karabash and Aleksey Kostenko
Department of Mathematics
Donetsk National University
Universitetskaja 24
83055 Donetsk, Ukraine
e-mail: karabashi@yahoo.com; karabashi@mail.ru
e-mail: duzer@skif.net; duzer80@mail.ru

From (1.2), one easily obtains the matrix identity

$$M^{[*]} = H^{-1}M^*H. \tag{1.3}$$

A matrix $M \in \mathbb{C}^{n\times n}$ is called *H-selfadjoint*, *H-skewadjoint*, or *H-unitary*, respectively, if $M^{[*]} = M$, $M^{[*]} = -M$, or $M^{[*]} = M^{-1}$, respectively. Using (1.3), we obtain the matrix identities

$$M^*H = HM, \quad M^*H + HM = 0, \quad \text{or } M^*HM = H \tag{1.4}$$

for H-selfadjoint, H-skewadjoint, or H-unitary matrices, respectively. These three types of matrices have been widely discussed in the literature, both in terms of theory and numerical analysis. Extensive lists of references can be found in [1, 9, 13, 21].

H-selfadjoint, H-skewadjoint, and H-unitary matrices are special cases of H-normal matrices. A matrix $M \in \mathbb{C}^{n\times n}$ is called H-normal if M commutes with its H-adjoint, i.e., if $MM^{[*]} = M^{[*]}M$, or, in other words, if and only if

$$MH^{-1}M^*H = H^{-1}M^*HM, \quad \text{or,} \quad HMH^{-1}M^*H = M^*HM. \tag{1.5}$$

In recent years, there has been great interest in H-normal matrices, see [8, 9, 10, 17, 18] and the references therein.

Spaces with degenerate inner products, that is, H is singular, are less familiar, although this case does appear in applications [14]. Some works here, primarily concerning infinite-dimensional degenerate Pontryagin spaces, include [23], [11] (and references there), [2], [12], and parts of the book [3]. The main problem in the context of degenerate inner products is that there is no straightforward definition of an H-adjoint. Indeed, if H is singular, then for a matrix $M \in \mathbb{C}^{n\times n}$ an H-adjoint, i.e., a matrix $N \in \mathbb{C}^{n\times n}$ satisfying $[x, My] = [Nx, y]$ for all $x, y \in \mathbb{C}^n$ need not exist, and if it exists, it need not be unique.

Example 1.1. Let $a, b \in \mathbb{C}$ and

$$H = \begin{pmatrix} 1 & 0 \\ 0 & 0 \end{pmatrix}, \quad M_1 = \begin{pmatrix} 1 & 0 \\ 0 & 1 \end{pmatrix}, \quad M_2 = \begin{pmatrix} 1 & 1 \\ 0 & 1 \end{pmatrix}, \quad N(a,b) = \begin{pmatrix} 1 & 0 \\ a & b \end{pmatrix}.$$

Then for all possible choices $a, b \in \mathbb{C}$, $N(a,b)$ satisfies $[x, M_1y] = [N(a,b)x, y]$ for all $x, y \in \mathbb{C}^n$. On the other hand, there is no matrix $N \in \mathbb{C}^{2\times 2}$ such that $[x, M_2y] = [Nx, y]$ for all $x, y \in \mathbb{C}^n$.

Despite the lack of the notion of an H-adjoint, one can define H-selfadjoint, H-skewadjoint and H-unitary for the case of singular H by the matrix identities (1.4) and this definition has been used in many sources, see, e.g., [20] and the references therein. The corresponding matrix identities (1.5) for H-normal matrices, however, require an inverse of H. A standard approach to circumvent this difficulty is the use of some generalized inverse, in particular the Moore-Penrose generalized inverse $H^\dagger$, instead. However, it seems that the use of the Moore-Penrose inverse leads to some inconsistencies in the theory of degenerate inner products. We illustrate this by an example.

Example 1.2. Let

$$H = \begin{pmatrix} 1 & 0 \\ 0 & 0 \end{pmatrix}, \quad A = \begin{pmatrix} 1 & 0 \\ 1 & 1 \end{pmatrix}, \quad N = \begin{pmatrix} 0 & 1 \\ 1 & 0 \end{pmatrix}.$$

Then we have $H^\dagger = H$. Observe that $A^*H = HA$, but $A \neq H^\dagger A^* H$. So, should A be considered as an H-selfadjoint matrix or not? On the other hand, note that $NH^\dagger N^* H = H^\dagger N^* HN$, but $HNH^\dagger N^* H \neq N^* HN$. So, should N be considered as an H-normal matrix or not?

In [16] H-normal matrices have been defined as matrices M satisfying

$$HMH^\dagger M^* H = M^* HM \tag{1.6}$$

and this definition has later been taken up in [4, 17]. (This way minimizes the number of times the Moore-Penrose generalized inverse of H appears in the defining equation of H-normal matrices.) In this paper, we will call matrices M satisfying (1.6) *Moore-Penrose H-normal matrices* in order to highlight the occurrence of the Moore-Penrose generalized inverse in the definition.

Although it is possible to prove some interesting results for Moore-Penrose H-normal matrices (e.g., concerning existence of invariant maximal semidefinite subspaces), there is an unpleasant mismatch between H-selfadjoint, H-skewadjoint, and H-unitary matrices on the one hand and Moore-Penrose H-normal matrices on the other hand. It is easy to check (see also [20]) that the kernel of H is always an invariant subspace for H-selfadjoint, H-skewadjoint, and H-unitary matrices (defined as in (1.4)). However, it has been shown in [17, Example 6.1] that there exist Moore-Penrose H-normal matrices A such that $\ker H$ is not A-invariant.

It is exactly the fact that $\ker H$ need not be invariant which makes the investigation of Moore-Penrose H-normal matrices challenging. For example, let us consider the problem of existence of semidefinite invariant subspaces. Recall that a subspace $\mathcal{M} \subseteq \mathbb{C}^n$ is called H-nonnegative if $[x, x] \geq 0$ for every $x \in \mathcal{M}$, *H-positive* if $[x, x] > 0$ for every nonzero $x \in \mathcal{M}$, and *H-neutral* if $[x, x] = 0$ for every $x \in \mathcal{M}$. An H-nonnegative subspace is said to be *maximal H-nonnegative* if it is not properly contained in any larger H-nonnegative subspace. It is easy to see that an H-nonnegative subspace is maximal if and only if its dimension is equal to $\nu_+(H) + \nu_0(H)$, where $\nu_+(H)$ and $\nu_0(H)$ denote the number (counted with multiplicities) of positive and zero eigenvalues of H, respectively. It has been shown in [17, Theorem 6.6] that any Moore-Penrose H-normal matrix has a maximal H-nonnegative invariant subspace, but the problem which H-nonnegative, H-positive, or H-neutral invariant subspaces of Moore-Penrose H-normal matrices can be extended to a maximal H-nonnegative invariant subspaces has only been solved for the case of invertible H so far [18, 19].

It is the aim of this paper to propose a different definition of H-normal matrices in degenerate inner product spaces that is based on a generalization of the H-adjoint $A^{[*]_H}$ of a matrix A for singular H. This generalization is obtained by dropping the assumption that the H-adjoint of a matrix is a matrix itself. Instead, the H-adjoint $A^{[*]_H}$ is defined to be a *linear relation in* $\mathbb{C}^n$, i.e., a linear subspace of $\mathbb{C}^{2n}$. For basic facts on linear relations and further references see, e.g., [5, 6, 7, 22]. Throughout this paper we identify a matrix $A \in \mathbb{C}^{n\times n}$ with its graph

$$\left\{ \begin{pmatrix} x \\ Ax \end{pmatrix} : x \in \mathbb{C}^n \right\} \subseteq \mathbb{C}^{2n}$$

which is a linear relation, also denoted by A. If $H \in \mathbb{C}^{n\times n}$ is invertible, then, by (1.2), we obtain that $A^{[*]_H}$ coincides with the linear relation

$$A^{[*]_H} = \left\{ \begin{pmatrix} y \\ z \end{pmatrix} \in \mathbb{C}^{2n} : [y, Ax]_H = [z, x]_H \text{ for all } x \in \mathbb{C}^n \right\}.$$

This representation allows a direct generalization of the concept of adjoint to the case of degenerate inner products and even to the case of starting with a linear relation rather than a matrix A, see [22].

Definition 1.3. Let $H \in \mathbb{C}^{n\times n}$ be Hermitian and let A be a linear relation in $\mathbb{C}^n$. Then the linear relation

$$A^{[*]_H} = \left\{ \begin{pmatrix} y \\ z \end{pmatrix} \in \mathbb{C}^{2n} : [y, w]_H = [z, x]_H \text{ for all } \begin{pmatrix} x \\ w \end{pmatrix} \in A \right\}$$

is called the H-*adjoint* of A.

Again, if there is no risk of ambiguity, we will suppress the subscript H in the notation. In this setting, we obtain (see Proposition 2.6 below) that

$$A^{[*]} = H^{-1}A^*H,$$

where H^{-1} is the inverse of H in the sense of linear relations (see Section 2). Observe that this coincides with (1.3) if H is invertible. Hence, the H-adjoint in degenerate inner product spaces is a natural generalization of the H-adjoint in nondegenerate inner product spaces.

The remainder of the paper is organized as follows. In Section 2, we discuss basic properties of the H-adjoint. In Section 3 we use the H-adjoint to define H-symmetric and H-isometric linear relations and we show that in the case of matrices these definitions coincide with the definitions of H-selfadjoint and H-unitary matrices via the identities (1.4). In Section 4, we present a new definition for H-normal matrices. We show that the set of H-normal matrices is a proper subset of the set of Moore-Penrose H-normal matrices and that H-normal matrices share the property with H-selfadjoint and H-unitary matrices that the kernel of H is always an invariant subspace. The latter fact allows us to obtain sufficient conditions for an H-positive invariant subspace of an H-normal matrix to be contained in a maximal H-nonnegative invariant subspace. This generalizes a result obtained in [18].

2. The adjoint in degenerate inner product spaces

We study linear relations in $\mathbb{C}^n$, i.e., linear subspaces of $\mathbb{C}^{2n}$. For the definitions of linear operations with relations and the inverse of relations we refer to [6]. We only mention the following. For linear relations $A, B \subseteq \mathbb{C}^{2n}$ we define

$$\operatorname{dom} A = \left\{ x \,:\, \begin{pmatrix} x \\ y \end{pmatrix} \in A \right\}, \qquad \text{the domain of } A,$$

$$\operatorname{ran} A = \left\{ y \,:\, \begin{pmatrix} x \\ y \end{pmatrix} \in A \right\}, \qquad \text{the range of } A,$$

$$\operatorname{mul} A = \left\{ y \,:\, \begin{pmatrix} 0 \\ y \end{pmatrix} \in A \right\}, \qquad \text{the multivalued part of } A,$$

$$A^{-1} = \left\{ \begin{pmatrix} y \\ x \end{pmatrix} \,:\, \begin{pmatrix} x \\ y \end{pmatrix} \in A \right\}, \qquad \text{the inverse of } A$$

and the product of A and B,

$$AB = \left\{ \begin{pmatrix} x \\ z \end{pmatrix} \,:\, \text{there exists a } y \in \mathbb{C}^n \text{ with } \begin{pmatrix} y \\ z \end{pmatrix} \in A, \begin{pmatrix} x \\ y \end{pmatrix} \in B \right\}.$$

In all cases, x, y, z are understood to be from $\mathbb{C}^n$. For a subset $M \subseteq \mathbb{C}^n$ we define

$$M^{[\perp]} = \{ x \,:\, [x, y] = 0 \text{ for all } y \in M \}.$$

The following lemma is needed for the proof of Proposition 2.2 below. It is contained in [15, proof of Lemma 2.2], but for the sake of completeness we give a separate proof.

Lemma 2.1. *Let $M \subseteq \mathbb{C}^n$ be a subspace. Then $(M^{[\perp]})^{[\perp]} = M + \ker H$.*

Proof. Obviously, we have $M + \ker H \subseteq (M^{[\perp]})^{[\perp]}$. If $u \notin M + \ker H$ then, as the quotient space $(\mathbb{C}^n / \ker H, [\cdot, \cdot]^\sim)$ with $[x + \ker H, y + \ker H]^\sim := [x, y]$, $x, y \in \mathbb{C}^n$, is nondegenerate, there exists an $v \in \mathbb{C}^n$ such that $[v, M] = [v + \ker H, M + \ker H]^\sim = \{0\}$ and $[v, u] = [v + \ker H, u + \ker H]^\sim \neq 0$. Therefore $u \notin (M^{[\perp]})^{[\perp]}$, which completes the proof. □

In the next proposition we collect some properties of the H-adjoint.

Proposition 2.2. *Let $A, B \subseteq \mathbb{C}^{2n}$ be linear relations. Then we have*

(i) $A \subseteq B \implies B^{[*]} \subseteq A^{[*]}$.

(ii) $\operatorname{mul} A^{[*]} = (\operatorname{dom} A)^{[\perp]}$. *If A is a matrix, we have* $\operatorname{mul} A^{[*]} = \ker H$.

(iii) $\ker A^{[*]} = (\operatorname{ran} A)^{[\perp]}$.

(iv) $(A^{[*]})^{[*]} = A + (\ker H \times \ker H)$.

Proof. Assertions (i), (ii), (iii) are easy consequences of Definition 1.3 and can be found, e.g., in [22].

In order to prove assertion (iv) we equip the space $\mathbb{C}^n \times \mathbb{C}^n$ with the inner product

$$\left[\!\left[\begin{pmatrix} y \\ z \end{pmatrix}, \begin{pmatrix} x \\ w \end{pmatrix} \right]\!\right] := i([y, w] - [z, x]) = \left\langle \begin{pmatrix} 0 & -iH \\ iH & 0 \end{pmatrix} \begin{pmatrix} y \\ z \end{pmatrix}, \begin{pmatrix} x \\ w \end{pmatrix} \right\rangle$$

where $x, y, w, z \in \mathbb{C}^n$. Then

$$A^{[*]} = A^{[\![\perp]\!]} \quad \text{and } (A^{[*]})^{[*]} = (A^{[\![\perp]\!]})^{[\![\perp]\!]}. \tag{2.1}$$

Lemma 2.1 remains true if we replace $\mathbb{C}^n$ by $\mathbb{C}^{2n}$, $[\perp]$ by $[\![\perp]\!]$ and $\ker H$ by $(\ker H \times \ker H)$. This and (2.1) shows assertion (iv) of Proposition 2.2. $\square$

If A is a matrix, then, by Proposition 2.2 (iv), we have that $y \in \operatorname{mul}(A^{[*]})^{[*]}$ if and only if there exists $f, g \in \ker H$ such that

$$\begin{pmatrix} 0 \\ y \end{pmatrix} = \begin{pmatrix} f \\ Af \end{pmatrix} + \begin{pmatrix} -f \\ g \end{pmatrix} = \begin{pmatrix} 0 \\ Af + g \end{pmatrix}.$$

This proves the following lemma.

Lemma 2.3. *Let $A \in \mathbb{C}^{n\times n}$ be a matrix. Then* $\ker H \subseteq \operatorname{mul}(A^{[*]})^{[*]}$. *If, in addition,* $\ker H$ *is A-invariant, then we have*

$$\operatorname{mul}(A^{[*]})^{[*]} = \ker H.$$

The following formula for the H-adjoint of matrices will be used frequently.

Lemma 2.4. *Let $A \in \mathbb{C}^{n\times n}$ be a matrix. Then*

$$A^{[*]} = \left\{ \begin{pmatrix} y \\ z \end{pmatrix} \in \mathbb{C}^{2n} \ : \ A^* H y = H z \right\}. \tag{2.2}$$

In particular, $A^{[*]}$ is a matrix if and only if H is invertible.

Proof. Clearly, since A is a matrix, we have

$$\begin{pmatrix} y \\ z \end{pmatrix} \in A^{[*]} \iff x^* A^* H y = x^* H z \text{ for all } x \in \mathbb{C}^n.$$

This implies (2.2). The remaining assertion of Lemma 2.4 follows from Proposition 2.2 (ii). $\square$

By Lemma 2.4 we have that $\operatorname{dom} A^{[*]} = \mathbb{C}^n$ if and only if $\operatorname{ran}(A^*H) \subseteq \operatorname{ran} H$.

Example 2.5. Using the notations of Example 1.1 we have

$$M_1^{[*]} = \operatorname{graph} I + (\{0\} \times \{0\} \times \{0\} \times \mathbb{C}) \quad \text{and } M_2^{[*]} = \{0\} \times \mathbb{C} \times \{0\} \times \mathbb{C},$$

where $+$ denotes the sum of linear subspaces. We then obtain

$$\begin{array}{ll} \operatorname{dom} M_1^{[*]} = \mathbb{C}^2, & \operatorname{mul} M_1^{[*]} = \{0\} \times \mathbb{C}, \\ \operatorname{dom} M_2^{[*]} = \{0\} \times \mathbb{C}, & \operatorname{mul} M_2^{[*]} = \{0\} \times \mathbb{C}. \end{array}$$

Let $A \subseteq \mathbb{C}^{2n}$ be a linear relation. Introducing a change of basis $x \mapsto Px$ on $\mathbb{C}^n$, where $P \in \mathbb{C}^{n\times n}$ is a nonsingular matrix, yields

$$P^{-1}AP = \left\{ \begin{pmatrix} P^{-1}x \\ P^{-1}w \end{pmatrix} : \begin{pmatrix} x \\ w \end{pmatrix} \in A \right\}. \tag{2.3}$$

Therefore, for

$$\begin{pmatrix} y \\ z \end{pmatrix} \in (P^{-1}AP)^{[*]_{P^*HP}} \quad \text{and arbitrary} \begin{pmatrix} x \\ w \end{pmatrix} \in A$$

we have $0 = [y, P^{-1}w]_{P^*HP} - [z, P^{-1}x]_{P^*HP} = [Py, w]_H - [Pz, x]_H$, i.e.,

$$\begin{pmatrix} Py \\ Pz \end{pmatrix} \in A^{[*]_H} \quad \text{or, equivalently,} \begin{pmatrix} y \\ z \end{pmatrix} \in P^{-1}A^{[*]_H}P.$$

This gives

$$(P^{-1}AP)^{[*]_{P^*HP}} = P^{-1}A^{[*]_H}P. \tag{2.4}$$

Moreover, with the help of (2.3), it is easily deduced that

$$(P^{-1}AP)^{-1} = P^{-1}A^{-1}P, \tag{2.5}$$

where the inverses are understood in the sense of linear relations.

Thus, if A is a matrix, then changing the basis of $\mathbb{C}^n$ accordingly, we may always assume that H and A have the forms

$$H = \begin{matrix} & m & n-m \\ m & H_1 & 0 \\ n-m & 0 & 0 \end{matrix} \quad \text{and} \quad A = \begin{matrix} & m & n-m \\ m & A_1 & A_2 \\ n-m & A_3 & A_4 \end{matrix} \in \mathbb{C}^{n\times n}, \tag{2.6}$$

where $H_1 \in \mathbb{C}^{m\times m}$ is nonsingular. When using these forms and identifying A with the linear relation $A \subseteq \mathbb{C}^{2n}$, then for the ease of simple notation we will usually omit the indication of dimensions of vectors if those are clear from the context. Thus, for example, we write

$$A = \left\{ \begin{pmatrix} x_1 \\ x_2 \\ A_1x_1 + A_2x_2 \\ A_3x_1 + A_4x_2 \end{pmatrix} : x_1 \in \mathbb{C}^m, x_2 \in \mathbb{C}^{n-m} \right\} = \left\{ \begin{pmatrix} x_1 \\ x_2 \\ A_1x_1 + A_2x_2 \\ A_3x_1 + A_4x_2 \end{pmatrix} \right\}.$$

Proposition 2.6. *Let $A \in \mathbb{C}^{n\times n}$ be a matrix. Then*

$$A^{[*]} = H^{-1}A^*H, \tag{2.7}$$

where H^{-1} is the inverse in the sense of linear relations. In particular, if H and A have the forms as in (2.6) then

$$A^{[*]} = \left\{ \begin{pmatrix} y_1 \\ y_2 \\ A_1^{[*]_{H_1}} y_1 \\ z_2 \end{pmatrix} : A_2^*H_1y_1 = 0 \right\}. \tag{2.8}$$

Moreover, we have $\operatorname{dom} A^{[*]} = \mathbb{C}^n$ *if and only if* $A_2 = 0$.

Similar to H-symmetric matrices, H-isometric matrices can be defined by passing to the concept of linear relations. (We mention that H-isometric relations in degenerate inner product spaces have been introduced in [22].)

Definition 3.6. A linear relation U in $\mathbb{C}^n$ is called H-isometric if $U^{-1} \subseteq U^{[*]}$.

We note that in the definition above U^{-1} is the inverse in the sense of linear relations. E.g., if $H = 0$, then every matrix is H-isometric. It follows from (2.4) and (2.5) that U is H-isometric if and only if $P^{-1}UP$ is P^*HP-isometric for any invertible $P \in \mathbb{C}^{n\times n}$.

Proposition 3.7. *Let $U \in \mathbb{C}^{n\times n}$ be a matrix. Then the following statements are equivalent.*

i) *U is H-isometric, i.e., $U^{-1} \subseteq U^{[*]}$.*

ii) *$U^*HU = H$.*

If one of the conditions is satisfied, then $\ker H$ *is U-invariant. In particular, if H and U have the forms as in (2.6), i.e.,*

$$U = \begin{pmatrix} U_1 & U_2 \\ U_3 & U_4 \end{pmatrix}, \quad H = \begin{pmatrix} H_1 & 0 \\ 0 & 0 \end{pmatrix}, \tag{3.1}$$

where H_1 is invertible, then U is H-isometric if and only if U_1 is H_1-unitary and $U_2 = 0$. Moreover, we have

$$(\operatorname{ran} U)^{[\perp]} = \ker H. \tag{3.2}$$

Proof. By Lemma 2.4, we have

$$U^{-1} \subseteq U^{[*]} \iff \begin{pmatrix} Uy \\ y \end{pmatrix} \in U^{[*]} \text{ for all } y \in \mathbb{C}^n \iff U^*HUy = Hy \text{ for all } y \in \mathbb{C}^n.$$

This shows the equivalence of the two statements in Proposition 3.7.

For the remainder of the proof, let H and U be in the forms as in (3.1) and assume that U is H-isometric. Then we obtain from the identity $U^*HU = H$ that

$$\begin{pmatrix} U_1^*H_1U_1 & U_1^*H_1U_2 \\ U_2^*H_1U_1 & U_2^*H_1U_2 \end{pmatrix} = \begin{pmatrix} H_1 & 0 \\ 0 & 0 \end{pmatrix}$$

This immediately implies $U_1^*H_1U_1 = H_1$, i.e., U_1 is H_1-unitary. In particular, with H_1 also U_1 must be invertible. This finally yields $U_2 = 0$ and (3.2). □

If we adjoin the inclusion $U^{-1} \subseteq U^{[*]}$ once more we get, similar to Proposition 3.4, the following characterization of H-isometric matrices.

Proposition 3.8. *Let $U \in \mathbb{C}^{n\times n}$ be a matrix. Then U is H-isometric if and only if*

$$(U^{-1})^{[*]} = (U^{[*]})^{[*]}$$

Proof. Let U be H-isometric, i.e., $U^{-1} \subseteq U^{[*]}$. Then, by Proposition 2.2 (i), we have $(U^{[*]})^{[*]} \subseteq (U^{-1})^{[*]}$. For the other inclusion, observe that, by Proposition 2.2 (iv), we have

$$\operatorname{dom}(U^{[*]})^{[*]} = \mathbb{C}^n.$$

Thus, using $(U^{[*]})^{[*]} \subseteq (U^{-1})^{[*]}$, we obtain that

$$\operatorname{dom}(U^{[*]})^{[*]} = \operatorname{dom}(U^{-1})^{[*]} = \mathbb{C}^n.$$

Proposition 2.2 (ii) and (3.2) imply

$$\operatorname{mul}(U^{-1})^{[*]} = (\operatorname{dom} U^{-1})^{[\perp]} = (\operatorname{ran} U)^{[\perp]} = \ker H$$

and, with Lemma 2.3 and the U-invariance of $\ker H$ (Proposition 3.7), we conclude

$$\operatorname{mul}(U^{[*]})^{[*]} = \ker H = \operatorname{mul}(U^{-1})^{[*]}.$$

Hence $(U^{-1})^{[*]} = (U^{[*]})^{[*]}$. The contrary follows from Proposition 2.2 and

$$U^{-1} \subset ((U^{-1})^{[*]})^{[*]} = ((U^{[*]})^{[*]})^{[*]} \subset U^{[*]}. \qquad \square$$

4. H-normal matrices

Recall that in the case of invertible H, a matrix A is called H-normal if and only if $AA^{[*]} = A^{[*]}A$. For the case that H is singular and that H and A are given in the forms as in (2.6), a straightforward computation reveals

$$AA^{[*]} = \left\{ \begin{pmatrix} y_1 \\ y_2 \\ A_1A_1^{[*]}y_1 + A_2z_2 \\ A_3A_1^{[*]}y_1 + A_4z_2 \end{pmatrix} \;:\; A_2^*H_1y_1 = 0 \right\} \tag{4.1}$$

and

$$A^{[*]}A = \left\{ \begin{pmatrix} y_1 \\ y_2 \\ A_1^{[*]}A_1y_1 + A_1^{[*]}A_2y_2 \\ z_2 \end{pmatrix} \;:\; A_2^*H_1A_1y_1 + A_2^*H_1A_2y_2 = 0 \right\}. \tag{4.2}$$

However, even in the case that A is H-symmetric (i.e., in the identities (4.1) and (4.2) we have $A_1A_1^{[*]} = A_1^{[*]}A_1 = A_1^2$ and $A_2 = 0$), we only obtain the inclusion $AA^{[*]} \subseteq A^{[*]}A$, while the other inclusion $A^{[*]}A \subseteq AA^{[*]}$ is only satisfied if A_4 is invertible. This motivates the following definition.

Definition 4.1. A relation A in $\mathbb{C}^n$ is called H-normal if $AA^{[*]} \subseteq A^{[*]}A$.

As for the case of H-isometric and H-symmetric matrices, we obtain that the kernel of H is always an invariant subspace for H-normal matrices.

Proposition 4.2. *Let $A \in \mathbb{C}^{n\times n}$ be an H-normal matrix. Then* $\ker H$ *is A-invariant. In particular, if A and H are in the forms as in* (2.6)*, then A is H-normal if and only if A_1 is H_1-normal and $A_2 = 0$.*

Proof. Without loss of generality, we may assume that A and H are in the forms as in (2.6). Clearly, if A_1 is H_1-normal and $A_2 = 0$ then it follows directly from (4.1) and (4.2) that $AA^{[*]} \subseteq A^{[*]}A$, i.e., A is H-normal. For the converse, assume that

A is H-normal. Then $AA^{[*]} \subseteq A^{[*]}A$ implies in particular that

$$\left(AA^{[*]} \cap (\{0\} \times \mathbb{C}^n)\right) \subseteq \left(A^{[*]}A \cap (\{0\} \times \mathbb{C}^n)\right)$$

Comparing the third block components of (4.1) and (4.2) this reduces to $A_2 z = 0$ for all $z_2 \in \mathbb{C}^{n-m}$ and this is only possible if $A_2 = 0$. But then $AA^{[*]} \subseteq A^{[*]}A$ implies $A_1 A_1^{[*]} y_1 = A_1^{[*]} A_1 y_1$ for all $y_1 \in \mathbb{C}^m$ and we obtain that A_1 is H_1-normal. Clearly, $\ker H$ is A-invariant, because of $A_2 = 0$. This concludes the proof. □

With Propositions 3.3 and 3.7 we immediately obtain the following corollary.

Corollary 4.3. *H-symmetric and H-isometric matrices are H-normal.*

The question arises, if we obtain a different characterization of H-normality in the style of Propositions 3.4 and 3.8 by the identity

$$A^{[*]}(A^{[*]})^{[*]} = (A^{[*]})^{[*]}A^{[*]}. \tag{4.3}$$

Let us investigate this question in detail. Without loss of generality assume that the matrix $A \in \mathbb{C}^{n\times n}$ and H are given in the forms (2.6). Then using Proposition 2.2 (iv), we obtain

$$(A^{[*]})^{[*]} = \left\{ \begin{pmatrix} x_1 \\ x_2 \\ A_1x_1 + A_2x_2 \\ A_3x_1 + A_4x_2 \end{pmatrix} + \begin{pmatrix} 0 \\ w_2 \\ 0 \\ z_2 \end{pmatrix} \right\} = \left\{ \begin{pmatrix} x_1 \\ w_2 \\ A_1x_1 + A_2x_2 \\ z_2 \end{pmatrix} \right\}.$$

Together with (2.8) this implies

$$A^{[*]}(A^{[*]})^{[*]} = \left\{ \begin{pmatrix} x_1 \\ w_2 \\ A_1^{[*]}A_1x_1 + A_1^{[*]}A_2x_2 \\ z_2 \end{pmatrix} : A_2^* H_1(A_1x_1 + A_2x_2) = 0 \right\} \tag{4.4}$$

and

$$(A^{[*]})^{[*]}A^{[*]} = \left\{ \begin{pmatrix} y_1 \\ y_2 \\ A_1A_1^{[*]}y_1 + A_2x_2 \\ z_2 \end{pmatrix} : A_2^* H_1 y_1 = 0 \right\}. \tag{4.5}$$

With the help of these formulas, we obtain that (4.3) may be satisfied even if the matrix $A \in \mathbb{C}^{n\times n}$ is not H-normal, see Example 4.4 below.

Example 4.4. Let A and H be given as

$$A = \left(\begin{array}{c|c} A_1 & A_2 \\ \hline A_3 & A_4 \end{array}\right) := \left(\begin{array}{cc|c} 1 & 0 & 1 \\ 0 & 1 & 0 \\ \hline 0 & 0 & 0 \end{array}\right), \quad H = \left(\begin{array}{c|c} H_1 & 0 \\ \hline 0 & 0 \end{array}\right) := \left(\begin{array}{cc|c} 0 & 1 & 0 \\ 1 & 0 & 0 \\ \hline 0 & 0 & 0 \end{array}\right).$$

Then A is not H-normal, because $\ker H$ is not A-invariant.

However, from (4.4) and (4.5), we immediately obtain

$$A^{[*]}(A^{[*]})^{[*]} = \left\{ \begin{pmatrix} x_1 \\ w_2 \\ x_1 + A_2 x_2 \\ z_2 \end{pmatrix} : A_2^* H_1 x_1 = 0 \right\} = (A^{[*]})^{[*]} A^{[*]}.$$

The following result shows that the set of matrices satisfying (4.3) contains the set of H-normal matrices.

Proposition 4.5. *Let $A \in \mathbb{C}^{n\times n}$ be a matrix. If A is H-normal, then*

$$A^{[*]}(A^{[*]})^{[*]} = (A^{[*]})^{[*]} A^{[*]}.$$

Proof. Without loss of generality let A and H be in the forms (2.6). Then we obtain by Proposition 4.2 that $A_2 = 0$. The identities (4.4) and (4.5) imply

$$(A^{[*]})^{[*]} A^{[*]} = \left\{ \begin{pmatrix} y_1 \\ y_2 \\ A_1 A_1^{[*]} y_1 \\ w_2 \end{pmatrix} \right\} \quad \text{and} \quad A^{[*]}(A^{[*]})^{[*]} = \left\{ \begin{pmatrix} x_1 \\ x_2 \\ A_1^{[*]} A_1 x_1 \\ z_2 \end{pmatrix} \right\}.$$

Indeed, these two sets are equal because of the H_1-normality of A_1 which is guaranteed by Proposition 4.2. □

Next, let us compare H-normal matrices with Moore-Penrose H-normal matrices. We obtain the following result.

Proposition 4.6. *Let $A \in \mathbb{C}^{n\times n}$ be a matrix. Then the following statements are equivalent.*

i) *A is H-normal, i.e., $AA^{[*]} \subseteq A^{[*]}A$.*
ii) *A is Moore-Penrose H-normal and $A^{[*]}(A^{[*]})^{[*]} = (A^{[*]})^{[*]} A^{[*]}$.*

Proof. Without loss of generality, let A and H have the forms (2.6). Then the Moore-Penrose generalized inverse of H is given by

$$H^{\dagger} = \begin{pmatrix} H_1^{-1} & 0 \\ 0 & 0 \end{pmatrix}$$

and the matrix A is Moore-Penrose H-normal if and only if

$$\begin{pmatrix} A_1^* H_1 A_1 & A_1^* H_1 A_2 \\ A_2^* H_1 A_1 & A_2^* H_1 A_2 \end{pmatrix} = \begin{pmatrix} H_1 A_1 H_1^{-1} A_1^* H_1 & 0 \\ 0 & 0 \end{pmatrix}. \tag{4.6}$$

"i) $\Rightarrow$ ii)": If A is H-normal, then by Proposition 4.2 we have that A_1 is H_1-normal and that $A_2 = 0$. Then (4.6) is satisfied and the remainder follows from Proposition 4.5.

"ii) $\Rightarrow$ i)": Let A be Moore-Penrose H-normal and let $A^{[*]}(A^{[*]})^{[*]} = (A^{[*]})^{[*]} A^{[*]}$. Then by (4.6) we have $A_2^* H_1 A_1 = 0$ and $A_2^* H_1 A_2 = 0$. Comparing this with (4.4), we find that $\operatorname{dom} A^{[*]}(A^{[*]})^{[*]} = \mathbb{C}^n$. But then, we must have $\operatorname{dom}(A^{[*]})^{[*]} A^{[*]} = \mathbb{C}^n$ as well which, using (4.5), implies $A_2^* H_1 y_1 = 0$ for all $y_1 \in \mathbb{C}^m$. From this, we obtain $A_2 = 0$. But then $A^{[*]}(A^{[*]})^{[*]} = (A^{[*]})^{[*]} A^{[*]}$ reduces to $A_1^{[*]} A_1 x_1 = A_1 A_1^{[*]} x_1$ for all $x_1 \in \mathbb{C}^m$ which implies H_1-normality of A_1. From this and Proposition 4.2, we finally obtain that A is H-normal. □

As a consequence, we obtain that the set of H-normal matrices is a strict subset of the set of Moore-Penrose H-normal matrices, because it has been shown in [17, Example 6.1] that there exist Moore-Penrose H-normal matrices A such that $\ker H$ is not A-invariant.

The fact that the kernel of H is invariant for H-normal matrices allows the generalization of extension results for H-semidefinite invariant subspaces of normal matrices for invertible H to the case of singular H. For example, if A is H-normal and H is invertible, then any H-nonnegative subspace that is invariant for both A and $A^{[*]}$ can be extended to an A-invariant maximal H-nonnegative subspace, see [15]. This result now easily generalizes to the case of singular H. Here, an invariant subspace $\mathcal{U} \subseteq \mathbb{C}^n$ of a linear relation A in $\mathbb{C}^n$ is defined by the implication

$$x \in \mathcal{U} \quad \text{and} \quad \begin{pmatrix} x \\ y \end{pmatrix} \in A \quad \Longrightarrow \quad y \in \mathcal{U}.$$

Theorem 4.7. *Let $A \in \mathbb{C}^{n\times n}$ be H-normal, and let $\mathcal{M}_0$ be an H-nonnegative A-invariant subspace that is also invariant for $A^{[*]}$. Then there exists an A-invariant maximal H-nonnegative subspace $\mathcal{M}$ containing $\mathcal{M}_0$ that is also invariant for $A^{[*]}$.*

Proof. Without loss of generality assume that A and H are in the forms (2.6) and that $\mathcal{M}_0$ can be written as a direct sum $\mathcal{M}_0 = \mathcal{M}_1 \dot{+} \mathcal{M}_2$, where $\mathcal{M}_2 \subseteq \ker H$ and

$$\mathcal{M}_1 = \left\{ \begin{pmatrix} x_1 \\ 0 \end{pmatrix} : x_1 \in \widetilde{\mathcal{M}}_1 \right\}$$

for some subspace $\widetilde{\mathcal{M}}_1 \subseteq \mathbb{C}^m$. It is easy to verify that the A- and $A^{[*]}$-invariance of $\mathcal{M}_0$ imply that $\widetilde{\mathcal{M}}_1$ is A_1-invariant as well as $A_1^{[*]}$-invariant. Thus, by [15] there exists an A_1-invariant maximal H-nonnegative subspace $\mathcal{M}_{\max}$ that is also invariant for $A_1^{[*]}$. Setting $\mathcal{M} = \mathcal{M}_{\max} \dot{+} \ker H$, we obtain that $\mathcal{M}$ contains $\mathcal{M}_0$ and is A-invariant and maximal H-nonnegative. It is easy to check that $\mathcal{M}$ is also $A^{[*]}$-invariant. □

If we drop the assumption that $\mathcal{M}_0$ is $A^{[*]}$-invariant, then extension results are not as immediate. Indeed, it has been shown in [18] that there exist H-normal matrices (in the case of invertible H) that have an invariant H-nonnegative subspace that cannot be extended to an invariant maximal H-nonnegative subspace. (There still exist such counterexamples if one restricts the subspace to be H-positive rather than H-nonnegative.) Thus, stronger conditions have to be imposed on an H-normal matrix such that extension of semidefinite invariant subspaces can be guaranteed, see [18, 19].

We conclude the paper by generalizing a result concerning the extension of H-positive invariant subspaces of H-normal matrices obtained in [18] to the case of singular H. The fact that $\ker H$ is always an invariant subspace for H-normal matrices plays a key role in this proof.

Theorem 4.8. *Let $A \in \mathbb{C}^{n\times n}$ be H-normal, and let $\mathcal{M}_0$ be an H-positive A-invariant subspace. Let $\mathcal{M}_{\text{com}}$ be a direct complement of* $\ker H$ *in $\mathcal{M}_0^{[\perp]}$, that is, $\mathcal{M}_0^{[\perp]} = \mathcal{M}_{\text{com}} \dot{+} \ker H$. Define*

$$A_{22} := PA|_{\mathcal{M}_{\text{com}}} : \mathcal{M}_{\text{com}} \to \mathcal{M}_{\text{com}},$$

where P is the projection onto $\mathcal{M}_{\text{com}}$ along $\mathcal{M}_0 \dot{+} \ker H$. Then $\mathcal{M}_{\text{com}}$ is nondegenerate. Equip $\mathcal{M}_{\text{com}}$ with the indefinite inner product induced by H. Assume that

$$\sigma(A_{22} + A_{22}^{[*]}) \subseteq \mathbb{R} \quad \text{or} \quad \sigma(A_{22} - A_{22}^{[*]}) \subseteq i\mathbb{R}. \tag{4.7}$$

Then there exists an A-invariant maximal H-nonnegative subspace $\mathcal{M}$ that contains $\mathcal{M}_0$ and that is also $A^{[*]}$-invariant.

The condition (4.7) *is independent of the particular choice of a direct complement $\mathcal{M}_{\text{com}}$ of* $\ker H$ *in $\mathcal{M}_0^{[\perp]}$.*

Proof. Since $\mathcal{M}_0$ is H-positive, we have that $\mathcal{M}_0^{[\perp]}$ is a direct complement of $\mathcal{M}_0$. Moreover, it is clear that a complement of $\ker H$ is nondegenerate. (By default, i.e., nonexistence of H-neutral vectors, $\{0\}$ is a nondegenerate subspace.) Thus, without loss of generality, we may assume that $\mathcal{M}_0 = \text{span}(e_1, \ldots, e_k)$ and $\mathcal{M}_{\text{com}} = \text{span}(e_{k+1}, \ldots, e_{k+l})$, $l \geq 0$, and $\mathcal{M}_0^{[\perp]} = \text{span}(e_{k+1}, \ldots, e_n)$, where e_j denotes the jth unit vector. Then A and H have the corresponding block forms

$$A = \begin{pmatrix} A_{11} & A_{12} & A_{13} \\ 0 & A_{22} & A_{23} \\ 0 & A_{32} & A_{33} \end{pmatrix}, \quad H = \begin{pmatrix} I_k & 0 & 0 \\ 0 & H_{22} & 0 \\ 0 & 0 & 0 \end{pmatrix},$$

where H_{22} is invertible. In this representation of A, we have

$$A_1 = \begin{pmatrix} A_{11} & A_{12} \\ 0 & A_{22} \end{pmatrix}, \quad \text{and} \quad H_1 = \begin{pmatrix} I_k & 0 \\ 0 & H_{22} \end{pmatrix},$$

where A_1 and H_1 are defined in analogy to the decomposition (2.6). By Proposition 4.2 we have that A_1 is H_1-normal and that $\ker H$ is A-invariant which implies $A_{13} = 0$ and $A_{23} = 0$. Setting

$$\widetilde{\mathcal{M}}_0 = \text{ran} \begin{pmatrix} I_k \\ 0 \end{pmatrix} \subseteq \mathbb{C}^{k+l}$$

which is an A_1-invariant H_1-positive subspace, we obtain, given the condition on A_{22}, that by [18, Theorem 4.4] there exists an A_1-invariant maximal H_1-nonnegative subspace $\widehat{\mathcal{M}}_0$ of dimension $\nu_+(H) = \nu_+(H_1)$ containing $\widetilde{\mathcal{M}}_0$. We choose appropriate matrices M_{11}, M_{12} and write $\widehat{\mathcal{M}}_0$ in the following way

$$\widehat{\mathcal{M}}_0 = \text{ran} \begin{pmatrix} I_k & M_{11} \\ 0 & M_{12} \end{pmatrix}.$$

Let

$$\mathcal{M} := \text{ran} \begin{pmatrix} I_k & M_{11} & 0 \\ 0 & M_{12} & 0 \\ 0 & 0 & I_{n-k-l} \end{pmatrix}.$$

Then it is straightforward to check that $\mathcal{M}$ is an A-invariant, H-nonnegative subspace of dimension $\nu_+(H) + \nu_0(H)$ containing $\mathcal{M}_0$. Clearly, $\mathcal{M}$ is also invariant for $A^{[*]}$.

It remains to show that the condition (4.7) is independent of the particular choice of the direct complement $\mathcal{M}_{\rm com}$. But choosing a different direct complement $\mathcal{M}_{\rm new}$ of $\ker H$ in $\mathcal{M}_0^{[\perp]} = \operatorname{span}(e_{k+1}, \ldots, e_n)$ amounts to a change of basis given by a matrix of the form

$$S = \begin{pmatrix} I_k & 0 & 0 \\ 0 & S_{22} & 0 \\ 0 & S_{32} & I_{n-k-l} \end{pmatrix}$$

with S_{22} invertible. With respect to the decomposition $\mathbb{C}^n = \mathcal{M}_0 \dot{+} \mathcal{M}_{\rm new} \dot{+} \ker H$ and the new basis, A and H take the forms

$$\tilde{A} = S^{-1}AS = \begin{pmatrix} A_{11} & * & 0 \\ 0 & S_{22}^{-1}A_{22}S_{22} & 0 \\ 0 & * & A_{33} \end{pmatrix}, \quad \tilde{H} = S^*HS = \begin{pmatrix} I_k & 0 & 0 \\ 0 & S_{22}^*H_{22}S_{22} & 0 \\ 0 & 0 & 0 \end{pmatrix}.$$

The compressions of $\tilde{A}$ and $\tilde{H}$ to $\mathcal{M}_{new}$ are $S_{22}^{-1}A_{22}S_{22}$ and $S_{22}^*H_{22}S_{22}$. Clearly, condition (4.7) is satisfied for these compressions if and only if (4.7) is satisfied for A_{22} and H_{22}. $\square$

Acknowledgment

We thank Heinz Langer for asking a question that initiated this research and Peter Jonas for fruitful comments on an earlier version of this paper.

References

[1] G. Ammar, C. Mehl, and V. Mehrmann, *Schur-like Forms for Matrix Lie Groups, Lie Algebras and Jordan Algebras.* Linear Algebra Appl. **287** (1999), 11–39.

[2] T.Ya. Azizov, *Completely Continuous Operators that are Selfadjoint with Respect to a Degenerate Indefinite Metric* (Russian). Mat. Issled. **7** (1972), 237–240, 259.

[3] T.Ya. Azizov and I.S. Iohvidov, *Linear Operators in Spaces with an Indefinite Metric.* John Wiley and Sons, Ltd., Chichester, 1989. (Translated from Russian.)

[4] V. Bolotnikov, C.K. Li, P. Meade, C. Mehl, and L. Rodman, *Shells of Matrices in Indefinite Inner Product Spaces.* Electron. J. Linear Algebra **9** (2002), 67–92.

[5] R. Cross, *Multivalued Linear Operators.* Marcel Dekker Inc., 1998.

[6] A. Dijksma and H.S.V. de Snoo, *Symmetric and Selfadjoint Relations in Krein Spaces I.* Oper. Theory Adv. Appl. **24** (1987), 145–166.

[7] A. Dijksma and H.S.V. de Snoo, *Symmetric and Selfadjoint Relations in Krein Spaces II.* Ann. Acad. Sci. Fenn. Math. **12** (1987), 199–216.

[8] I. Gohberg, P. Lancaster, and L. Rodman, *Matrices and Indefinite Scalar Products.* Birkhäuser, 1983.

[9] I. Gohberg, P. Lancaster, and L. Rodman, *Indefinite Linear Algebra.* Birkhäuser, 2005.

[10] I. Gohberg and B. Reichstein, *On Classification of Normal Matrices in an Indefinite Scalar Product.* Integral Equations Operator Theory **13** (1990), 364–394.

[11] M. Kaltenbäck and H. Woracek, *Selfadjoint Extensions of Symmetric operators in Degenerated Inner Product Spaces*, Integral Equations Operator Theory **28** (1997), 289–320.

[12] P. Lancaster, A.S. Markus, and P. Zizler, *The Order of Neutrality for Linear Operators on Inner Product Spaces.* Linear Algebra Appl. **259** (1997), 25–29.

[13] P. Lancaster and L. Rodman, *Algebraic Riccati Equations.* Clarendon Press, 1995.

[14] H. Langer, R. Mennicken, and C. Tretter, *A Self-Adjoint Linear Pencil* $Q - \lambda P$ *of Ordinary Differential Operators.* Methods Funct. Anal. Topology **2** (1996), 38–54.

[15] H. Langer, *Invariante Teilräume definisierbarer* J*-selbstadjungierter Operatoren.* Ann. Acad. Sci. Fenn. Ser. A.I. Math. **475** (1971), 1–23

[16] C.K. Li, N.K. Tsing, and F. Uhlig, *Numerical Ranges of an Operator on an Indefinite Inner Product Space.* Electron. J. Linear Algebra **1** (1996), 1–17.

[17] C. Mehl, A. Ran, and L. Rodman, *Semidefinite Invariant Subspaces: Degenerate Inner Products.* Oper. Theory Adv. Appl. **149** (2004), 467–486.

[18] C. Mehl, A. Ran, and L. Rodman, *Hyponormal Matrices and Semidefinite Invariant Subspaces in Indefinite Inner Products.* Electron. J. Linear Algebra **11** (2004), 192–204.

[19] C. Mehl, A. Ran, and L. Rodman, *Extension to Maximal Semidefinite Invariant Subspaces for Hyponormal Matrices in Indefinite Inner Products.* Linear Algebra Appl. **421** (2007), 110–116.

[20] C. Mehl and L. Rodman, *Symmetric Matrices with Respect to Sesquilinear Forms.* Linear Algebra Appl. **349** (2002), 55–75.

[21] V. Mehrmann, *Existence, Uniqueness, and Stability of Solutions to Singular Linear Quadratic Optimal Control Problems.* Linear Algebra Appl. **121** (1989), 291–331.

[22] B. C. Ritsner, *The Theory of Linear Relations.* (Russian) Voronezh, Dep. VINITI, No. 846-82, 1982.

[23] H. Woracek, *Resolvent Matrices in Degenerated Inner Product Spaces.* Math. Nachr. **213** (2000), 155–175.

Christian Mehl
Institut für Mathematik, MA 4-5
Technische Universität Berlin
Str. des 17. Juni 136
D-10623 Berlin, Germany
e-mail: `mehl@math.tu-berlin.de`

Carsten Trunk
Institut für Mathematik, MA 6-3
Technische Universität Berlin
Str. des 17. Juni 136
D-10623 Berlin, Germany
e-mail: `trunk@math.tu-berlin.de`

Operator Theory:
Advances and Applications, Vol. 175, 211–224

Symmetric Hermite-Biehler Polynomials with Defect

Vyacheslav Pivovarchik

Abstract. The polynomial $\omega = P(\lambda) + iQ(\lambda)$ with real $P(\lambda)$ and $Q(\lambda)$ which belongs to Hermite-Biehler class (all its zeros lie in the open upper half-plane) and is symmetric ($\omega(-\overline{\lambda}) = \overline{\omega(\lambda)}$) is modified as follows

$$\omega_c(\lambda) = \tilde{P}(\lambda^2 + c) + i\lambda\tilde{\tilde{Q}}(\lambda^2 + c), \quad c > 0.$$

Here $\tilde{P}(\lambda^2) = P(\lambda)$, $\tilde{\tilde{Q}}(\lambda^2) = \lambda^{-1}Q(\lambda)$ and

$$P(\lambda) = \frac{\omega(\lambda) + \omega(-\lambda)}{2}, \quad Q(\lambda) = \frac{\omega(\lambda) - \omega(-\lambda)}{2i}.$$

The conditions are obtained necessary and sufficient for a set of complex numbers to be the zeros of a polynomial of the form $\omega_c(\lambda)$.

Mathematics Subject Classification (2000). Primary 12D10; Secondary 26C10, 30C15.

Keywords. Hermite-Biehler polynomial, Hurwitz polynomials, zeros in the lower half-plane.

1. Introduction

In [1] the following so-called generalized Regge boundary problem was considered:

$$y'' - q(x)y = \lambda^2 y, \quad x \in (0, a), \tag{1.1}$$

$$y(0) = y'(a) + (i\lambda\alpha + \beta)y(a) = 0, \quad \alpha > 0, \quad \beta \in R.$$

It was shown that the spectrum of this problem coincides with the set of zeros of the characteristic function

$$\varphi(\lambda) = s'(\lambda, a) + (i\alpha\lambda + \beta)s(\lambda, a), \tag{1.2}$$

where $s(\lambda, x)$ is the solution of (1.1) which satisfies the conditions $s(\lambda, 0) = s'(\lambda, 0) - 1 = 0$. The function $\varphi(\lambda)$ is symmetric with respect to the imaginary axis, i.e.,

$$\varphi(-\overline{\lambda}) = \overline{\varphi(\lambda)}, \quad \lambda \in C.$$

For a symmetric polynomial $\omega(\lambda)$ the following is valid

$$\omega(\lambda) = P(\lambda) + iQ(\lambda) = P(\lambda) + i\lambda\hat{Q}(\lambda) = \tilde{P}(\lambda^2) + i\lambda\tilde{\hat{Q}}(\lambda^2), \tag{2.2}$$

where $P(\lambda)$ and $\hat{Q}(\lambda)$ are real even function. Here

$$\tilde{P}(\lambda^2) = P(\lambda), \tag{2.3}$$

$$\tilde{\hat{Q}}(\lambda^2) = \hat{Q}(\lambda). \tag{2.4}$$

Lemma 2.5. *Let the polynomial $\omega(\lambda)$ of the form (2.2) belong to SHB. Then the polynomial*

$$\tilde{P}(\lambda) + i\tilde{\hat{Q}}(\lambda),$$

where $\tilde{P}$ and $\tilde{\hat{Q}}$ are given by (2.3) and (2.4) belongs to HB.

Proof. Due to Theorem 2.3

$$P(\lambda) = \prod_1^n (\lambda^2 - a_k^2) \qquad \text{and} \qquad Q(\lambda) = (-1)^{n+n'}\lambda \prod_1^{n'} (\lambda^2 - b_k^2),$$

where $n' = n$ or $n' = n-1$, $a_{-k} \overset{\text{def}}{=} -a_k$, $b_{-k} \overset{\text{def}}{=} -b_k$, and the zeros a_k and b_k interlace as follows:

$$\cdots < b_{-1} < a_{-1} < 0 = b_0 < a_1 < b_1 < a_2 < b_2 < \cdots .$$

We assume in above formulae $\prod_1^0 (\lambda^2 - b_k^2) \equiv 1$ by definition.

Then

$$\tilde{P}(\lambda) = \prod_1^n (\lambda - a_k^2) \qquad \text{and} \qquad \tilde{\hat{Q}}(\lambda) = (-1)^{n+n'} \prod_1^{n'} (\lambda - b_k^2).$$

Here

$$0 < a_1^2 < b_1^2 < a_2^2 < b_2^2 < \cdots , \tag{2.5}$$

i.e., the zeros of $\tilde{P}(\lambda)$ and $\tilde{\hat{Q}}(\lambda)$ interlace and therefore, condition (1) of Theorem 2.3 is satisfied. Let us prove that condition (2) is satisfied too. It is clear that

$$\tilde{P}(0) = (-1)^n \prod_1^n a_k^2, \qquad \tilde{\hat{Q}}(0) = (-1)^{n+2n'} \prod_1^{n'} b_k^2,$$

$$\tilde{P}'(0) = (-1)^{n-1} \sum_{k'=1}^{n} \prod_{k=1, k\neq k'}^{n} a_k^2, \qquad \tilde{\hat{Q}}'(0) = (-1)^{n+2n'-1} \sum_{k'=1}^{n'} \prod_{k=1, k\neq k'}^{n'} b_k^2.$$

Using the above expressions we obtain

$$\tilde{\tilde{Q}}'(0)\tilde{P}(0) - \tilde{\tilde{Q}}(0)\tilde{\tilde{P}}'(0) = -\left(\sum_{k'=1}^{n'} \prod_{k=1,k\neq k'}^{n'} b_k^2 \prod_{k=1}^{n} a_k^2 - \sum_{k'=1}^{n'} \prod_{k=1,k\neq k'}^{n} a_k^2 \prod_{k=1}^{n'} b_k^2\right)$$

$$= \prod_{k=1}^{n'} b_k^2 \prod_{k=1}^{n} a_k^2 \left(\sum_{k=1}^{n} a_k^{-2} - \sum_{k=1}^{n'} b_k^{-2}\right). \qquad (2.6)$$

Taking into account (2.5) and the inequality $n' \leq n$ we have

$$\sum_{k=1}^{n} a_k^{-2} - \sum_{k=1}^{n'} b_k^{-2} > 0. \qquad (2.7)$$

Now (2.6) and (2.7) imply

$$\tilde{\tilde{Q}}'(0)\tilde{P}(0) - \tilde{\tilde{Q}}(0)\tilde{\tilde{P}}'(0) > 0.$$

Using again Theorem 2.3 we finish the proof. □

Corollary 2.6. *Let the polynomial $\omega(\lambda)$ of the form (2.2) belong to SHB. Then the polynomial*

$$\tilde{P}(\lambda + c) + i\tilde{\tilde{Q}}(\lambda + c)$$

belongs to HB for every real c.

Definition 2.7. *Two real polynomials $P(\lambda)$ and $Q(\lambda)$ are said to compose a real pair, if they have no common zeros and if any linear combination $\mu P(\lambda) + \nu Q(\lambda)$ of them with real coefficients μ and ν has no complex zeros. (We refer to not lying on the real axis as complex zeros).*

By Theorem 4 ([7], Chap. 7, Sec. 2, p. 315) we have the following statement.

Corollary 2.8. *The polynomials $\tilde{P}(\lambda)$ and $\tilde{\tilde{Q}}(\lambda)$ form a real pair and*

$$\tilde{P}(\lambda + c)\frac{\partial}{\partial\lambda}\tilde{\tilde{Q}}(\lambda + c) - \tilde{\tilde{Q}}(\lambda + c)\frac{\partial}{\partial\lambda}\tilde{P}(\lambda + c) > 0, \ \text{for} \lambda \in R, c \in R. \qquad (2.8)$$

To describe the sets of zeros of polynomials of the form (2.2) let us make use of the following definitions introduced in [1].

Definition 2.9. *A finite or infinite sequence of complex numbers $\{\zeta_k\}_{k\in Z}$ or $\{\zeta_k\}_{0\neq k\in Z}$ is said to be properly enumerated if*

1) $\operatorname{Re}\zeta_k \geq \operatorname{Re}\zeta_p$ *for all* $k > p$.
2) $\zeta_{-k} = -\overline{\zeta_k}$ *for all* ζ_k *not purely imaginary.*
3) *A certain complex number appears in the sequence at most finitely many times.*

Definition 2.10. *Let κ be a nonnegative integer. Then the properly enumerated sequence $\{\zeta_k\}_{k=-n}^{n}$ ($\{\zeta_k\}_{k=-n,\ k\neq 0}^{n}$) is said to have the SHB_κ^+ (resp., SHB_κ^-) property if*

1) *All but κ terms of the sequence lie in the open upper half-plane.*
2) *All terms in the closed lower half-plane are purely imaginary and occur only once. If $\kappa \geq 1$, we denote them as $\zeta_{-j} = -i|\zeta_{-j}|$ ($j = 1, \ldots, \kappa$). We assume that $|\zeta_{-j}| < |\zeta_{-(j+1)}|$ ($j = 1, \ldots, \kappa - 1$).*
3) *If $\kappa \geq 1$, the numbers $i|\zeta_{-j}|$ ($j = 1, \ldots, \kappa$) (with the exception of ζ_{-1} if it equals zero) are not terms of the sequence.*
4) *If $\kappa \geq 2$, then the number of terms in the intervals $(i|\zeta_{-j}|, i|\zeta_{-(j+1)}|)$ ($j = 1, \ldots, \kappa - 1$) is odd.*
5) *If $|\zeta_{-1}| > 0$, then the interval $(0, i|\zeta_{-1}|)$ contains no terms at all or an even number of terms.*
6) *If $\kappa \geq 1$, then the interval $(i|\zeta_{-\kappa}|, i\infty)$ contains a nonzero even (in the case of SHB_κ^+) or odd (in the case of SHB_κ^-) number of terms.*
7) *If $\kappa = 0$, then the sequence has an odd (in the case of SHB_κ^+) or even or zero (in the case of SHB_κ^-) number of positive imaginary terms.*

Now we introduce the class of polynomials to which this paper is devoted.

Definition 2.11. *Let the polynomial $\omega(\lambda) \in SHB$. Then the polynomial*

$$\omega(\lambda, c) = \tilde{P}(\lambda^2 + c) + i\lambda\tilde{\hat{Q}}(\lambda^2 + c),$$

where $\tilde{P}$ and $\tilde{\hat{Q}}$ are given by (2.3)–(2.4) is said to belong to the class of shifted symmetric Hermite-Biehler polynomials or in other words to the symmetric Hermite-Biehler class with a defect (SHB_c).

Lemma 2.12. *If the polynomial $\omega(\lambda) \in SHB_c$, then it has no real zeros except of possible zero at $\lambda = 0$.*

Proof. If

$$\tilde{P}(\lambda^2 + c) + i\lambda\tilde{\hat{Q}}(\lambda^2 + c) = 0$$

for real $\lambda \neq 0$, then

$$\tilde{P}(\lambda^2 + c) = \tilde{\hat{Q}}(\lambda^2 + c) = 0,$$

what contradicts Corollary 2.6. □

Lemma 2.13. *If $-i\tau$ ($\tau \geq 0$) is a zero of $\omega(\lambda, c) \in SHB_c$ at $c > 0$ fixed, then this zero is simple.*

Proof. If $-i\tau$ ($\tau > 0$) is a multiple zero of $\omega(\lambda, c)$, then

$$\tilde{P}(-\tau^2 + c) + \tau\tilde{\hat{Q}}(-\tau^2 + c) = 0, \tag{2.9}$$

$$-2i\tau\tilde{P}'(\lambda)\Big|_{\lambda=-\tau^2+c} + i\tilde{\hat{Q}}(-\tau^2 + c) - 2i\tau^2\,\tilde{\hat{Q}}'(\lambda)\Big|_{\lambda=-\tau^2+c} = 0. \tag{2.10}$$

Combining (2.9) with (2.10) we obtain

$$2\tilde{P}(-\tau^2+c)\left(\tilde{P}'(\lambda)\Big|_{\lambda=-\tau^2+c}\tilde{\hat{Q}}(-\tau^2+c)-\tilde{P}(-\tau^2+c)\,\tilde{\hat{Q}}'(\lambda)\Big|_{\lambda=-\tau^2+c}\right)$$
$$+\tilde{\hat{Q}}^3(-\tau^2+c)=0 \tag{2.11}$$

$$\tilde{\hat{Q}}(-\tau^2+c)\left(-2\tau\left(\tilde{P}'(\lambda)\Big|_{\lambda=-\tau^2+c}\tilde{\hat{Q}}(-\tau^2+c)-\tilde{P}(-\tau^2+c)\,\tilde{\hat{Q}}'(\lambda)\Big|_{\lambda=-\tau^2+c}\right)\cdot\right.$$
$$\left.+\tilde{\hat{Q}}^2(-\tau^2+c)\right)=0 \tag{2.12}$$

As the zeros of $\tilde{P}(\lambda)$ and $\tilde{\hat{Q}}(\lambda)$ interlace we obtain from (2.9) that $\tilde{\hat{Q}}(-\tau^2+c)\neq 0$ for $\tau>0$. Then (2.12) implies

$$-2\tau\left(\tilde{P}'(\lambda)\Big|_{\lambda=-\tau^2+c}\tilde{\hat{Q}}(-\tau^2+c)-\tilde{P}(-\tau^2+c)\,\tilde{\hat{Q}}'(\lambda)\Big|_{\lambda=-\tau^2+c}\right)+\tilde{\hat{Q}}^2(-\tau^2+c)=0,$$

what is impossible for $\tau>0$ due to Corollary 2.8.

Let now $\tau=0$. Then from (2.9) we obtain

$$\tilde{P}(c)=0. \tag{2.13}$$

Suppose $\lambda=0$ is a multiple zero. Then (2.11) implies

$$\tilde{\hat{Q}}(c)=0$$

what together with (2.13) contradicts the interlacing conditions. □

Now we are going to prove that the set of zeros of any polynomial from SHB_c belongs to SHB_κ^+ or SHB_κ^-. First we consider the zeros in the lower half-plane.

Lemma 2.14. *Let $\omega(\lambda,c)\in SHB_c$. Then all the zeros of $\omega(\lambda,c)$ in the closed lower half-plane are purely imaginary.*

Proof. Let $\lambda_0=-i\tau$, $(\tau>0)$ be a zero of $\omega(\lambda,c_0)$, where $c_0>0$. The polynomial $\omega(\lambda,c)$ is analytic in λ for c fixed and analytic in c for λ fixed. Therefore (see for example [8]), the zeros $\lambda_j(c)$ of ω are continuous and piece-wise analytic functions of c, i.e., in some neighborhood of c_0 the following is true

$$\lambda_j(c)=\lambda_0+\sum_{k=1}^{\infty}\beta_k\left((c-c_0)_j^{1/r}\right)^k,\qquad (j=\overline{1,r}), \tag{2.14}$$

where $(c-c_0)_j^{1/r}$ $(j=\overline{1,r})$ means the set of all branches of the root. Here due to the symmetry of the function each of the β_j is either real or purely imaginary. When c changes from 0 to $c_1>0$ the zeros of $\omega(\lambda,c)$ can come into the lower half-plane from infinity or crossing the real axis. The order of the polynomial $\omega(\lambda,c)$ does not depend on c. Thus, the zeros cannot come to the lower half-plane from infinity. Due to Lemma 2.12 they can cross the real axis only at the origin. Moving along the negative imaginary half-axis the zeros do not collide because at the collision a multiple zero should appear what would contradict Lemma 2.13. The zeros remain purely imaginary due to the symmetry of the problem. □

Now we are ready to state the main theorem of this section.

Theorem 2.15. *Let the polynomial $\omega(\lambda, c)$ belong to SHB_c. Then the set of its zeros belongs to SHB_κ^+ (if the order of $\omega(\lambda, c)$ is odd) or to SHB_κ^- (if the order of $\omega(\lambda, c)$ is even) with κ equal to the number of zeros of $P(\lambda)$ lying on $(0, \sqrt{c}]$ if $c > 0$ and $\kappa = 0$ if $c \leq 0$.*

Proof. Let us consider the following polynomial of two variables λ and α (c is fixed):

$$\Omega(\lambda, \alpha) = \tilde{P}(\lambda^2 + c) + i\alpha\lambda\tilde{\hat{Q}}(\lambda^2 + c).$$

It is clear that $\Omega(\lambda, 0) = \tilde{P}(\lambda^2 + c)$ and $\Omega(\lambda, 1) = \omega(\lambda, c)$.

To continue the proof of Theorem 2.15 we need the following four lemmas.

Lemma 2.16. *If $\Omega(\lambda, 1) \in SHB_c$, then $\Omega(\lambda, \alpha) \in SHB_c$ for all $\alpha > 0$.*

Proof. It is clear that if $P(\lambda)$ and $Q(\lambda)$ satisfy conditions (1) and (2) of Theorem 2.3, then $P(\lambda)$ and $\alpha Q(\lambda)$ also satisfy (1) and (2) for all $\alpha > 0$. The assertion of Lemma 2.16 follows. □

Lemma 2.17.

1) *If $\Omega(0, 0) \neq 0$ then $\Omega(0, \alpha) \neq 0$ for all $\alpha \in R$.*
2) *If $\Omega(0, 0) = 0$ then $\lambda = 0$ is a double zero of $\Omega(\lambda, 0)$ and a simple zero of $\Omega(\lambda, \alpha)$ for all $\alpha > 0$.*

Proof. The first statement follows from the identity $\Omega(0, \alpha) = \tilde{P}(c)$.

If $\Omega(0, 0) = 0$ then $\tilde{P}(c) = 0$. This zero is double because the function $\tilde{P}(\lambda^2 + c)$ is even. The simplicity of $\lambda = 0$ as the zero of $\Omega(\lambda, \alpha)$ for all $\alpha > 0$ follows from Lemma 2.13. □

Lemma 2.18. *The number of zeros of $\Omega(\lambda, \alpha)$ in the closed lower half-plane (all the zeros are simple and purely imaginary) does not depend on α for $\alpha \in (0, \infty)$.*

Proof. Let $\lambda_k(\alpha)$ be a purely imaginary zero of $\Omega(\lambda, \alpha)$. Then differentiating

$$\tilde{P}(\lambda_k^2(\alpha) + c) + i\alpha\lambda_k(\alpha)\tilde{\hat{Q}}(\lambda_k^2(\alpha) + c) = 0 \tag{2.15}$$

with respect to α we obtain

$$\lambda_k'(\alpha) = \frac{-i\lambda_k(\alpha)\tilde{\hat{Q}}(\lambda_k^2(\alpha) + c)}{2\lambda_k(\alpha)\tilde{P}'(\lambda_k^2(\alpha) + c) + i\alpha\tilde{\hat{Q}}(\lambda_k^2(\alpha) + c) + 2i\alpha\lambda_k^2(\alpha)\tilde{\hat{Q}}'(\lambda_k^2(\alpha) + c)}. \tag{2.16}$$

From (2.15) we express

$$\lambda_k(\alpha) = i\frac{\tilde{P}(\lambda_k^2(\alpha) + c)}{\alpha\tilde{\hat{Q}}(\lambda_k^2(\alpha) + c)} \tag{2.17}$$

and substitute it into (2.16) and obtain

$$\lambda_k'(\alpha) = (-i\lambda_k(\alpha)\tilde{\hat{Q}}^2(\lambda_k^2(\alpha)+c))$$
$$\Big(2\lambda_k(\alpha)(\tilde{\hat{Q}}(\lambda_k^2(\alpha)+c)\tilde{P}'(\lambda_k^2(\alpha)+c) - \tilde{P}(\lambda_k^2(\alpha)+c)\tilde{\hat{Q}}'(\lambda_k^2(\alpha)+c))$$
$$+i\alpha\tilde{\hat{Q}}^2(\lambda_k^2(\alpha)+c)\Big)^{-1}. \tag{2.18}$$

Due to Corollary 2.8

$$\tilde{\hat{Q}}(\lambda_k^2(\alpha)+c)\tilde{P}'(\lambda_k^2(\alpha)+c) - \tilde{P}(\lambda_k^2(\alpha)+c)\tilde{\hat{Q}}'(\lambda_k^2(\alpha)+c) < 0.. \tag{2.19}$$

Using (2.19) we obtain from (2.18) that if $\mathrm{Re}\lambda_k(\alpha) = 0$, $\mathrm{Im}\lambda_k(\alpha) < 0$ (we keep $\alpha > 0$), then Re $\lambda_k'(\alpha) = 0$ and Im $\lambda_k'(\alpha) > 0$. Combining this result with Lemmas 2.13, 2.16 and 2.17 we finish the proof of Lemma 2.18. □

Let us continue the proof of Theorem 2.15. If $\lambda_k(0) \neq 0$ then from (2.18) we obtain

$$\lambda_k'(0) = \frac{-i\tilde{\hat{Q}}^2(\lambda_k^2(0)+c)}{2(\tilde{\hat{Q}}(\lambda_k^2(0)+c)\tilde{P}'(\lambda_k^2(0)+c) - \tilde{P}(\lambda_k^2(0)+c)\tilde{\hat{Q}}'(\lambda_k^2(0)+c)}. \tag{2.20}$$

Due to (2.19) equation (2.20) implies Re $\lambda_k'(0) = 0$ and Im $\lambda_k'(0) > 0$.

Remark 2.19. One can prove Lemmas 2.13, 2.14 and 2.18 using the results of [9] on quadratic operator pencils.

Remark 2.20. If $\alpha > 0$ then the denominator in (2.18) is equal to zero if and only if the corresponding purely imaginary zero $\lambda_k(\alpha)$ is multiple (see the proof of Lemma 2.13).

All the other (simple) purely imaginary zeros can be classified according to the sign of the mentioned denominator. An analogue of such a classification for eigenvalues of quadratic operator pencils can be found in [10].

Definition 2.21. *The purely imaginary zero λ_k of a symmetric polynomial is said to be of the first (second) kind if*

$$-2i\lambda_k(\tilde{\hat{Q}}(\lambda_k^2+c)\tilde{P}'(\lambda_k^2+c) - \tilde{P}(\lambda_k^2+c)\tilde{\hat{Q}}'(\lambda_k^2+c)) + \tilde{\hat{Q}}^2(\lambda_k^2+c) > 0 \quad (< 0).$$

Remark 2.22. The purely imaginary zeros in the open lower half-plane are all of the first kind and according to (2.18) and (2.19) they all move upwards when $\alpha > 0$ grows. The purely imaginary zeros of the first (second) kind in the open upper half-plane move downwards (upwards).

Lemma 2.23. *If $-i\tau$ $(\tau > 0)$ is a zero of $\omega(\lambda, c) \in SHB_c$ then $i\tau$ is not a zero of it.*

Proof. Suppose the both $-i\tau$ and $i\tau$ to be zeros of $\omega(\lambda, c)$. Then

$$\tilde{P}(-\tau^2+c) + \tau\tilde{\hat{Q}}(-\tau^2+c) = 0 \quad \text{and} \quad \tilde{P}(-\tau^2+c) - \tau\tilde{\hat{Q}}(-\tau^2+c) = 0$$

and, consequently, $\tilde{P}(-\tau^2+c) = \tilde{\hat{Q}}(-\tau^2+c) = 0$, what is impossible. □

Let us continue proving Theorem 2.15. Taking into account the symmetry of the problem on reflection with respect to the imaginary axis, we have $\lambda_{-k}(\alpha) = -\overline{\lambda_k(\alpha)}$ for all not purely imaginary $\lambda_{-k}(\alpha)$ with $\alpha \geq 0$, and hence new zeros can appear on the imaginary axis in pairs, which implies the statements of Theorem 2.15.

3. Inverse problem

Let us consider the inverse theorem, i.e., we are going to prove that the condition of Theorem 2.15 is not only sufficient but also necessary.

Theorem 3.1. *Let the set* $\{\lambda_k\}_{k=-n,\ k\neq 0}^{n}$ $(\{\lambda_k\}_{k=-n}^{n})$ *belong to* SHB_κ^- (SHB_κ^+) *with* $\kappa \geq 0$ *Then the polynomial*

$$\varphi(\lambda) = \prod_{k=-n,\ k\neq 0}^{n} (\lambda - \lambda_k) \left(\varphi(\lambda) = \prod_{k=-n}^{n} (\lambda - \lambda_k)\right) \tag{3.1}$$

belongs to SHB_c, *with some* $c > 0$.

Proof. Let $\{\lambda_k\}_{k=-n,\ k\neq 0}^{n} \in SHB_\kappa^-$. Consider the auxiliary polynomial

$$\varphi_0(\lambda) = \prod_{k=1}^{n} (\lambda - \lambda_k^{(0)})(\lambda - \lambda_{-k}^{(0)}),$$

where $\lambda_k^{(0)} = k + i\epsilon$, $\epsilon > 0$. It is clear that $\varphi_0 \in SHB$.

Put

$$P_0(\lambda) = \frac{\varphi_0(\lambda) + \varphi_0(-\lambda)}{2}, \tag{3.2}$$

$$\hat{Q}_0(\lambda) = \frac{\varphi_0(\lambda) - \varphi_0(-\lambda)}{2i\lambda}. \tag{3.3}$$

Then $P_0(\lambda)$ and $\hat{Q}_0(\lambda)$ are both even polynomials. Let us introduce the following real polynomials

$$\tilde{P}_0(\lambda) = P_0(\sqrt{\lambda}), \tag{3.4}$$

$$\tilde{\hat{Q}}_0(\lambda) = \hat{Q}_0(\sqrt{\lambda}). \tag{3.5}$$

Since the zeros $\{a_k^{(0)}\}_{k=-n,\ k\neq 0}^{n}$ of $P_0(\lambda)$ and the zeros $\{b_k^{(0)}\}_{k=-n,\ k\neq 0}^{n}$ of $\hat{Q}_0(\lambda)$ interlace, we have

$$\cdots < b_{-2}^{(0)} < a_{-2}^{(0)} < b_{-1}^{(0)} < a_{-1}^{(0)} < 0 < a_1^{(0)} < b_1^{(0)} < a_2^{(0)} < b_2^{(0)} < \cdots. \tag{3.6}$$

Now we need the following proposition.

Proposition 3.2. *There exists a set of continuous and piecewise analytic functions* $\{\lambda_k(t)\}_{k=-n,\ k\neq 0}^{n}$ *such that* $\lambda_k(0) = \lambda_k^{(0)}$ *and* $\lambda_k(1) = \lambda_k$, $(k = \pm 1, \pm 2, \ldots, \pm n)$ *and*

$$\{\lambda_k(t)\}_{k=-n,\ k\neq 0}^{n} \in SHB_{\kappa(t)}^-$$

for any fixed $t \in [0,1]$ *and* $\kappa(t) \geq 0$ *depending on* t *stepwise.*

Proof. If there are not purely imaginary λ_k then there exists $\ell \in N \cup \{0\}$ such that $\operatorname{Re}\lambda_s > 0$ for $s \geq \ell+1$ and $\operatorname{Re}\lambda_s < 0$ for $s \leq -(\ell+1)$. For such λ_s we define

$$\lambda_s(t) = \lambda_s^{(0)} + (\lambda_s - \lambda_s^{(0)})t.$$

So it remains to construct the piecewise analytic functions involving the purely imaginary eigenvalues λ_s $(0 < |s| \leq \ell)$.

Now let λ_{-j} be a (purely imaginary) term in the closed lower half-plane. Consider the interval $(i|\lambda_{-j}|, i|\lambda_{-(j+1)}|)$. Then the number of terms on this interval is odd. Let us choose one of them and denote it by λ_{+j}, thus resulting in the positive imaginary, so-called principal, eigenvalues $\lambda_{+1}, \ldots, \lambda_{+\kappa}$. The remaining positive imaginary eigenvalues, which are even in number when multiplicities are taken into account, are grouped into pairs $\{\lambda_{+s}, \lambda_{-s}\}$ $(s = \kappa+1, \ldots, \ell)$ such that each of the intervals $(i|\lambda_{-j}|, i|\lambda_{-(j+1)}|)$ contains an integer number of such pairs.

Let us move the points $\lambda_s^{(0)}$ $(0 < |s| \leq \ell)$ in various steps. First, we define for $0 \leq t \leq (1/(\kappa+2))$

$$\lambda_j(t) = \begin{cases} \lambda_j^{(0)}, & 0 < |j| < |\kappa|, \\ \lambda_{\pm\kappa}^{(0)} + (\kappa+2)t\left(\frac{\lambda_{-\kappa}+\lambda_{+\kappa}}{2} - \lambda_{\pm\kappa}^{(0)}\right), & j = \pm\kappa, \\ \lambda_j^{(0)}, & |\kappa| < |j| \leq \ell. \end{cases}$$

Next, for $s = 0, \ldots, \kappa-2$ and $t \in [\frac{1+s}{\kappa+2}, \frac{2+s}{\kappa+2}]$ we define

$$\lambda_j(t) = \begin{cases} \frac{\lambda_j+\lambda_{-j}}{2} + ((\kappa+2)t - s - 1)\frac{\lambda_j-\lambda_{-j}}{2}, & j = \pm(\kappa-s), \\ \lambda_j^{(0)} + ((\kappa+2)t - s - 1)\left(\frac{\lambda_j+\lambda_{-j}}{2} - \lambda_j^{(0)}\right), & j = \pm(\kappa-s-1), \\ \lambda_j^{(0)}, & 0 < |j| < \kappa-s-1, \\ \lambda_j, & \kappa - s < |j| \leq \ell. \end{cases}$$

Next, for $t \in [\frac{\kappa}{\kappa+2}, \frac{\kappa+1}{\kappa+2}]$ we define

$$\lambda_j(t) = \begin{cases} \frac{\lambda_1+\lambda_{-1}}{2} + ((\kappa+2)t - \kappa)\frac{\lambda_{\pm1}-\lambda_{\mp1}}{2}, & j = \pm1, \\ \lambda_j, & 1 < |j| \leq \kappa, \\ \lambda_j^{(0)} + ((\kappa+2)t - \kappa)\left(\frac{\lambda_j+\lambda_{-j}}{2} - \lambda_j^{(0)}\right), & \kappa < |j| \leq \ell. \end{cases}$$

Finally, for $t \in [\frac{\kappa+1}{\kappa+2}, 1]$ we define

$$\lambda_j(t) = \begin{cases} \lambda_j, & 0 < |j| \leq \kappa, \\ \frac{\lambda_j+\lambda_{-j}}{2} + ((\kappa+2)t - \kappa - 1)\frac{\lambda_j-\lambda_{-j}}{2}, & \kappa < |j| \leq \ell, \end{cases}$$

which completes the proof. □

Continuing the proof of Theorem 3.1, let us construct the function

$$\varphi(\lambda, t) = \prod_{k=-n,\ k\neq0}^{n} (\lambda - \lambda_k(t)). \tag{3.7}$$

Then

$$\begin{aligned}\varphi(\lambda,0) &= \prod_{k=-n,\ k\neq 0}^{n} (\lambda-\lambda_k^{(0)}),\\ \varphi(\lambda) := \varphi(\lambda,1) &= \prod_{k=-n,\ k\neq 0}^{n} (\lambda-\lambda_k).\end{aligned} \tag{3.8}$$

Put

$$P(\lambda,t) = \frac{\varphi(\lambda,t)+\varphi(-\lambda,t)}{2}, \tag{3.9}$$

$$\hat{Q}(\lambda,t) = \frac{\varphi(\lambda,t)-\varphi(-\lambda,t)}{2i\lambda}, \tag{3.10}$$

and then define $P(\lambda) = P(\lambda,1)$ and $\hat{Q}(\lambda) = \hat{Q}(\lambda,1)$. Next, put

$$\tilde{P}(\lambda,t) = P(\sqrt{\lambda},t),$$

$$\tilde{\hat{Q}}(\lambda,t) = \hat{Q}(\sqrt{\lambda},t),$$

and then define $\tilde{P}(\lambda) = \tilde{P}(\lambda,1)$ and $\tilde{\hat{Q}}(\lambda) = \tilde{\hat{Q}}(\lambda,1)$.

Denote by $\{a_k(t)\}_{k=-n,\ k\neq 0}^{n}$ the set of zeros of $P(\lambda,t)$ and by $\{b_k(t)\}_{k=-n,\ k\neq 0}^{n}$ the set of zeros of $\hat{Q}(\lambda,t)$. Then $\{a_k(t)^2\}_{k=1}^{n}$ are the zeros of $\tilde{P}(\lambda,t)$ and $\{b_k(t)^2\}_{k=1}^{n}$ are the zeros of $\tilde{\hat{Q}}(\lambda,t)$.

Proposition 3.3. *For any fixed $t\in[0,1]$, the sets of zeros $\{a_k(t)^2\}_{k=1}^{n}$ and $\{b_k(t)^2\}_{k=1}^{n}$ interlace, i.e.,*

$$-\infty < a_1(t)^2 < b_1(t)^2 < a_2(t)^2 < b_2(t)^2 < \cdots < b_n(t)^2. \tag{3.11}$$

Proof of Proposition 3.3. Due to (3.6), this statement is true for $t=0$. The functions $\tilde{P}(\lambda,t)$ and $\tilde{\hat{Q}}(\lambda,t)$ are entire functions of λ for every $t\in[0,1]$ and continuous functions of $t\in[0,1]$ for every $\lambda\in C$. This means that their zeros are continuous functions of t. Therefore, inequalities (3.11) can be violated only if for some $t_1\in[0,1]$ and some k we have $b_k(t_1)^2 = a_k(t_1)^2$ or $b_k(t_1)^2 = a_{k+1}(t_1)^2$. But each of the two identities implies

$$\tilde{P}(b_k(t_1)^2,t_1) = \tilde{\hat{Q}}(b_k(t_1)^2,t_1) = 0,$$

and, consequently,

$$\varphi(b_k(t_1),t_1) = \varphi(-b_k(t_1),t_1) = 0,$$

where $b_k(t_1)$ is real or purely imaginary. Since Lemma 2.12 implies that $\varphi(\lambda,t)$ can have a real zero only at the origin for any $t\in[0,1]$, we conclude that $b_k(t_1)$ is purely imaginary or zero. Since $P(\lambda,t)$ and $\hat{Q}(\lambda,t)$ are even functions of λ, we have

$$P(b_k(t_1),t_1) = P(-b_k(t_1),t_1) = 0$$

and

$$\hat{Q}(b_k(t_1),t_1) = \hat{Q}(-b_k(t_1),t_1) = 0,$$

and consequently $\varphi(b_k(t_1),t_1)=\varphi(-b_k(t_1),t_1)=0$, which contradicts Proposition 3.2 if $b_k(t_1)\neq 0$. On the other hand, if $b_k(t_1)=0$, then it is a double zero, because it is a zero of the even functions $\tilde{P}(\lambda,t_1)$ and $\tilde{\hat{Q}}(\lambda,t_1)$ and consequently a zero of $\omega(\lambda,t_1)$, which contradicts Proposition 3.2. □

Let us continue proving Theorem 3.1. Let us choose c such that

$$a_1(1)^2>c. \tag{3.12}$$

Let us consider the polynomial

$$\omega(\lambda,c)=\tilde{P}(\lambda^2+c,1)+i\lambda\tilde{\hat{Q}}(\lambda^2+c,1).$$

The zeros of $\tilde{P}(\lambda^2+c,1)$ and of $\tilde{\hat{Q}}(\lambda^2+c,1)$ are $\{\sqrt{a_k^2(1)-c}\}_{k=-n,\ k\neq 0}^{n}$ and $\{\sqrt{a_k^2(1)-c}\}_{k=-n,k\neq 0}^{n}$, respectively. They interlace as follows

$$-\sqrt{b_{-n}^2(1)-c}<-\sqrt{a_{-n}^2(1)-c}<-\sqrt{b_{-n+1}^2(1)-c}<\cdots<-\sqrt{a_{-1}^2(1)-c}$$
$$<0<\sqrt{a_1^2(1)-c}<\cdots<\sqrt{b_n^2(1)-c}. \tag{3.13}$$

Thus, $\omega(\lambda,c)$ fulfils condition (1) of Theorem 2.3. To show the validity of condition (2) for $\omega(\lambda,c)$ we notice that

$$\left.\frac{\partial}{\partial\lambda}\lambda\tilde{\hat{Q}}(\lambda^2+c,1)\right|_{\lambda=0}=\tilde{\hat{Q}}(c,1),$$
$$\left.\lambda\tilde{\hat{Q}}(\lambda^2+c,1)\right|_{\lambda=0}=0.$$

Using these equations we obtain

$$\left.\left(\tilde{P}(\lambda^2+c,1)\frac{\partial}{\partial\lambda}\lambda\tilde{\hat{Q}}(\lambda^2+c,1)-\lambda\tilde{\hat{Q}}(\lambda^2+c,1)\frac{\partial}{\partial\lambda}\tilde{P}(\lambda^2+c,1)\right)\right|_{\lambda=0}$$
$$=\tilde{\hat{Q}}(c,1)\tilde{P}(c,1).$$

Using (3.12) and (3.13) we obtain

$$\tilde{\hat{Q}}(c,1)\tilde{P}(c,1)>0$$

and therefore

$$\left.\left(\tilde{P}(\lambda^2+c,1)\frac{\partial}{\partial\lambda}\lambda\tilde{\hat{Q}}(\lambda^2+c,1)-\lambda\tilde{\hat{Q}}(\lambda^2+c,1)\frac{\partial}{\partial\lambda}\tilde{P}(\lambda^2+c,1)\right)\right|_{\lambda=0}>0.$$

That means condition (2) of Theorem 2.3 is satisfied too. The first statement of Theorem 3.1 follows. In the same way one can prove the second statement, i.e., that if $\{\lambda_k\}_{-n}^{n}\in SHB_\kappa^+$, then

$$\varphi(\lambda)=\prod_{k=-n,\ k\neq 0}^{n}(\lambda-\lambda_k)\in SHB_c^+.$$

□

Acknowledgment

This work is partially supported by Grants UM1-2567-OD-03 and UKM2-2811-OD-06 of Civil Research and Development Foundation.

References

[1] V. Pivovarchik, C. van der Mee, *The inverse generalized Regge problem.* Inverse Problems, **17** (2000), 1831–1845.

[2] C. van der Mee, V. Pivovarchik, *A Sturm-Liouville Spectral Problem with Boundary Conditions Depending on the Spectral Parameter.* Functional Analysis and Its Applications, **36**, No. 4 (2002), 315–317.

[3] C. van der Mee, V. Pivovarchik, *The Sturm-Liouville inverse spectral problem with boundary conditions depending on the spectral parameter.* To appear in Opuscula Mathematica.

[4] M. Möller, V. Pivovarchik, *Spectral properties of a forth order differential equation.* To appear in Zeitschrift für Analysis und ihre Anwendungen.

[5] C. Hermite, *Extract d'une lettre de M. Ch. Hermite de Paris à Mr. Borchardt de Berlin sur le nombre des racines d'une équation algébrique comprises entre des limites données.* J. Reine Angew. Math. **52** (1856), 39–51; reprinted in his Oeuvres, Vol. 1, Gauthier-Villars, Paris, 1905, 397–414.

[6] F.R. Gantmaher, *Theory of Matrices (in Russian).* Nauka, Moscow, 1988.

[7] B.Ja. Levin, *Distribution of Zeros of Entire Functions, Trans. Math. Monographs.* **5**, Amer. Math. Soc., Providence, RI, 1980.

[8] A.I. Markushevich, *Theory of Analytic Functions, (in Russian)*, **1**, Nauka, Moscow, 1968.

[9] V. Pivovarchik. *On Positive Spectra of One Class of Polynomial Operator Pencils.* Integral Equations and Operator Theory, **19** (1994), 314–326.

[10] A.G. Kostyuchenko, M.B. Orazov, *The problem of oscillations of an elastic half-cylinder and related self-adjoint quadratic pencil (in Russian).* Trudy Sem. Petrovsk., N6 (1981), 97–147.

Vyacheslav Pivovarchik
Preobrazhenskaya str. 59/61, a.17
65045 Odessa, Ukraine
e-mail: v.pivovarchik@paco.net

Operator Theory:
Advances and Applications, Vol. 175, 225–229

A Note on Indefinite Douglas' Lemma

Leiba Rodman

Abstract. The Douglas lemma on majorization and factorization of Hilbert space operators is extended to the setting of Krein space operators.

Mathematics Subject Classification (2000). 47A68, 47A63, 46C20.

Keywords. Majorization, factorization, Hilbert space operators, Krein spaces.

1. Introduction and main results

We let $\mathcal{E}$, $\mathcal{F}$, $\mathcal{G}$ be (complex) Hilbert spaces, with the inner products $(x,y)_{\mathcal{E}}$, $(x,y)_{\mathcal{F}}$, and $(x,y)_{\mathcal{G}}$, respectively, and let $\mathcal{L}(\mathcal{E},\mathcal{F})$ denote the Banach space of all linear bounded operators acting from $\mathcal{E}$ into $\mathcal{F}$.

The celebrated Douglas lemma, or more precisely one of its several versions, states that if $H \in \mathcal{L}(\mathcal{E},\mathcal{F})$ and $G \in \mathcal{L}(\mathcal{E},\mathcal{G})$ are such that $H^*H \geq G^*G$, in other words, $H^*H - G^*G$ is positive semidefinite, then there exists an operator $F \in \mathcal{L}(\mathcal{F},\mathcal{G})$ satisfying $Gx = FHx$ for every $x \in \mathcal{F}$. Although generally speaking the operator F is not unique, it can be always chosen so that $\|F\| \leq 1$. The original reference for the Douglas lemma is [2]. It has been extensively used in operator theory, in particular in studies of division and quotients of operators (see [8, 4]), operator range inclusions [3], and operator inequalities [5]; see, e.g., [6] for connections with lifting and Leech's theorem. A survey of results around the Douglas lemma in the context of Banach spaces is given in a recent paper [1].

In this note we extend the Douglas lemma to the setting of Krein space operators. Yu. L. Šmul'jan [7, 9] was the first who obtained some interesting results in this direction; in particular, he pointed out that generally the Douglas lemma does not hold verbatim in Krein spaces, even the finite-dimensional ones.

In what follows, for a bounded selfadjoint operator X, we denote by $i_-(X)$ the dimension (finite or infinite) of the spectral X-invariant subspace corresponding to the negative part of $\sigma(X)$.

Let $J_{\mathcal{E}} \in \mathcal{L}(\mathcal{E},\mathcal{E})$, $J_{\mathcal{F}} \in \mathcal{L}(\mathcal{F},\mathcal{F})$, $J_{\mathcal{G}} \in \mathcal{L}(\mathcal{G},\mathcal{G})$ be operators that are simultaneously selfadjoint and unitary, and such that $i_-(J_{\mathcal{G}}) < \infty$. The operators $J_{\mathcal{E}}$, $J_{\mathcal{F}}$,

Now define $F = F_0$ on the closure of $\{Hy \,|\, y \in \mathcal{N}\}$, and define F as in (2.7) on $\{Hy \,|\, y \in \text{Span}\,\{\xi_1, \dots, \xi_p\}\}$. The definition is correct, because, as one checks easily,

$$\overline{\{Hy \,|\, y \in \mathcal{N}\}} \cap \{Hy \,|\, y \in \text{Span}\,\{\xi_1, \dots, \xi_p\}\} = \{0\}.$$

It follows that we have $G = FH$ on $\mathcal{E}$. □

3. Proofs of Theorems 1.1 and 1.2.

We start with Theorem 1.1. Using the easily verifiable equalities

$$H^{*J_{\mathcal{F}},J_{\mathcal{E}}} = J_{\mathcal{E}}^{-1} H^* J_{\mathcal{F}}, \quad G^{*J_{\mathcal{G}},J_{\mathcal{E}}} = J_{\mathcal{E}}^{-1} G^* J_{\mathcal{G}},$$

the condition that the operator Q given by the left-hand side of (1.1) is $J_{\mathcal{E}}$-nonnegative takes the form

$$H^* J_{\mathcal{F}} H - G^* J_{\mathcal{G}} G + J_{\mathcal{E}} V \geq 0.$$

Rewrite this inequality:

$$H^* H + G^* G \geq -J_{\mathcal{E}} V + H^*(I - J_{\mathcal{F}})H - G^*(I - J_{\mathcal{G}})G.$$

Note that $H^*(I - J_{\mathcal{F}})H \geq 0$ and

$$i_-(-J_{\mathcal{E}} V - G^*(I - J_{\mathcal{G}})G) \leq \operatorname{rank} V + i_-(G^*(J_{\mathcal{G}} - I)G) \leq \operatorname{rank} V + i_-(J_{\mathcal{G}}).$$

Thus,

$$i_-(H^* H + G^* G) \leq \operatorname{rank} V + i_-(J_{\mathcal{G}}),$$

and it remains to apply Lemma 2.1. □

The proof of Theorem 1.2 follows the same line of argument, using Lemma 2.2.

Acknowledgments

The author is grateful to V. Bolotnikov for constructive discussions concerning the subject matter of this note, and to T. Azizov for useful suggestions and for pointing out the references [7, 9]. The research is partially supported by NSF Grant DMS-0456625.

References

[1] A.B. Barnes. Majorization, range inclusion, and factorization for bounded linear operators. *Proc. Amer. Math. Soc.* 133 (2005), 155–162.

[2] R.G. Douglas. On majorization, factorization, and range inclusion of operators on Hilbert space. *Proc. Amer. Math. Soc.* 17 (1966), 413–415.

[3] P.A. Fillmore and J.P. Williams. On operator ranges. *Advances in Math.* 7 (1971), 254–281.

[4] S. Izumino. Quotients of bounded operators. *Proc. Amer. Math. Soc.* 106 (1989), 427–435.

[5] C.-S. Lin. On Douglas's majorization and factorization theorem with applications. *Int. J. Math. Sci.* 3 (2004), 1–11.

[6] M. Rosenblum and J. Rovnyak. *Hardy classes and operator theory.* Oxford University Press, New York, 1985.

[7] Ju.L. Šmul'jan. Non-expanding operators in a finite-dimensional space with an indefinite metric. *Uspehi Mat. Nauk*, 18 (1963), 225–230. (Russian.)

[8] Ju.L. Šmul'jan. Two-sided division in the ring of operators. *Mat. Zametki*, 1 (1967), 605–610. (Russian).

[9] Ju.L. Šmul'jan. Division in the class of J-expansive operators. *Matem. Sbornik (N.S.)*, 74 (116) (1967), 516–525. (Russian.)

Leiba Rodman
Department of Mathematics
The College of William and Mary
Williamsburg VA 23187-8795, USA
e-mail: `lxrodm@math.wm.edu`

Operator Theory:
Advances and Applications, Vol. 175, 231–240

Some Basic Properties of Polynomials in a Linear Relation in Linear Spaces

Adrian Sandovici

Abstract. The behavior of the domain, the range, the kernel and the multivalued part of a polynomial in a linear relation is analyzed, respectively.

Mathematics Subject Classification (2000). Primary 47A05; Secondary 47A06.

Keywords. Polynomial, linear space, linear relation, domain, range, kernel, multivalued part.

1. Introduction

Let S be a linear relation in a linear space $\mathfrak{H}$ over the field $\mathbb{K}$ of real or complex numbers, let n and m_i, $1 \leq i \leq n$ be some positive integers, and let $\lambda_i \in \mathbb{K}$, $1 \leq i \leq n$ be some distinct constants. Then, the polynomial p in S given by

$$p(S) = \prod_{i=1}^{n} (S - \lambda_i)^{m_i} \tag{1.1}$$

is a linear relation in $\mathfrak{H}$ too. Certain properties of polynomials in a linear relation have been obtained long time ago (see for instance [1, 4, 5]). However, it seems that the behavior of the domain, the range, the kernel and the multivalued part of $p(S)$ has not yet been described, cf. [3]. It is the goal of this technical note to cover this gap. For instance, if α and β are two constants and p and q are two positive integers, then the results in this note show that

$$\operatorname{dom}(S-\alpha)^p(S-\beta)^q = \operatorname{dom} S^{p+q}, \tag{1.2}$$

and

$$\operatorname{mul}(S-\alpha)^p(S-\beta)^q = \operatorname{mul} S^{p+q}. \tag{1.3}$$

If additionally, $\alpha \neq \beta$ then also

$$\operatorname{ran}(S-\alpha)^p(S-\beta)^q = \operatorname{ran}(S-\alpha)^p \cap \operatorname{ran}(S-\beta)^q, \tag{1.4}$$

3. Main results

This section contains the main results of this note. They represent the generalization of (1.2)–(1.5) to the case of a polynomial in a linear relation S in a linear space $\mathfrak{H}$.

Lemma 3.1. *Let S be a linear relation in a linear space $\mathfrak{H}$ and let $\lambda \in \mathbb{K}$. Then*

$$\operatorname{dom}(S-\lambda)^n = \operatorname{dom} S^n \quad \text{for all} \quad n \in \mathbb{N}. \tag{3.1}$$

Proof. If $\lambda = 0$ then (3.1) is obvious. Assume that $\lambda \neq 0$ and proceed by induction. The case $n = 1$ is clear. Assume now that $n > 1$ and let $x_0 \in \operatorname{dom}(S-\lambda)^n$, so that $\{x_i, x_{i+1}\} \in S - \lambda$ for some $x_i \in \mathfrak{H}$, $0 \leq i \leq n-1$. Define the vectors y_p, $0 \leq p \leq n$ by $y_p = \sum_{j=1}^{p} \binom{p}{j} \lambda_j x_{p-j}$, $\quad 0 \leq p \leq n$, so that

$$\{y_p, y_{p+1}\} = \sum_{j=0}^{p} \binom{p}{j} \lambda^{p-j} \{x_j, x_{j+1} + \lambda x_j\} \in S, \quad 0 \leq p \leq n,$$

which implies that $\{y_0, y_n\} \in S^n$. Hence, $x_0 = y_0 \in \operatorname{dom} S^n$, which shows that $\operatorname{dom}(S-\lambda)^n \subset \operatorname{dom} S^n$. The converse inclusion follows using similar arguments. □

Theorem 3.2. *Let S be a linear relation in a linear space $\mathfrak{H}$, let $n \in \mathbb{N}$, $\lambda_j \in \mathbb{K}$, $m_j \in \mathbb{N}$, $1 \leq j \leq n$. Assume that λ_j, $1 \leq j \leq n$ are distinct. Then*

$$\operatorname{dom} p(S) = \operatorname{dom} S^{\sum_{j=1}^{n} m_j}. \tag{3.2}$$

Proof. Proceed by induction. The case $n = 1$ follows from Lemma 3.1. Assume that (3.2) holds true for some $n = k \geq 1$ and denote $M_k = \sum_{j=1}^{k} m_j$. Let $x_0 \in \operatorname{dom} \prod_{j=1}^{k+1} (S-\lambda_j)^{m_j}$, so that $\{x_0, y_0\} \in (S-\lambda_{k+1})^{m_{k+1}}$ and $\{y_0, z_0\} \in \prod_{j=1}^{k} (S-\lambda_j)^{m_j}$ for some y_0, $z_0 \in \mathfrak{H}$. By induction hypothesis it follows that $y_0 \in \operatorname{dom} S^{M_k}$, so that $\{y_j, y_{j+1}\} \in S$, $0 \leq j \leq M_k - 1$ for some $y_i \in \mathfrak{H}$, $1 \leq j \leq M_k$. It follows from $\{x_0, y_0\} \in (S-\lambda_{k+1})^{m_{k+1}}$ that

$$\{x_p, x_{p+1}\} \in S - \lambda_{k+1}, \quad 0 \leq p \leq m_{k+1} - 1$$

for some $x_p \in \mathfrak{H}$, $1 \leq p \leq m_{k+1}$, with $x_{m_{k+1}} = y_0$. This implies that

$$\{x_p, x_{p+1} + \lambda_{k+1} x_p\} \in S, \quad 0 \leq p \leq m_{k+1} - 1.$$

Furthermore, define the vectors $w_p \in \mathfrak{H}$, $0 \leq p \leq m_{k+1}$ by

$$w_p = \sum_{j=0}^{p} \binom{p}{j} \lambda_{k+1}^{j} x_{p-j}, \quad 0 \leq p \leq m_{k+1},$$

so that

$$\{w_p, w_{p+1}\} = \sum_{j=0}^{p} \binom{p}{j} \lambda_{k+1}^{p-j} \{x_j, x_{j+1} + \lambda_{k+1} x_j\} \in S, \quad 0 \leq p \leq m_{k+1} - 1.$$

Define also the vectors $w_{m_{k+1}+q} \in \mathfrak{H}$, $1 \leq q \leq M_k$ by

$$w_{m_{k+1}+q} = \sum_{j=0}^{m_{k+1}-1} \binom{m_{k+1}+q}{j} \lambda_{k+1}^{m_{k+1}+q-j} x_j + \sum_{i=0}^{q} \binom{m_{k+1}-1+q-i}{m_{k+1}-1} \lambda_{k+1}^{q-i} y_i,$$

so that

$$\begin{aligned}\{w_{m_{k+1}+q}, w_{m_{k+1}+q+1}\} = &\sum_{j=0}^{m_{k+1}-1} \binom{m_{k+1}+q}{j} \lambda_{k+1}^{m_{k+1}+q-j} \{x_j, x_{j+1}+\lambda_{k+1}x_j\} \\ &+ \sum_{i=0}^{q} \binom{m_{k+1}-1+q-i}{m_{k+1}-1} \lambda_{k+1}^{q-i} \{y_i, y_{i+1}\} \in S\end{aligned}$$

for all $0 \leq q \leq M_k - 1$. Hence, $\{w_0, w_{m_{k+1}+M_k}\} \in S^{m_{k+1}+M_k}$, which leads to $x_0 = w_0 \in \operatorname{dom} S^{\sum_{j=1}^{k+1} m_k}$. This shows that

$$\operatorname{dom} \prod_{j=1}^{k+1} (S-\lambda_j)^{m_j} \subset \operatorname{dom} S^{\sum_{j=1}^{k+1} m_j}.$$

Thus, the induction is complete. The converse inclusion will be next proved. Denote $M = \sum_{j=1}^{n} m_j$ and let μ_j, $1 \leq j \leq M$ be the λ_i's counted with their multiplicities, so that $\prod_{j=1}^{M}(S-\mu_j) = \prod_{j=1}^{n}(S-\lambda_j)^{m_j}$. Define the constants $\eta_p^{q_p}$, $0 \leq p \leq M$, $0 \leq q_p \leq p$ by $\eta_0^0 = 1$ for $p = 0$, and

$$\eta_p^p = 1, \quad \eta_p^{p-q_p} = (-1)^{q_p} \sum_{1 \leq i_1 < \cdots < i_{q_p} \leq p} \mu_{i_1} \cdots \mu_{i_{q_p}}, \quad 1 \leq q_p \leq p$$

for $p \geq 1$. Let now $x_0 \in \operatorname{dom} S^M$, so that $\{x_p, x_{p+1}\} \in S$, $0 \leq p \leq M-1$ for some vectors $x_p \in \mathfrak{H}$, $1 \leq p \leq M$. Define the vectors $y_p \in \mathfrak{H}$, $0 \leq p \leq M$ by $y_p = \sum_{j=0}^{p} \eta_p^j x_j$, so that $y_0 = x_0$. Let p be arbitrarily chosen between 0 and $M-1$. Now it is easily seen that $\{y_p, y_{p+1}\} \in S - \mu_{p+1}$, $0 \leq p \leq M-1$, which implies that $\{y_0, y_M\} \in \prod_{j=1}^{M}(S-\mu_j)$. Therefore $x_0 = y_0 \in \operatorname{dom} \prod_{j=1}^{M}(S-\mu_j) = \operatorname{dom} \prod_{i=1}^{n}(S-\lambda_i)^{m_i}$, which completes the proof. □

Theorem 3.3. *Let S be a linear relation in a linear space $\mathfrak{H}$, let $n \in \mathbb{N}$, $\lambda_j \in \mathbb{K}$, $m_j \in \mathbb{K}$, $1 \leq j \leq n$. Assume that λ_j, $1 \leq j \leq n$ are distinct. Then*

$$\operatorname{ran} p(S) = \bigcap_{i=1}^{n} \operatorname{ran}(S-\lambda_i)^{m_i}. \tag{3.3}$$

Proof. In order to prove that

$$\operatorname{ran} \prod_{i=i}^{n} (S-\lambda_i)^{m_i} \subset \bigcap_{i=1}^{n} \operatorname{ran}(S-\lambda_i)^{m_i}, \tag{3.4}$$

let $y \in \operatorname{ran} \prod_{i=i}^{n}(S-\lambda_i)^{m_i}$, and let j be arbitrarily chosen between 1 and n. Applying Corollary 2.1 it follows that $y \in \operatorname{ran}(S-\lambda_j)^{m_j} \prod_{i=1,\, i\neq j}^{n}(S-\lambda_i)^{m_i}$. By (2.2) it follows that $y \in \operatorname{ran}(S-\lambda_j)^{m_j}$ for all $1 \leq j \leq n$. Thus, (3.4) holds true.

The converse inclusion in (3.3) will be next proved by induction on n. The case $n = 1$ is obvious. Assume that the inclusion

$$\bigcap_{i=1}^{n} \mathrm{ran}\,(S-\lambda_i)^{m_i} \subset \mathrm{ran} \prod_{i=i}^{n} (S-\lambda_i)^{m_i} \tag{3.5}$$

holds for some $n = k$ and it will be proved for $n = k+1$. For simplicity denote $\lambda = \lambda_{k+1}$ and $m = m_{k+1}$. Assume that $y \in \bigcap_{i=1}^{k+1} \mathrm{ran}\,(S-\lambda_i)^{m_i}$, so that

$$y \in \bigcap_{i=1}^{k} \mathrm{ran}\,(S-\lambda_i)^{m_i} \bigcap \mathrm{ran}\,(S-\lambda)^{m} \subset \mathrm{ran} \prod_{i=i}^{k} (S-\lambda_i)^{m_i} \bigcap \mathrm{ran}\,(S-\lambda)^{m}.$$

Since $y \in \mathrm{ran}\,(S-\lambda)^m$ it follows that

$$\{z_{i+1}, z_i\} \in S-\lambda, \quad 0 \le i \le m-1$$

for some vectors $z_i \in \mathfrak{H}$, $0 \le i \le m$ with $z_0 = y$. Also, $y \in \mathrm{ran} \prod_{i=1}^{k} (S-\lambda_i)^{m_i}$ implies that

$$\{x_i, x_{i+1}\} \in (S-\lambda_i)^{m_i}, \quad 1 \le i \le k, \tag{3.6}$$

for some vectors $x_i \in \mathfrak{H}$, $1 \le i \le k+1$, with $x_{k+1} = y$. Furthermore, it follows from (3.6) that

$$\{x_{p,q_p}, x_{p,q_p-1}\} \in S-\lambda_p, \quad 1 \le p \le k, \quad 1 \le q_p \le m_p \tag{3.7}$$

for some vectors $x_{p,q_p} \in \mathfrak{H}$, with $x_{k,0} = y$, $x_{p,m_p} = x_p$, $1 \le p \le k$, and $x_{p-1,0} = x_{p,m_p}$, $2 \le p \le k$. Define the constants ξ_t^{p,q_p}, $1 \le p \le k$, $1 \le q_p \le m_p$, $1 \le t \le m$ by

$$\xi_t^{1,q_1} = (-1)^t \binom{m_1 - q_1 + t - 1}{t-1} (\lambda - \lambda_1)^{-(m_1 - q_1 + t)},$$

for $p = 1$, and

$$\xi_t^{p,q_p} = (-1)^t \prod_{j=1}^{p-1} (\lambda-\lambda_j)^{-m_j} (\lambda-\lambda_p)^{-(m_p+1-q_p)} \sum_{l=0}^{t-1} \binom{m_p - q_p + l}{l} \psi_{t,l}^{p} (\lambda-\lambda_p)^{-l}$$

for $p > 1$, where

$$\psi_{t,l}^{p} = \sum_{\sum_{h=1}^{p-1} r_h = t-1-l} \prod_{j=1}^{p-1} \binom{m_j - 1 + r_j}{r_j} (\lambda-\lambda_j)^{-r_j}, \quad 0 \le l \le t-1.$$

Define also the constants η_t^s, $1 \le s \le t \le m$ by

$$\eta_t^s = (-1)^{t+s} \prod_{j=1}^{k} (\lambda-\lambda_j)^{-m_j} \sum_{\sum_{h=1}^{k} r_h = t-s} \prod_{j=1}^{k} \binom{m_j - 1 + r_j}{r_j} (\lambda-\lambda_j)^{-r_j},$$

and define the vectors $w_t \in \mathfrak{H}$, $0 \le t \le m$ by $w_0 = x_{1,m_1}$, and

$$w_t = \sum_{p=1}^{k} \sum_{q_p=1}^{m_p} \xi_t^{p,q_p} x_{p,q_p} + \sum_{s=1}^{t} \eta_t^s z_s, \quad 1 \le t \le m.$$

By a straightforward computation it follows that

$$\{w_t, w_{t-1}\} = \sum_{p=1}^{k} \sum_{q_p=1}^{m_p} \xi_t^{p,q_p} \{x_{p,q_p}, x_{p,q_p-1} - (\lambda - \lambda_p) x_{p,q_p}\} + \sum_{s=1}^{t} \eta_t^s \{z_s, z_{s-1}\} \in S - \lambda$$

for all t between 1 and m. Therefore, $\{w_m, w_0\} \in (S-\lambda)^m$, so that $\{w_m, x_{1,m_1}\} \in (S-\lambda)^m$. Since $\{x_{1,m_1}, y\} \in \prod_{i=1}^{k}(S-\lambda_i)^{m_i}$ it follows that

$$\{w_m, y\} \in \left(\prod_{i=1}^{k}(S-\lambda_i)^{m_i}\right)(S-\lambda)^m.$$

Thus, $y \in \operatorname{ran} \prod_{i=1}^{k+1}(S-\lambda_i)^{m_i}$ and the proof is now complete. □

Theorem 3.4. *Let S be a linear relation in a linear space $\mathfrak{H}$, let $n \in \mathbb{N}$, $\lambda_j \in \mathbb{K}$, $m_j \in \mathbb{K}$, $1 \le j \le n$. Assume that λ_j, $1 \le j \le n$ are distinct. Then*

$$\ker p(S) = \sum_{i=i}^{n} \ker (S-\lambda_i)^{m_i}. \tag{3.8}$$

Proof. To prove the inclusion "$\supset$" in (3.8) let $x \in \sum_{i=i}^{n} \ker (S-\lambda_i)^{m_i}$, so that $x = \sum_{i=1}^{n} x_i$, $1 \le i \le n$ for some $x_i \in \ker (S-\lambda_i)^{m_i}$. It follows using (2.3) that

$$x_i \in \ker (S-\lambda_i)^{m_i} \prod_{j=1,\, j\neq i}^{n} (S-\lambda_j)^{m_j} = \ker \prod_{j=1}^{n}(S-\lambda_j)^{m_j}.$$

Hence $x = \sum_{i=1}^{n} x_i \in \ker \prod_{j=1}^{n}(S-\lambda_j)^{m_j}$.

The inclusion "$\subset$" in (3.8) will be next proved by induction on n. The case $n=1$ is obvious. Assume that the inclusion holds true for some $n = k \ge 1$ and let $x_0 \in \ker \prod_{i=1}^{k+1}(S-\lambda_i)^{m_i}$. Without lose of generality assume that $m_i \le m_{k+1}$ for all $1 \le i \le k$. For simplicity denote $m = m_{k+1}$ and $\lambda = \lambda_{k+1}$. Since

$$\{x_0, 0\} \in \prod_{i=1}^{k+1}(S-\lambda_i)^{m_i} = \left(\prod_{i=1}^{k}(S-\lambda_i)^{m_i}\right)(S-\lambda)^m,$$

it follows that $\{x_0, y_0\} \in (S-\lambda)^m$, and $\{y_0, 0\} \in \prod_{i=1}^{k}(S-\lambda_i)^{m_i}$ for some vector $y_0 \in \mathfrak{H}$, which further implies that $\{x_i, x_{i+1}\} \in S-\lambda$, $0 \le i \le m-1$ for some vectors $x_i \in \mathfrak{H}$, $1 \le i \le m$, with $x_m = y_0$. Since $y_0 \in \ker \prod_{i=1}^{k}(S-\lambda_i)^{m_i}$ it follows that $y_0 = \sum_{i=1}^{k} y_{i0}$, $1 \le i \le k$ for some vectors $y_{i0} \in \ker (S-\lambda_i)^{m_i}$, $1 \le i \le k$. Since $m_i \le m$, it follows that $\ker (S-\lambda_i)^{m_i} \subset \ker (S-\lambda_i)^m$, which implies that

$$\{y_{i,j_i}, y_{i,j_i+1}\} \in S-\lambda_i, \quad 1 \le i \le k, \quad 0 \le j_i \le m-1$$

for some vectors $y_{i,j_i} \in \mathfrak{H}$ with $y_{i,m} = 0$. Thus

$$\{y_{i,j_i}, y_{i,j_i+1} - (\lambda-\lambda_i) y_{i,j_i}\} \in S-\lambda, \quad 1 \le i \le k, \quad 0 \le j_i \le m-1.$$

Define now the vectors $w_p \in \mathfrak{H}$, $0 \le p \le m-1$ by

$$w_p = \sum_{j=p}^{m-1} x_j + \sum_{i=1}^{k}\sum_{j=0}^{m-1}\sum_{q=0}^{m-p-1} (-1)^q \binom{j+g}{q} (\lambda - \lambda_i)^{-(j+q+1)} y_{ij},$$

and define $w_m = 0$. By a straightforward computation it follows that

$$\begin{aligned}\{w_p, w_{p+1}\} &= \sum_{j=p}^{m-1} \{x_j, x_{j+1}\} \\ &+ \sum_{i=1}^{k}\sum_{j=0}^{m-1}\sum_{q=0}^{m-p-1} (-1)^q \binom{j+g}{q} (\lambda - \lambda_i)^{-(j+q+1)} \{y_{ij}, y_{i,j_i+1} - (\lambda - \lambda_i) y_{i,j_i}\},\end{aligned}$$

so that $\{w_p, w_{p+1}\} \in S - \lambda$ for all $0 \le p \le m-1$. Since $w_m = 0$ it follows that $w_p \in \ker\,(S-\lambda)^{m-p} \subset \ker\,(S-\lambda)^m$ for all $0 \le p \le m-1$, which implies inductively that $x_p \in \ker\,(S-\lambda)^{m-p} + \sum_{i=1}^{k} \ker\,(S-\lambda_i)^{m_i}$. Therefore,

$$x_0 \in \ker\,(S-\lambda)^m + \sum_{i=1}^{k} \ker\,(S-\lambda_i)^{m_i} = \sum_{i=1}^{k+1} \ker\,(S-\lambda_i)^{m_i},$$

which shows that the induction's step holds true. This completes the proof. □

Lemma 3.5. *Let S be a linear relation in a linear space $\mathfrak{H}$, and let $\lambda \in \mathbb{K}$. Then*

$$\operatorname{mul}(S-\lambda)^n = \operatorname{mul} S^n, \quad n \in \mathbb{N}. \tag{3.9}$$

Proof. The case $\lambda = 0$ is trivial so assume that $\lambda \neq 0$. The proof is done by induction.

If $n = 1$ let $x \in \operatorname{mul}(S-\lambda)$, so that $\{0, x\} \in S - \lambda$, which implies that $\{0, x\} \in S$ and then $\operatorname{mul}(S-\lambda) \subset \operatorname{mul} S$. The converse inclusion, i.e., $\operatorname{mul} S \subset \operatorname{mul}(S-\lambda)$ follows using a similar argument, so that the case $n = 1$ is clear.

Assume next that (3.9) holds for some $n = k \ge 1$ and let $x \in \operatorname{mul}(S-\lambda)^{k+1}$, so that $\{0, x\} \in (S-\lambda)^{k+1}$. Then $\{0, y\} \in (S-\lambda)^k$ and $\{y, x\} \in S - \lambda$ for some $y \in \mathfrak{H}$. By induction hypothesis it follows that $y \in \operatorname{mul} S^k$. Furthermore, it follows from $\{y, x\} \in S - \lambda$ that $\{y, x + \lambda y\} \in S$, which together with the fact that $y \in \operatorname{mul} S^k$ implies that $x + \lambda y \in \operatorname{mul} S^{k+1}$. Hence,

$$x = (x + \lambda y) - \lambda y \in \operatorname{mul} S^{k+1} + \operatorname{mul} S^k = \operatorname{mul} S^{k+1}.$$

Therefore, $\operatorname{mul}(S-\lambda)^{k+1} \subset \operatorname{mul} S^{k+1}$. Similar arguments show that the inclusion $\operatorname{mul} S^{k+1} \subset \operatorname{mul}(S-\lambda)^{k+1}$ holds also true. This completes the proof. □

Theorem 3.6. *Let S be a linear relation in a linear space $\mathfrak{H}$, let $n \in \mathbb{N}$, $\lambda_j \in \mathbb{K}$, $m_j \in \mathbb{N}$, $1 \le j \le n$. Assume that λ_j, $1 \le j \le n$ are distinct. Then*

$$\operatorname{mul} p(S) = \operatorname{mul} S^{\sum_{j=1}^{n} m_j}. \tag{3.10}$$

Proof. The proof is done by induction. The case $n = 1$ follows from Lemma 3.5. Assume now that (3.10) holds for some $n = k \geq 1$ and let $x_0 \in \operatorname{mul} \prod_{j=1}^{k+1}(S - \lambda_j)^{m_j}$, so that

$$\{0, y_0\} \in \prod_{j=1}^{k}(S - \lambda_j)^{m_j}, \quad \{y_0, x_0\} \in (S - \lambda_{k+1})^{m_{k+1}} \tag{3.11}$$

for some $y_0 \in \mathfrak{H}$. By induction hypothesis it follows that $y_0 \in \operatorname{mul} S^{M_k}$ where $M_k = \sum_{j=1}^{k} m_j$. Furthermore, the second relation in (3.11) implies that there exist the vectors $y_j \in \mathfrak{H}$, $1 \leq j \leq m_{k+1}$, with $y_{m_{k+1}} = x_0$, such that

$$\{y_j, y_{j+1}\} \in S - \lambda_{k+1}, \quad 0 \leq j \leq m_{k+1} - 1,$$

which implies that

$$\{y_j, y_{j+1} + \lambda_{k+1} y_j\} \in S, \quad 0 \leq j \leq m_{k+1} - 1. \tag{3.12}$$

Define the vectors $z_p \in \mathfrak{H}$, $0 \leq p \leq m_{k+1}$ by

$$z_p = \sum_{j=0}^{p} \binom{p}{j} \lambda_{k+1}^{j} y_{p-j}, \quad 0 \leq p \leq m_{k+1}. \tag{3.13}$$

It follows from (3.12) that $z_0 = y_0$ and also that

$$\{z_p, z_{p+1}\} = \sum_{j=0}^{p} \binom{p}{j} \lambda_{k+1}^{p-j} \{y_j, y_{j+1} + \lambda_{k+1} y_j\} \in S, \quad 0 \leq p \leq m_{k+1} - 1,$$

so that $z_p \in \operatorname{mul} S^{M_k+p}$ for all $0 \leq p \leq m_{k+1}$, which implies that $y_p \in \operatorname{mul} S^{M_k+p}$ for all $0 \leq p \leq m_{k+1}$. In particular, $x_0 = y_{m_{k+1}} \in \operatorname{mul} S^{\sum_{j=i}^{k+1} m_j}$. Therefore,

$$\operatorname{mul} \prod_{j=1}^{k+1}(S - \lambda_j)^{m_j} \subset \operatorname{mul} S^{\sum_{j=1}^{k+1} m_j}.$$

Conversely, let $x_0 \in \operatorname{mul} \prod_{j=1}^{k+1} S^{m_j}$, so that

$$\{0, u_0\} \in \prod_{j=1}^{k} S^{m_j}, \quad \{u_0, x_0\} \in S^{m_{k+1}}, \tag{3.14}$$

for some $u_0 \in \mathfrak{H}$. By induction hypothesis it follows that $u_0 \in \operatorname{mul} \prod_{j=1}^{k}(S-\lambda_j)^{m_j}$. Furthermore, the second relation in (3.14) implies that there exist the vectors $u_j \in \mathfrak{H}$, $1 \leq j \leq m_{k+1}$, with $u_{m_{k+1}} = x_0$, such that

$$\{u_j, u_{j+1}\} \in S, \quad 0 \leq j \leq m_{k+1} - 1,$$

which implies that

$$\{u_j, u_{j+1} - \lambda_{k+1} y_j\} \in S - \lambda_{k+1}, \quad 0 \leq j \leq m_{k+1} - 1. \tag{3.15}$$

Define the vectors $w_p \in \mathfrak{H}$, $0 \leq p \leq m_{k+1}$ by

$$w_p = \sum_{j=0}^{p} \binom{p}{j} (-\lambda_{k+1})^j y_{p-j}, \quad 0 \leq p \leq m_{k+1}. \tag{3.16}$$

It follows from (3.15) that $w_0 = u_0$ and also that

$$\{w_p, w_{p+1}\} = \sum_{j=0}^{p} \binom{p}{j} (-\lambda_{k+1})^{p-j} \{u_j, u_{j+1} - \lambda_{k+1} u_j\} \in S - \lambda_{k+1}$$

for all $0 \leq p \leq m_{k+1} - 1$, so that $w_p \in \operatorname{mul}(S - \lambda_{k+1})^p \prod_{j=1}^{k} (S - \lambda_j)$, for all $0 \leq p \leq m_{k+1}$, which implies by (2.3) that $u_p \in \operatorname{mul}(S - \lambda_{k+1})^p \prod_{j=1}^{k} (S - \lambda_j)$, for all $0 \leq p \leq m_{k+1}$. In particular, $x_0 = u_{m_{k+1}} \in \operatorname{mul} \prod_{j=1}^{k+1} (S - \lambda_j)^{m_j}$, so that

$$\operatorname{mul} S^{\sum_{j=1}^{k+1} m_j} \subset \operatorname{mul} \prod_{j=1}^{k+1} (S - \lambda_j)^{m_j}.$$

This completes the proof. □

Remark 3.7. Let S be a linear relation in a linear space $\mathfrak{H}$, let $\lambda \in \mathbb{K}$ and let $p, k \in \mathbb{N} \cup \{0\}$. A combination of (2.6), (3.1) and (3.9) leads to

$$\ker (A - \lambda)^p \subset \operatorname{dom} A^k, \quad \operatorname{mul} A^p \subset \operatorname{ran} (A - \lambda)^k.$$

References

[1] R. Arens, *Operational calculus of linear relations.* Pacific J. Math. **11** (1961), 9–23.

[2] T. Azizov, B. Curgus and A. Dijksma, *Standard symmetric operators in Pontryagin spaces: a generalized von Neumann formula and minimality of boundary coefficients.* J. Funct. Anal. **198** (2003), 361–412.

[3] R. Cross, *Multivalued linear operators.* Marcel Dekker, New York, 1998.

[4] M.J. Kascic, *Polynomials in linear relations.* Pacific J. Math. **24** (1968), 291–295.

[5] M.J. Kascic, *Polynomials in linear relations. II.* Pacific J. Math. **29** (1969), 593–607.

[6] A. Sandovici, *Standard symmetric linear relations in Pontryagin spaces* (in preparation).

Adrian Sandovici
Department of Mathematics
University of Groningen
P.O. Box 800
9700 AV Groningen, The Netherland
e-mail: v.a.sandovici@rug.nl